RNA Polymerases as Molecular Motors

Edited by

Henri Buc
CIS, Institut Pasteur, Paris, France

Terence Strick
Centre National de la Recherche Scientifique,
Institut Jacques Monod, and University of Paris Diderot-Paris 7,
Paris, France

RSCPublishing

ISBN: 978-0-85404-134-3

A catalogue record for this book is available from the British Library

© The Royal Society of Chemistry 2009

All rights reserved

Apart from fair dealing for the purposes of research for non-commercial purposes or for private study, criticism or review, as permitted under the Copyright, Designs and Patents Act 1988 and the Copyright and Related Rights Regulations 2003, this publication may not be reproduced, stored or transmitted, in any form or by any means, without the prior permission in writing of The Royal Society of Chemistry or the copyright owner, or in the case of reproduction in accordance with the terms of licences issued by the Copyright Licensing Agency in the UK, or in accordance with the terms of the licences issued by the appropriate Reproduction Rights Organization outside the UK. Enquiries concerning reproduction outside the terms stated here should be sent to The Royal Society of Chemistry at the address printed on this page.

Published by The Royal Society of Chemistry,
Thomas Graham House, Science Park, Milton Road,
Cambridge CB4 0WF, UK

Registered Charity Number 207890

For further information see our web site at www.rsc.org

RSC Biomolecular Sciences

Editorial Board:

Professor Stephen Neidle (Chairman), *The School of Pharmacy, University of London, UK*
Dr Marius Clore, *National Institutes of Health, USA*
Professor Roderick E Hubbard, *University of York and Vernalis, Cambridge, UK*
Professor David M J Lilley FRS, *University of Dundee, UK*

This Series is devoted to coverage of the interface between the chemical and biological science especially structural biology, chemical biology, bio- and chemo-informatics, drug discovery development, chemical enzymology and biophysical chemistry. Ideal as reference and state-art guides at the graduate and post-graduate level.

Titles in the Series:

Biophysical and Structural Aspects of Bioenergetics
Edited by Mårten Wikström

Computational and Structural Approaches to Drug Discovery: Ligand-Protein Interactions
Edited by Robert M Stroud and Janet Finer-Moore

Exploiting Chemical Diversity for Drug Discovery
Edited by Paul A. Bartlett and Michael Entzeroth

Metabolomics, Metabonomics and Metabolite Profiling
Edited by William J. Griffiths

Nucleic Acid-Metal Ion Interactions
Edited by Nicholas V. Hud

Oxidative Folding of Peptides and Proteins
Edited by Johannes Buchner and Luis Moroder

Protein-Carbohydrate Interactions in Infectious Disease
Edited by Carole A. Bewley

Protein Folding, Misfolding and Aggregation: Classical Themes and Novel Approaches
Edited by Victor Muñoz

Protein-Nucleic Acid Interactions: Structural Biology
Edited by Phoebe A. Rice and Carl C. Correll

Quadruplex Nucleic Acids
Edited by Stephen Neidle and Shankar Balasubramanian

Quantum Tunnelling in Enzyme-Catalysed Reactions
Edited by Nigel S. Scrutton and Rudolf K. Allemann

Ribozymes and RNA Catalysis
Edited by David MJ Lilley FRS and Fritz Eckstein

RNA Polymerases as Molecular Motors
Edited by Henri Buc and Terence Strick

Sequence-specific DNA Binding Agents
Edited by Michael Waring

Structural Biology of Membrane Proteins
Edited by Reinhard Grisshammer and Susan K. Buchanan

Structure-based Drug Discovery: An Overview
Edited by Roderick E. Hubbard

Therapeutic Oligonucleotides
Edited by Jens Kurreck

Visit our website on www.rsc.org/biomolecularsciences

For further information please contact:
Sales and Customer Care, Royal Society of Chemistry, Thomas Graham House, Science Park, Milton Road, Cambridge, CB4 0WF, UK
Telephone: +44 (0)1223 432360, Fax: +44 (0)1223 426017, Email: sales@rsc.org

RNA Polymerases as Molecular Motors

This
sel

Foreword

The process of making RNA obliges the DNA-dependent RNA polymerases to perform a remarkable number of separate tasks:

- They scan duplex DNA to find the sites for initiating transcription.
- At these sites, the promoters, they select the DNA strand that will serve as the transcription template and expose it for base-paired binding of their ribonucleoside triphosphate substrates.
- They polymerize RNA chains through distinctive processes of initiation, promoter escape and elongation.
- They discriminate against the incorporation of deoxyribonucleotides into RNA.
- They recognize specific DNA sites for disassembling the elongating transcription complex, releasing the nascent transcript and disengaging from the DNA template.

They also do many things that are less obviously intrinsic to a nucleic acid polymerase:

- They abortively and repetitively initiate transcription at the promoter, generating short ribo-oligonucleotides.
- They elongate RNA chains at non-uniform rates, pausing at characteristic sites.
- They detect incorrect nucleotide addition to the elongating transcript and increase the accuracy of transcription by proofreading.
- They set intrinsically unequal rates of transcription at different promoters by DNA sequence-determined variations of binding affinity, rates of promoter opening and ease of escape from the promoter.
- They exert force on transcription-obstructing DNA-bound proteins, clearing them out of their path as they elongate RNA chains.
- They exert force on their DNA templates, with diverse consequences for the architecture of the eukaryotic nucleus and the prokaryotic nucleoid/chromoid.
- They monitor their DNA templates for damage.

- They recruit accessory proteins for several of these activities, particularly transcript elongation, termination and DNA repair, and they also interact with other molecular machines in order to couple transcription with subsequent RNA processing.

This is an exciting time for understanding and admiring transcription in its machine-like and mechanistic terms. Within the past approximately 10 years, the determination of the structure of the multisubunit RNA polymerases has profoundly transformed the way in which ideas about transcription mechanisms are formulated and tests of these ideas are designed. The determination of the structure of the eukaryotic RNA polymerase II from budding yeast and, most recently, of an archaeal RNA polymerase has made the common evolutionary roots of the multisubunit enzymes vividly apparent. At the same time, comparison of structures of single-subunit nucleic acid polymerases with multisubunit RNA polymerases establishes the existence of catalytic mechanisms that are common to all nucleic acid polymerizations.

During the same period, the spectacular development of methods for examining single molecules of RNA polymerase in action has opened up entirely new possibilities for probing the mechanical and motor-like aspects of RNA polymerases. The process of directly observing RNA synthesis one molecule at a time reveals insights that are difficult to retrieve from, or are entirely obscured in, observations of ensembles in bulk solution. At the same time, technical advances have significantly increased the power of longer-established analytical approaches (*e.g.*, fast reaction kinetics).

This book presents a synthesis of these streams of endeavor. Overview chapters that focus on the mechanism–structure interface and the structure–machine interface introduce the two sections of the book, while individual chapters within each section concentrate more specifically on particular processes – kinetic analysis, single-molecule spectroscopy, and termination of transcription, for example. Seen from the perspective of (nearly) 50 years ago, the detail in which every step of transcription is currently understood is remarkable, and the ways in which that detail illuminates every aspect of gene regulation is enormously satisfying. From the current perspective, the dominant sense is of unanswered questions, of experiments addressing key aspects of mechanism that remain open to competing interpretation, of further technical development that would yield insights currently just out of reach, of molecular computations that have not yet been done – *i.e.*, of a work in progress and a field of activity urgently engaged in finding fascinating new questions to answer.

The current understanding of the mechanistic and machine-like aspects of transcription has been formed primarily through work with the bacterial RNA polymerases. While the common evolutionary and mechanistic basis of all transcription can now be appreciated at the structural level, research on the eukaryotes has been dominated by the challenge of enumerating and understanding elaborations of the core transcription machinery with extrinsic initiation, elongation and termination factors and complexes (the core transcription initiation factors of budding yeast alone comprise nearly 50 polypeptide chains with an aggregate mass of more than 2.5×10^6 MDa), the dominant role of chromatin structure and modification in regulation of transcription, direct coupling of transcription with post-transcriptional RNA processing and, recently, the role of small RNAs in these processes. Many questions about the RNA polymerase machine that are specific to eukaryotic transcription, especially transcription of nucleosomal chromatin and the important role of elongation factors, are open to lines of experimentation and analysis that are described here for the bacterial enzymes. For these lines of inquiry, the work on bacterial transcription that is presented here points to, and lights up, the path.

A Foreword is a good place to relate how and where this very large endeavor started. The activity of DNA-dependent RNA polymerase was discovered at the University of Chicago's Argonne Cancer Research Hospital in 1959 and announced in a brief note in the *Journal of the American Chemical Society* in August of that year by S.B. Weiss and his assistant L. Gladstone. They showed that incorporation of ^{32}P-labeled CTP into acid-insoluble (*i.e.*, polymeric) form in a preparation of rat liver nuclei required all four ribo NTPs, ATP, GTP, CTP and UTP. The product of their synthesis was degraded by pancreatic ribonuclease but not by deoxyribonuclease I. Moreover, degradation with alkali of the radioactive product made with CTP ^{32}P-labeled in the α position distributed radioactivity to all four ribonucleoside 2′ and 3′ monophosphates. This implied the synthesis of RNA polymers of complex sequence, as opposed to the mere addition of CMP to the ends of nucleic acid chains, either adventitiously to DNA or as the matured CCA adduct of tRNAs. RNA synthesis was strongly inhibited by pyrophosphate but indifferent to orthophosphate, distinguishing the enzyme from polynucleotide phosphorylase. The role of DNA in RNA synthesis was, however, not resolved.

A year later, A. Stevens, then a postdoctoral fellow at NIH, and J. Hurwitz, A. Bresler and R. Diringer at the NYU School of Medicine separately announced the existence of a comparable activity in extracts of *Escherichia coli*. RNA synthesis by the abundantly active bacterial extracts was readily shown to be profoundly dependent on DNA. In contrast, Weiss could not separate the mammalian RNA polymerase activity from DNA and also turned to work with a bacterial enzyme (from *Micrococcus luteus*). With the bacterial preparations, J.J. Furth, Hurwitz and M. Goldman and also A. Stevens, followed by Weiss and T. Nakamoto soon showed the correspondence of the relative incorporation of (AMP+UMP) to (GMP+CMP) into synthesized RNA with the guanine-cytosine content of added DNA. Weiss and Nakamoto extended the analysis of the DNA template–RNA product relationship to the level of nearest neighbor nucleotide pairs by essentially copying an elegant analytical strategy devised for DNA polymerase (by J. Josse, A.D. Kaiser and A. Kornberg) that had been published just months before.

Using CsCl density centrifugation, B.D. Hall and S. Spiegelman had just provided the definitive proof that RNA made in phage T2-infected *E. coli* was specific to the infecting virus by showing that it was able to form DNA–RNA hybrid duplexes with phage DNA. Weiss, Nakamoto and I adopted this approach to show that the RNA synthesized *in vitro* by the bacterial RNA polymerase generated a polymeric product that was fully complementary to the eliciting double-stranded T2 phage DNA and corresponded, in that sense, to RNA made in the phage-infected cell. This (simple) experiment showed that DNA, which was still widely referred to as the "primer" of RNA synthesis, in fact was its template. The same series of experiments yielded the additional information that the newly synthesized RNA was released from its template and that transcription did not separate template DNA strands. However, both strands of T2 DNA were transcribed, yielding RNA that was self-complementary. In similar experiments with the RNA polymerase activity from *E. coli* and phage φX174 DNA, M.N. Hayashi and S. Spiegelman at the University of Illinois, as well as M. Chamberlin and P. Berg at Stanford, also found both strands of their phage DNA templates transcribed. In contrast, the RNA isolated from cells infected with diverse phages was soon found to be DNA-strand-selected, or "asymmetric," which is consistent with the requirements of messenger RNA function in instructing protein synthesis. Was something missing from the *in vitro* RNA synthesis system, or was it conceivable that mRNA might have to be selected after transcription by an additional process, with unusable transcripts rapidly disposed of? The first alternative implied an ability to select specific sites on DNA for starting transcription; the alternative hypothesis clearly failed on two counts: it was too elaborate and inelegant to be plausible, and it implied that transcription yields its functional products at the cost of an energy-consuming futile cycle. Thus, the hunt for DNA strand-selective "asymmetric" transcription was on. Of course, it can now be appreciated that the dichotomy was, to some extent, false. The human genome is pervasively transcribed, with both complementary DNA strands frequently serving as transcription templates. Moreover, cellular processes for very fast disposal of nonfunctional/unusable transcripts do exist.

In any case, the question of template strand selection in transcription was soon answered by J. Marmur's group at Brandeis, Hayashi and Spiegelman, as well as G.P. Tocchini-Valentini and co-workers at Chicago and the International Laboratory of Genetics and Biophysics in Naples, with bacterial RNA polymerase preparations that yielded strand-selective transcription, implying the ability of the enzyme to select specific DNA sites for production of RNA *in vitro*. In hindsight, the something missing from, or inactivated in, the polymerase preparations used for the initial experiments must have been σ, the initiation-specific subunit of bacterial RNA polymerases, discovered several years later by R.R. Burgess and A.A. Travers at Harvard, and E.K.F. Bautz and J.J. Dunn at Rutgers. Within the next year, partially purified bacterial RNA polymerase had been prepared in several laboratories, including those already referred to, W. Zillig's group and others. Those first experiments also provided what we now understand to have been a straightforward and simple demonstration of the existence of genes under positive transcriptional control: "asymmetric" transcription of phage T4 DNA, and the DNA of large-tailed

Foreword

phages infecting *Bacillus subtilis*, selectively yielded transcripts that correspond to RNA produced at the outset of phage infection, the so-called early RNA. This is, in a sense, the historical baseline for the "modern synthesis" represented by this book.

E.P. Geiduschek
Division of Biological Sciences, UCSD, La Jolla

Preface

It is now fashionable to view the cell as an ensemble of molecular machines or motors working in concert. What exactly do we mean by employing these terms normally reserved for industry and engineering? Up to what point is the analogy drawn between man-made motors and Nature's molecular motors justified? When does this analogy become counterproductive? If one considers RNA polymerase, the analogy is evidently useful for helping ask the right questions about its function. How does a given RNA polymerase molecule cope with the various biochemical and biomechanical tasks it has to accomplish – physically translocating along DNA, acting as a faithful chemical replicator, and properly targeting biological "start" and "stop" signals along a given DNA sequence? These are some of the naïve questions posed and addressed within the ten chapters of this book.

Tackling these challenges and concisely reporting the present achievements and bottlenecks encountered in this venture requires both clarity and humility. We are very grateful to all the authors for their illuminating contributions, and for their compliance with our numerous requests.

We are also extremely grateful to David Lilley for having offered us the opportunity to publish this book in the prestigious collection of the RSC Biomolecular Science series. It has been a pleasure working with Janet Freshwater and Annie Jacob of the Royal Society of Chemistry and we thank them for their support, patience and humor.

In Dante's Inferno there must exist a tiny, poorly-lit room where pale fellows frantically scratch at thick piles of papers while still receiving "final" versions, totally ruining their previous efforts. We have left Tantalus and Sisyphus there, thanks to our saviors, Marie-Hélène and Cary.

Henri Buc and Terence Strick

Contents

There and Back Again: A Structural Atlas of RNAP 1
Seth Darst

Part I From Promoter Recognition to Promoter Escape

Chapter 1 Where it all Begins: An Overview of Promoter Recognition and Open Complex Formation
Stephen Busby, Annie Kolb and Henri Buc

	1.1	Gene Expression as a Driver of Life	13
	1.2	*Escherichia coli* RNA Polymerase	14
	1.3	Promoters and Core Promoter Elements	17
	1.4	Biochemistry: It Works with RNA Polymerase!	19
	1.5	Biochemistry of Promoter Regulation	22
	1.6	A Word about the Intracellular Environment	26
	1.7	Coupling Transcription to Changes in a Complex Environment	27
	1.8	A Global View of the RNA Polymerase Economy	30
	1.9	The Real World, Emergency Procedures and RNA Polymerase	32
	1.10	This is just the Beginning!	33
		References	34

Chapter 2 Opening the DNA at the Promoter; The Energetic Challenge
Bianca Sclavi

2.1	Introduction	38
2.2	Structural Characterization	42
	2.2.1 Crystal Structure of the Holoenzyme	42
	2.2.2 Crystal Structure of the Holoenzyme with Fork-junction DNA	43
	2.2.3 Structural Model of the Open Complex	45
2.3	Physical Characterization and Structure of the Intermediates	47
2.4	Finding the Promoter. Induced fit and Indirect Sequence Recognition	48
2.5	Formation of the Closed Complex	49
	2.5.1 On a Unique Structure of the Closed Complex	49
	2.5.2 Role of Upstream Contacts for the Stability of the Closed Complex and in Leading the Complex Towards Subsequent Isomerization	50
2.6	The First Isomerization Step. The role of Sigma, Formation of Specific Interactions	51
	2.6.1 Upstream Contacts	51
	2.6.2 Nucleation of the Single Stranded Region and its Propagation	52
	2.6.3 Phasing of −10 and −35 Regions, the Role of the Spacer	54
	2.6.4 Probing Possible Sequential Linear Pathways by the use of Temperature	54
	2.6.5 Specific Protein Domains Destabilize the Intermediates in the Pathway	57
	2.6.6 Overstabilization is sometimes used as a Regulatory Mechanism	59
2.7	Formation of Transcriptionally Active Open Complex and the Rate-limiting Step: Protein Conformational Changes or DNA Melting?	60
2.8	A Rugged Energy Landscape	61
2.9	Summary and Conclusions	62
	Acknowledgements	63
	References	63

Chapter 3 Intrinsic *In vivo* Modulators: Negative Supercoiling and the Constituents of the Bacterial Nucleoid
Georgi Muskhelishvili and Andrew Travers

3.1	Introduction	69
3.2	DNA Superhelicity – Structures and Implications	69

	3.3	Structure of the Bacterial Nucleoid	72
	3.4	Supercoiling Utilization	75
	3.5	Promoter Structure and DNA Supercoiling	78
	3.6	Role of RNA Polymerase Composition	80
		3.6.1 Exchange of σ Factors	81
		3.6.2 Auxiliary Subunits	82
		3.6.3 Role of ppGpp	82
	3.7	Model and Implications	82
	3.8	Causality	85
	3.9	Cooperation with Nucleoid Associated Proteins	86
	3.10	Conversion of Supercoil Energy into Genomic Transcript Patterns	87
	3.11	Conclusions	88
	References		88

Chapter 4 **Transcription by RNA Polymerases: From Initiation to Elongation, Translocation and Strand Separation**
Thomas A Steitz

	4.1	Introduction	96
	4.2	Transition from the Initiation to the Elongation Phase	98
		4.2.1 T7 RNA Polymerase	98
		4.2.2 Multi-subunit RNA Polymerases	103
	4.3	Translocation and Strand Separation	103
		4.3.1 T7 RNA Polymerase	103
		4.3.2 Multi-subunit Cellular RNAPs	108
	4.4	Additional Similarities between Single and Multi-subunit Polymerases	112
	Acknowledgements		113
	References		113

Chapter 5 **Single-molecule FRET Analysis of the Path from Transcription Initiation to Elongation**
Achillefs N. Kapanidis and Shimon Weiss

	5.1	Introduction	115
	5.2	Methodology: FRET and ALEX Spectroscopy	117
	5.3	Transcription Mechanisms Addressed using Single-molecule FRET and ALEX	124
	5.4	Fate of Initiation Factor σ^{70} in Elongation	126
	5.5	Mechanism of Initial Transcription	133

	5.6	Kinetic Analysis of Initial Transcription and Promoter Escape	141
	5.7	Comparison of FRET Approaches with Magnetic-trap Approaches	142
	5.8	Future Prospects	145
	5.9	Summary	147
	Acknowledgements		148
	References		148

Chapter 6 Real-time Detection of DNA Unwinding by *Escherichia coli* RNAP: From Transcription Initiation to Termination
Terence R. Strick and Andrey Revyakin

	6.1	Introduction	157
	6.2	Twist Deformations at the Promoter	158
	6.3	Magnetic Trapping and Supercoiling of a Single DNA Molecule	159
		6.3.1 General Features of the Magnetic Trap	159
		6.3.2 Calibrating the DNA Sensor	161
	6.4	Characterization of RPo at two Canonical Promoters	166
		6.4.1 Structural Characterization of RPo	167
		6.4.2 Kinetic Analysis of RPo	168
		6.4.3 Effect of Environmental Variables on Kinetics of RPo	171
	6.5	Promoter Escape by DNA Scrunching	172
		6.5.1 Characterization of DNA Scrunching during Abortive Initiation	173
		6.5.2 Characterization of DNA Scrunching during Promoter Escape	176
	6.6	Future Directions	182
	References		183

Part II Transcription Elongation and Termination

Interlude
The Engine and the Brake
Henri Buc and Terence Strick

	I.1	Introduction	191
	I.2	The Engine	193
		I.2.1 Mechano-chemical Coupling at the Catalytic Site	194
		I.2.2 Coupling between Translocation and Topology	200
	I.3	The Brake	201

Contents

	I.4	Conclusions	202
	References		204

Chapter 7 **Substrate Loading, Nucleotide Addition, and Translocation by RNA Polymerase**
Jinwei Zhang and Robert Landick

	7.1	Basic Mechanisms of Transcript Elongation by RNA Polymerase	206
		7.1.1 Active-site Features of an Elongation Complex	207
		7.1.2 The Nucleotide Addition Cycle	207
		7.1.3 Pyrophosphorolysis and Transcript Cleavage	208
		7.1.4 Regulation of Transcript Elongation by Pauses	211
	7.2	Structural Basis of NTP Loading and Nucleotide Addition	212
		7.2.1 Bridge-helix-centric Models of Nucleotide Addition and Translocation	213
		7.2.2 Central Role of the Trigger Loop in Nucleotide Addition and Pausing	216
		7.2.3 A Trigger-loop Centric Mechanism for Substrate Loading and Catalysis	217
	7.3	Models of Translocation: Power-stroke *versus* Brownian Ratchet	219
		7.3.1 Key Distinctions between Power-stroke and Brownian Ratchet Models	220
		7.3.2 Power-stroke Models	221
		7.3.3 Brownian Ratchet Models	221
		7.3.4 Technical Outlook in Detecting the Precise Translocation Register	222
	7.4	Kinetic Models of Nucleotide Addition	223
		7.4.1 Allosteric NTP Binding Model	223
		7.4.2 NTP-driven Translocation Model	226
		7.4.3 Two-pawl Ratchet Model	226
		7.4.4 Biophysical Models for Transcript Elongation	227
	7.5	Technological Advances in Studies of Transcript Elongation	228
	7.6	Concluding Remarks	228
	References		229

Chapter 8 **Regulation of RNA Polymerase through its Active Center**
Sergei Nechaev, Nikolay Zenkin and Konstantin Severinov

8.1	Introduction	236
8.2	Regulatory Checkpoints of the RNAP Active Center	237

		8.2.1	Versatility of the Active Center. How many Metals are Enough?	237

		8.2.2	Delivery of NTPs to the Active Center. How many Channels are Enough?	238
		8.2.3	Nucleotide Selection. How many Steps are Enough?	241
	8.3	Regulators that Target the RNAP Active Center		244
		8.3.1	Small-molecule Effectors of RNAP	244
		8.3.2	Regulation of RNAP by Proteins that Bind in the Secondary Channel	250
	8.4	Transcript Proofreading		254
		8.4.1	Transcriptional Proofreading through Pyrophosphorolysis	255
		8.4.2	Proofreading by Transcript Cleavage Factors	256
		8.4.3	Transcript-assisted Proofreading. A New Class of Ribozymes?	257
	8.5	Conclusions		258
	Acknowledgements			259
	References			259

Chapter 9   Kinetic Modeling of Transcription Elongation
Lu Bai, Alla Shundrovsky and Michelle D. Wang

	9.1	Introduction		263
	9.2	Background		265
	9.3	Mechano-chemical Coupling of Transcription		266
		9.3.1	NTP Incorporation Cycle	266
		9.3.2	NTP Incorporation Pathway in a Simple Brownian Ratchet Model	266
		9.3.3	NTP Incorporation Pathways in more Elaborate Brownian Ratchet Models	267
		9.3.4	NTP Incorporation Pathway in a Power-stroke Model	269
		9.3.5	Elongation Kinetics	270
		9.3.6	Force-dependent Elongation Kinetics	271
	9.4	Sequence-dependent RNAP Kinetics		274
		9.4.1	Thermodynamic Analysis of the TEC	274
		9.4.2	Sequence-dependent NTP Incorporation Kinetics in Brownian Ratchet Models	275
		9.4.3	Model Predictions of Pause Locations, Kinetics and Mechanisms	277
	Acknowledgements			278
	References			278

Chapter 10 Mechanics of Transcription Termination
Evgeny Nudler

10.1	Introduction	281
10.2	Structure/Function Overview of the Elongation Complex (EC)	282
10.3	Mechanism of Intrinsic Termination	283
	10.3.1 The Pausing Phase	285
	10.3.2 The Termination Phase	287
10.4	Mechanism of Rho Termination	294
10.5	Summary	295
10.6	Concluding Remarks	296
	References	296

Conclusion
Past, Present, and Future of Single-molecule Studies of Transcription
Carlos Bustamante and Jeffrey R. Moffitt

C.1	Introduction	302
C.2	RNA Polymerase as a Molecular Machine: Past and Present	303
C.3	Technical Developments in Optical Tweezers	307
C.4	A Look into the Future	309
	References	312

Subject Index 315

There and Back Again: A Structural Atlas of RNAP

SETH DARST

Rockefeller University, 1230 York Avenue, New York, NY 10021, USA

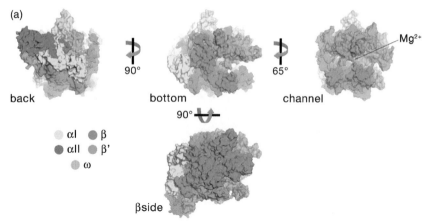

Figure A.1 Structural overview of bacterial core RNAP, σ, and holoenzyme. (a) Reference views of *Thermus aquaticus* core RNAP,[1] shown as a molecular surface with subunits colored as indicated. A Mg^{2+}-ion chelated at the RNAP active center is shown as a yellow sphere (barely visible in the channel view). A 295-residue non-conserved insert (β' residues 158–452) is not shown (see ref. 2). For parts (b)–(e), each of the reference views is expanded into four components: *Upper left*: Core RNAP [molecular surface colored as part (a)]. *Lower left*: $σ^A$ domains 2–4 [($σ_2$, $σ_3$, $σ_{3-4}$ loop, $σ_4$; see ref. 3); molecular surface colored orange] as they are seen in the RNAP holoenzyme structure,[4,5] along with the RNAP active center Mg^{2+} (purple sphere). *Upper right*: RNAP holoenzyme [refs 4. and 5; molecular surface colored as in part (a) and with $σ^A$ orange]. *Lower right*: RNAP holoenzyme (molecular surface with the core RNAP colored white and rendered transparent). Structural components, when visible, are labeled: **β1**: Upstream lobe of the β subunit.[1] β2: Downstream lobe of the β subunit.[1] **Clamp**: Structural element consisting of parts of both the β and β' subunits.[6] Undergoes large conformational change, "clamping" down on the RNA/DNA hybrid in the RNAP active site to stabilize the elongation complex. **Flap**: Structural element of the β subunit.[1] Occludes the RNA exit channel[1,7,8] and interacts with $σ_4$.[4,9] **Jaw**: Structural element of the β' subunit. **Lid**: Structural element of the β' subunit.[10] Interacts with the $σ_{3-4}$ loop[4] and is involved in nucleic acid interactions at the upstream edge of the RNA/DNA hybrid in the elongation complex. Believed to be involved in "peeling off" the RNA from its hybrid with the template DNA strand.[11] **Rudder**: Structural element of the β' subunit[1] involved in nucleic acid interactions at the upstream edge of the RNA/DNA hybrid in the elongation complex (Westover et al. 2004). Elongation complexes prepared with rudderless RNAP mutant displays reduced stability.[12] **Zbd**: Zinc-binding domain – Structural element of the β' subunit.[5] Involved in nucleic acid interactions at the upstream edge of the RNA/DNA hybrid in the elongation complex. Zbd mutants form stable TECs but are defective in termination and antitermination.[13] **Zipper**: Structural element of the β' subunit[10] involved in interactions at the upstream edge of the RNA/DNA hybrid in the elongation complex.

(b) **bottom view**

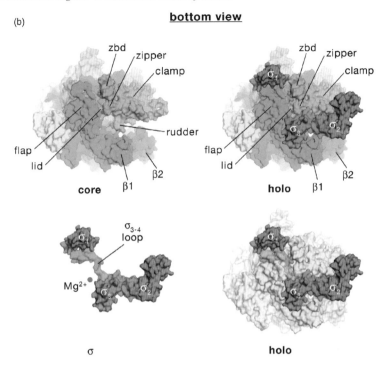

(c) **back view**

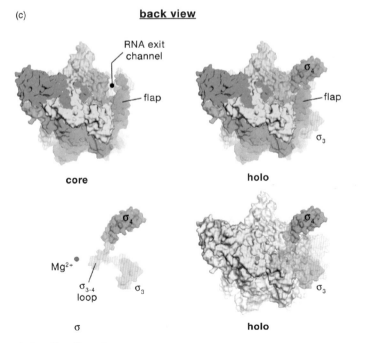

Figure A.1 Continued

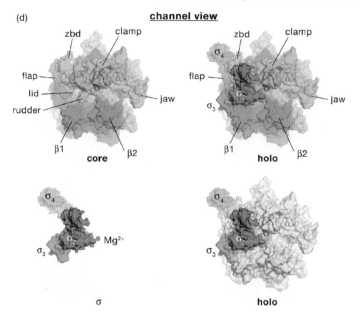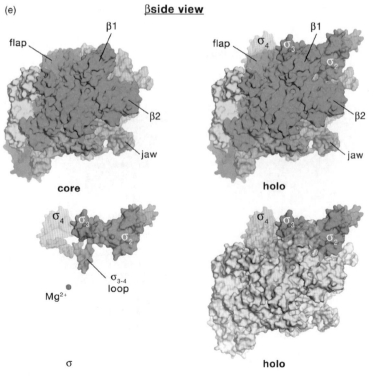

Figure A.1 Continued

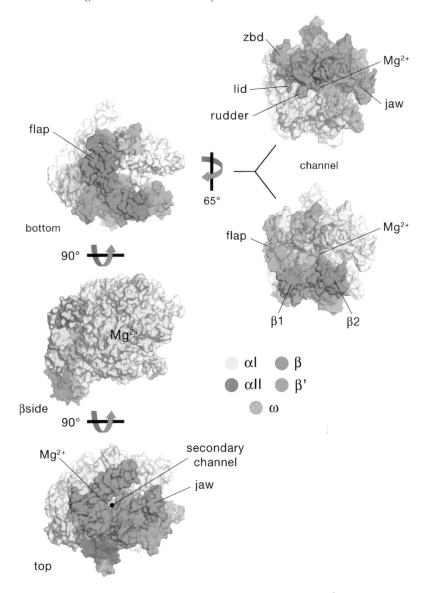

Figure A.2 Core RNAP subunits. Various views of core RNAP.[1] In each view, the RNAP is shown as a white, transparent molecular surface – except one subunit is shown non-transparent and colored (to show how the subunit fills up the space of the core RNAP). The active center Mg^{2+} is shown as a purple sphere. The bottom view highlights the β subunit. The β side view highlights the α subunit dimer. The top view highlights the β' subunit. There are two channel views, one highlighting β' (above), one highlighting β (below).

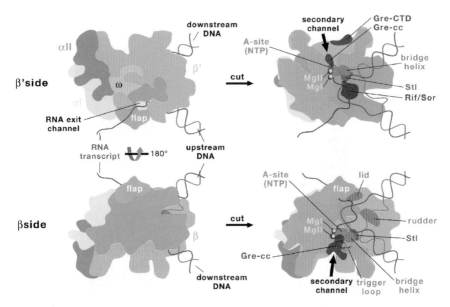

Figure A.3 Cartoon schematics of the RNAP elongation complex.[6–8,11,14–16] The top row shows the β' side view, the bottom row gives the β side view. The left-hand column shows the entire elongation complex structure, the right-hand column depicts cross-sections in which obscuring protein elements have been cut away to reveal the inside of the active site channel. The active center Mg^{2+} is shown as a yellow sphere labeled "MgI". Important structural elements of the RNAP are shown in magenta (bridge helix, lid, rudder, trigger loop). The trigger loop is shown in two positions, one pointing away from the active center, and one seen in structures of elongating complexes bound to the incoming NTP substrate (magenta dots).[17] (Vassylyev et al. 2007). In the cross-sectional views, binding sites for structurally characterized RNAP ligands or effectors are highlighted as follows: **A-site (NTP)**, green: Binding site for the incoming NTP substrate (Westover et al. 2004; Vassylyev et al. 2007), also sometimes called the insertion site (IS), or also the $i+1$ site. The incoming NTP substrate arrives in the A-site via the secondary channel,[1] (Westover et al. 2004) and chelated to a Mg^{2+} ion, which becomes an essential component of the active site (MgII, also sometimes called metal B; see ref. 18). **Gre**, red: Gre-factors (ref. 19) bind via their C-terminal domain (Gre-CTD) outside the secondary channel, and insert their coiled-coil finger (Gre-cc) into the secondary channel.[20] The tip of the Gre-cc contains absolutely conserved acidic residues that serve to stabilize the binding of MgII, which is required for the endonucleolytic cleavage of backtracked RNA.[21,22] **Rif/Sor**, blue: Rifamycins,[23] a key component of tuberculosis therapy,[24] bind in a pocket of the β subunit in the active site channel[25,26] and inhibit RNAP by blocking the path of the elongating RNA transcript.[25] Sor-angicin A, a chemically unrelated inhibitor,[27] binds in the identical site and likely inhibits RNAP by a similar mechanism.[28] **Stl**, orange: The bacterial RNAP inhibitor streptolydigin[29] binds to a site along the bridge helix.[30,31]

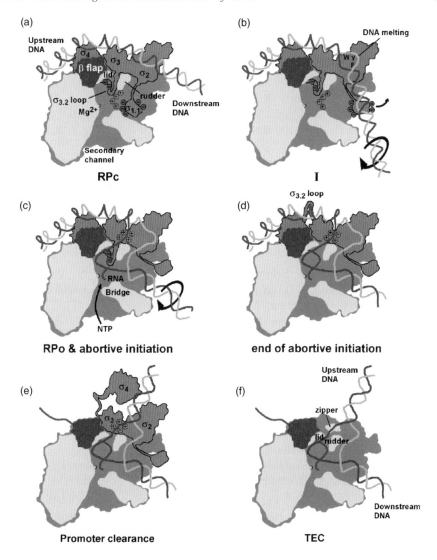

Figure A.4 Structural transitions during the steps of transcription initiation. Shown are cross-sectional views of the RNAP holoenzyme (β flap, blue; σ, orange; rest of RNAP, gray; catalytic Mg^{2+}, yellow sphere), promoter DNA (template strand, dark green; nontemplate strand, light green), and the RNA transcript (red) at (a) the RPc, (b) intermediate (I), (c) RPo and abortive initiation, (d) end of abortive initiation, (e) promoter clearance and (f) TEC stages of transcription initiation. The view is looking down on top of the β subunit, but with most of β removed, revealing the inside of the RNAP active site channel.

References

1. G. Zhang, E. A. Campbell, L. Minakhin, C. Richter, K. Severinov and S. A. Darst, Crystal structure of *Thermus aquaticus* core RNA polymerase at 3.3 Å resolution, *Cell*, 1999, **98**, 811–824.
2. M. Chlenov, S. Masuda, K. S. Murakami, V. Nikiforov, S. A. Darst and A. Mustaev, Structure and function of lineage-specific sequence insertions in the bacterial RNA polymerase b' subunit, *J. Mol. Biol.*, 2005, **353**, 138–154.
3. E. A. Campbell, O. Muzzin, M. Chlenov, J. L. Sun, C. A. Olson, O. Weinman, M. L. Trester-Zedlitz and S. A. Darst, Structure of the bacterial RNA polymerase promoter specificity sigma factor, *Mol. Cell*, 2002, **9**, 527–539.
4. K. Murakami, S. Masuda and S. A. Darst, Structural basis of transcription initiation: RNA polymerase holoenzyme at 4 Å resolution, *Science*, 2002, **296**, 1280–1284.
5. D. G. Vassylyev, S. Sekine, O. Laptenko, J. Lee, M. N. Vassylyeva, S. Borukhov and S. Yokoyama, Crystal structure of a bacterial RNA polymerase holoenzyme at 2.6 Å resolution, *Nature*, 2002, **417**, 712–719.
6. A. L. Gnatt, P. Cramer, J. Fu, D. A. Bushnell and R. D. Kornberg, Structural basis of transcription: An RNA polymerase II elongation complex at 3.3 Å resolution, *Science*, 2001, **292**, 1876–1882.
7. N. Korzheva, A. Mustaev, M. Kozlov, A. Malhotra, V. Nikiforov, A. Goldfarb and S. A. Darst, A structural model of transcription elongation, *Science*, 2000, **289**, 619–625.
8. D. G. Vassylyev, M. N. Vassylyeva, A. Perederina, T. H. Tahirov and I. Artsimovitch, Structural basis for transcription elongation by bacterial RNA polymerase, *Nature*, 2007, **448**, 157–162.
9. K. Kuznedelov, L. Minakhin, A. Niedziela-Majka, S. L. Dove, D. Rogulja, B. E. Nickels, A. Hochschild, T. Heyduk and K. Severinov, A role for interaction of the RNA polymerase flap domain with the sigma subunit in promoter recognition, *Science*, 2002, **295**, 855–857.
10. P. Cramer, D. A. Bushnell and R. D. Kornberg, Structural basis of transcription: RNA polymerase II at 2.8 Å resolution, *Science*, 2001, **292**, 1863–1876.
11. K. D. Westover, D. A. Bushnell and R. D. Kornberg, Structural basis of transcription: nucleotide selection by rotation in the RNA polymerase II active center, *Cell*, 2004, **119**(4), 481–9.
12. K. Kuznedelov, N. Korzheva, A. Mustaev and K. Severinov, Structure-based analysis of RNA polymerase function: the largest subunit's rudder contributes critically to elongation complex stability and is not involved in the maintenance of RNA-DNA hybrid length, *EMBO J.*, 2002, **21**, 1369–1378.
13. R. A. King, D. Markov, R. Sen, K. Severinov and R. A. Weisberg, A conserved zinc binding domain in the largest subunit of DNA-dependent RNA polymerase modulates intrinsic transcription termination and

antitermination but does not stabilize the elongation complex, *J. Mol. Biol.*, 2004, **342**, 1143–1154.
14. H. Kettenberger, K. -J. Armache and P. Cramer, Complete RNA polymerase II elongation complex structure and its interactions with NTP and TFIIS, *Mol. Cell*, 2004, **16**, 955–965.
15. K. D. Westover, D. A. Bushnell and R. D. Kornberg, Structural basis of transcription: Separation of RNA from DNA by RNA polymerase II, *Science*, 2004, **303**, 1014–1016.
16. D. G. Vassylyev, M. N. Vassylyeva, J. Zhang, M. Palangat, I. Artsimovitch and R. Landick, Structural basis for substrate loading in bacterial RNA polymerase, *Nature*, 2007, **448**, 163–168.
17. D. Wang, D. A. Bushnell, K. D. Westover, C. D. Kaplan and R. D. Kornberg, Structural basis of transcription: role of the trigger loop in substrate specificity and catalysis, *Cell*, 2006, **127**(5), 941–54.
18. T. A. Steitz, A mechanism for all polymerases, *Nature*, 1998, **391**, 231–232.
19. S. Borukhov, V. Sagitov and A. Goldfarb, Transcript cleavage factors from *E. coli*, *Cell*, 1993, **72**, 459–466.
20. N. Opalka, M. Chlenov, P. Chacon, W. J. Rice, W. Wriggers and S. A. Darst, Structure and function of the transcription elongation factor GreB bound to bacterial RNA polymerase, *Cell*, 2003, **114**, 335–345.
21. O. Laptenko, J. Lee, I. Lomakin and S. Borukhov, Transcript cleavage factors GreA and GreB act as transient catalytic components of RNA polymerase, *EMBO J.*, 2003, **23**, 6322–6334.
22. E. Sosunova, V. Sosunov, M. Kozlov, V. Nikiforov, A. Goldfarb and A. Mustaev, Donation of catalytic residues to RNA polymerase active center by transcription factor Gre. *Proc. Natl. Acad. Sci. USA*, 2003, **100**, 15469–15474.
23. P. Sensi, History of the development of rifampin, *Rev. Infect. Dis.*, 1983, **5**(Supp. 3), 402–406.
24. H. G. Floss and T. W. Yu, Rifamycin-mode of action, resistance, and biosynthesis, *Chem. Rev.*, 2005, **105**, 621–632.
25. E. A. Campbell, N. Korzheva, A. Mustaev, K. Murakami, S. Nair, A. Goldfarb and S. A. Darst, Structural mechanism for rifampicin inhibition of bacterial RNA polymerase, *Cell*, 2001, **104**, 901–912.
26. I. Artsimovitch, M. N. Vassylyeva, D. Svetlov, V. Svetlov, A. Perederina, N. Igarashi, N. Matsugaki, S. Wakatsuki, T. H. Tahirov and D. G. Vassylyev, Allosteric modulation of the RNA polymerase catalytic reaction is an essential component of transcription control by rifamycins, *Cell*, 2005, **122**, 351–363.
27. H. Irschik, R. Jansen, K. Gerth, G. Hofle and H. Reichenbach, The sorangicins, novel and powerful inhibitors of eubacterial RNA polymerase isolated from myxobacteria, *J. Antibiotics*, 1985, **40**, 7–13.
28. E. A. Campbell, O. Pavlova, N. Zenkin, F. Leon, H. Irschik, R. Jansen, K. Severinov and S. A. Darst, Structural, functional, and genetic analysis of sorangicin inhibition of bacterial RNA polymerase, *EMBO J.*, 2005, **24**, 674–682.

29. G. Cassani, R. R. Burgess, H. M. Goodman and L. Gold, Inhibition of RNA polymerase by streptolydigin, *Nature New Biol.*, 1971, **230**, 197–200.
30. D. Temiakov, N. Zenkin, M. N. Vassylyeva, A. Perederina, T. H. Tahirov, E. Kashkina, M. Savkina, S. Zorov, V. Nikiforov, N. Igarashi, N. Matsugaki, S. Wakatsuki, K. Severinov and D. G. Vassylyev, Structural basis of transcription inhibition by antibiotic streptolydigin, *Mol. Cell*, 2005, **19**, 655–666.
31. S. Tuske, S. G. Sarafianos, X. Wang, B. Hudson, E. Sineva, J. Mukhopadhyay, J. J. Birktoft, O. Leroy, S. Ismail, A. D. J. Clark, C. Dharia, A. Napoli, O. Laptenko, J. Lee, S. Borukhov, R. H. Ebright and E. Arnold, Inhibition of bacterial RNA polymerase by streptolydigin: Stabilization of a straight-bridge-helix active-center conformation, *Cell*, 2005, **122**, 541–552.

Part I From Promoter Recognition to Promoter Escape

CHAPTER 1
Where it all Begins: An Overview of Promoter Recognition and Open Complex Formation

STEPHEN BUSBY,[a] ANNIE KOLB[b] AND HENRI BUC[c]

[a] School of Biosciences, University of Birmingham, Birmingham B15 2TT, United Kingdom; [b] Institut Pasteur, Molecular Genetics Unit and CNRS URA 2172, 25 rue du Dr. Roux, 75724 Paris Cedex 15, France; [c] CIS Institut Pasteur, 75724 Paris Cedex 15, France.

1.1 Gene Expression as a Driver of Life

The importance of transcription, the process by which information encoded in DNA is copied into RNA, cannot be overstated. As soon as the dogma that DNA makes RNA makes protein was established, the hunt was on for the machinery that orchestrates transcription. Thus, in the late 1950s and early 1960s, classical methods of protein fractionation were used to identify DNA-dependent RNA polymerase activity. Remarkably, in parallel, primarily using *Escherichia coli* genetics, Jacob, Monod and their colleagues were discovering gene regulatory proteins and establishing the paradigm that gene transcription was the key point at which gene regulation is effected.[1] Thus, right from the start, *Escherichia coli* K-12 was established as the model system to use and, with the benefit of hindsight, it is easy to see now how 40 years of amazing progress was sparked by the fusion of two very different worlds, one populated by the biochemists and the other by the bacterial geneticists. Put very simply, the stories in this book expand on how the biochemistry explains the genetics and how the genetics gives reason to the biochemistry. The crucial discoveries that set the scene for these stories were made in the late 1960s: the characterization of the single multi-subunit RNA polymerase in *E. coli*, the discovery of promoters and terminators, and the realization that different genes are transcribed at widely differing frequencies. The pace accelerated with the arrival of cloning and DNA sequencing in the 1970s and in-depth studies of how different promoters are regulated exploiting increasingly sophisticated methodologies. The arrival of whole genome sequences in the late 1990s led to the complete catalogue of the different players and attempts to integrate our knowledge with systems biology approaches. And finally, the structural biologists have provided us with models of many of the major players, including the multi-subunit RNA polymerases, the principal topic of this book.

1.2 *Escherichia coli* RNA Polymerase

The view of bacterial RNA polymerase as a 500 kDa enzyme with subunit structure $\alpha_2\beta\beta'\omega\sigma$ had a long and slow birth, emerging from heroic biochemistry in both the USA and in Germany. It is easy to overlook the difficulties encountered by the pioneers in this field of proving the integrity and function of such a large multi-subunit complex. DNA cloning technologies had not yet arrived and early efforts to demonstrate specific DNA-directed transcription mostly had to exploit viral templates, notably bacteriophages. Perhaps the most influential single observation was the chance discovery by Dick Burgess and colleagues in 1969 that passage of the preparation of *E. coli* RNA polymerase through phosphocellulose led to loss of its ability to initiate specific transcripts and that this loss was due to the loss of the σ factor.[2] This led to the definition of two forms of RNA polymerase, the holo-enzyme with composition $\alpha_2\beta\beta'\omega\sigma$, and the core enzyme, $\alpha_2\beta\beta'\omega$, devoid of σ, and the notion of σ as the factor controlling transcript initiation. Another influential early finding came from Mike Chamberlin and colleagues, who showed that the transcriptionally competent complexes formed between the holoenzyme and DNA were resistant to heparin[3] (see also ref. 4). In these complexes, which could form in the absence of any nucleotides, the heparin resistance arises from the template DNA strands being locally unwound around the transcription start site. These observations gave birth to the idea of a pathway to transcription initiation, with the kinetically competent or open complex being preceded by a heparin-sensitive closed complex in which the DNA strands are not open.[5,6] Amazingly, the nature of closed complexes, the mechanics of the closed to open transition and the number of intermediates remain hot topics for study and debate today.[7,8]

One of the early proofs that *E. coli* contained a single core RNA polymerase was that RNA synthesis could be completely inhibited by the drug, rifampicin, but a single point mutation can confer complete resistance and normal RNA synthesis.[9] The location of these rif^R mutations led to the identification of the co-transcribed *rpoB* and *rpoC* genes, which encode the RNA polymerase large β and β′ subunits (1342 and 1407 amino acids respectively).[10] Subsequently, the genes encoding the other RNA polymerase subunits were identified at different locations on the *E. coli* chromosome, and the pathway of subunit assembly was established. The first step is the formation of a dimer of two 329 amino acid α subunits, which acts as a scaffold for the addition of first β and then β′/ω to give core enzyme. The holoenzyme is then formed by the addition of the σ subunit. This pathway was established by Akira Ishihama, who later showed that the C-terminal 100 amino acids of each α subunit are dispensable for RNA polymerase assembly.[11,12] The reason for this is that the RNA polymerase α subunit consists of two domains, with the 230 amino acid N-terminal domain being essential for enzyme assembly, whilst the C-terminal contains a separate independently folding domain that plays a key role at certain promoters. We now have detailed structures for both the core and holo enzymes, due largely to the efforts of Seth Darst, Dmitry Vassylyev and their coworkers using RNA polymerases from thermophilic bacteria.[13–15] The structures show the large β and β′ subunits assembled on the two α subunit N-terminal domains, with the β and β′ subunits forming a "crab claw" to accommodate DNA, with the catalytic centre of the enzyme right at the heart of the claw (Figure 1.1; full details are in Chapter 2). This organization is echoed in the structures of yeast RNA polymerase II that emerged from Roger Kornberg's laboratory at the same time, underlining its importance at all levels of life.[16]

Another major landmark in the study of bacterial RNA polymerases has been the realization that most bacteria contain multiple σ factors. This idea first emerged from studies by Rich Losick and others of the genes needed for spore formation by *Bacillus subtilis*. The products of some of these genes showed striking sequence similarities to already discovered σ factors. Since it was clear that σ factors were needed for both promoter specificity and open complex formation, Losick's proposal that the sporulation pathway was driven by the synthesis of new σ factors, which switch on new sets of genes, was soon accepted.[17] In fact, most bacteria contain one dominant σ factor and between 0 and 64 alternatives. The dominant σ, known as the "housekeeping σ" is an essential protein that is responsible for most transcription initiation. In *E. coli*, the predominant σ is σ^{70} with a molecular size of 70 kDa (613 amino acids) and it is this σ factor, encoded by the *rpoD* gene, that is found in most RNA polymerase preparations. The *E. coli* genome encodes six alternative σ factors (encoded by the *rpoS*, *rpoH*, *rpoE*, *rpoF*, *rpoN* and *fecI* genes), which are concerned with the management of different stresses.[18] An increase in the intracellular level of an alternative σ factor (*e.g.*, in response to a specific stress) results in the formation of a subpopulation of RNA polymerase holoenzyme molecules dedicated to initiate transcription at a particular subset of promoters.[19] As more bacterial genomes have been studied, the belief that alternative σ factors have evolved to drive programs of microbial adaptation and differentiation has been reinforced.

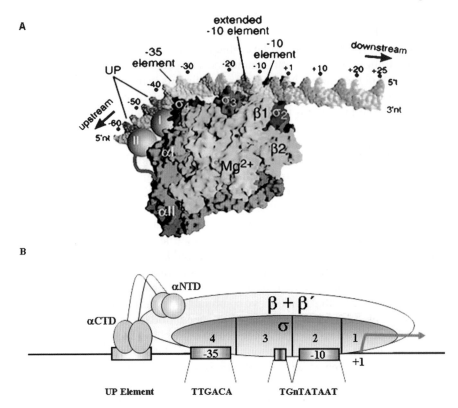

Figure 1.1 (A) Docking of a multi-subunit RNA polymerase to a promoter. The DNA double helix is shown in green, except for the −10 and −35 elements (in yellow) and for the UP element and the TGn extension (in red). The β and β′ subunits of RNA polymerase are in light green and pink respectively, the various domains of σ are in red and the two N-terminal domains of α in grey. Grey balls contacting the UP elements of the promoter represent the mobile C-terminal domains of α. They have not been located in any crystallographic study. A Mg^{2+} ion (magenta) locates the catalytic site, at the interior of the picture. (Reproduced with permission from the American Association for the Advancement of Sciences.[23]) (B) Simplified version of the model shown in (A), emphasizing the crucial protein contacts on the promoter sequence.

DNA sequencing then allowed comparison of the primary structure of various σ factors. Carol Gross and colleagues established that many of them contain four conserved regions of sequence that appear to fall in four protein domains.[20,21] With the notable exception of the RpoN (σ^{54}) family of σ factors,[22] this organization, or close variants, applies to most of the hundreds of σ factors that have now been characterized. In fact, biophysical studies have now confirmed the existence of four independent domains. In the RNA polymerase holoenzyme, Domains 2, 3 and 4 are arrayed on the surface, while Domain 1 slots in the crab claw DNA binding channel. The formation of a transcriptionally competent open complex involves recognition of different promoter elements by σ domain 2, and/or domain 3 and/or domain 4, expulsion of domain 1 from the crab claw channel, and entry of DNA surrounding the transcription startpoint[23] (full details are in Chapter 2).

1.3 Promoters and Core Promoter Elements

As soon as it became possible to measure the synthesis of individual RNAs in *E. coli*, it became clear that there was enormous selectivity in gene expression. Some genes are highly transcribed whilst others are rarely transcribed. The idea that individual transcripts start and finish at specific points, and that much of the selectivity was due to differences in rates of initiation was embraced with enthusiasm.[24] The rapid adoption of these concepts was probably due more to their simplicity than to the weight of experimental evidence, and, since they provided a convenient framework, pragmatism had the upper hand. What was incontrovertible was that specific mutations, located in front of the three genes that constitute the *E. coli* lactose (*lac*) operon, could reduce their simultaneous expression to near zero.[25,26] The simplest and, as it turned out, the correct explanation was that the mutations were removing something that was channeling RNA polymerase to make a specific transcript. This something became known as a promoter, and as it became possible directly to determine transcript start points, and base by base mutational analysis could be achieved, detailed understanding of the architecture of the *lac* promoter, and other promoters, could be built up.[27]

The picture that emerged of a typical promoter was that it contained two or more functionally important distinct sequence elements upstream of the transcription start site.[28] The first such elements to be established were the –10 and –35 hexamers, and early studies attempted, with mixed success, to score promoter strength on the basis of the correspondence of the actual sequence to the –10 consensus 5′-TATAAT-3′, the –35 consensus 5′-TTGACA-3′ and the separation between the two elements.[29] Crucially, most previously selected promoter "down" mutations fell in one or the other element and changed the sequence away from the consensus. Similarly, most known promoter "up" mutations also mapped to these elements and changed the promoter sequence towards the consensus. Subsequently, using suppression genetics,[30-32] it was shown that the –10 and –35 elements were recognized by determinants in Domain 2 and Domain 4 respectively of the RNA polymerase σ^{70} subunit during promoter recognition, and details of the molecular basis of this recognition are now understood (Chapter 2).[23,33] The simple model of the two principal promoter elements being recognized by two different domains of σ immediately provided an attractive explanation for how RNA polymerase containing alternative σ factors could recognize different promoters.

Although –10 and –35 elements play a major role in setting promoter activity, other elements that play important roles were soon discovered, and these can compensate for –10 and –35 elements that correspond poorly to the consensus.[29] The base sequences immediately upstream of the –10 hexamer constitute the "extended –10" element that is recognized by Domain 3 of σ^{70} and, according to promoter context, can increase promoter strength by up to 20-fold.[34] Similarly, the 20 base pair tract upstream of the –35 region provides a target for the two RNA polymerase α subunit C-terminal domains. This tract, known as the UP element, is found upstream of many strong promoters and functions by increasing the recruitment of RNA polymerase such that promoter strength can be increased up to 50-fold.[35] Other determinants of promoter activity are the spacing between the –10 and –35 elements, which determines their juxtaposition as they are recognized by the RNA polymerase σ subunit, and sequences downstream of the –10 hexamer, which determine the stability of the open complex and the facility with which RNA polymerase escapes from the open complex as the nascent RNA chain is elongated (Figure 1.2).[7]

When it comes to bacterial promoters, the keyword is variety: just as many combinations of small coins can make up a dollar, different combinations of elements can make a promoter! However, it is here that one encounters a problem, because if the contributing elements, acting together, are too strong, the RNA polymerase has difficulty escaping, since the forces that recruited it cannot be undone. This is because the energy to move RNA polymerase out of the initiation complex has to come from the free energy associated with RNA formation. This explains why promoters have evolved to have non-consensus

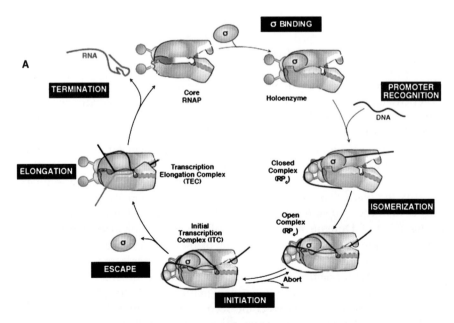

Figure 1.2 The transcription cycle according to a cartoon adapted from Geszvain and Landick. Note that the steps at which the σ subunit leaves and rebinds to the core enzyme are not necessarily the ones depicted here.[39]

elements and why some of the strongest promoters, which recruit RNA polymerase most efficiently, are constructed so that the open complex is unstable, and hence RNA polymerase can be cleared rapidly. In fact, as demonstrated by Hermann Bujard, a sequence having all the consensus elements would be useless as a promoter, as RNA polymerase would bind to it, but then would be trapped.[36]

Most of the research described in this book is concerned with RNA polymerase–DNA interactions, and focuses on simple promoters where RNA polymerase alone is sufficient for transcript initiation. However, in bacteria, the expression of most genes is regulated by the growth conditions and this involves different proteins that interact at promoters to activate or repress the process of transcript initiation. Thus, the activity of many promoters is set, not only by the different promoter elements but by these proteins.[37] Some of these proteins control changes in transcription initiation in response to specific signals in the environment, *e.g.*, the *lac* repressor responds to lactose, whilst others set the chromosomal context in which the promoter functions. Some transcription factors respond to global metabolic signals (*e.g.*, carbon, nitrogen or oxygen supply) and interact at scores of different promoters. For example, initiation of transcription at the *lac* promoter, one of the first *E. coli* promoters to be studied in detail, is almost completely dependent on the cyclic AMP receptor protein (CRP), a global regulator that is often regarded as a paradigm for transcription factors. To understand fully the transactions that take place at bacterial promoters, an appreciation of how different proteins influence open complex formation is essential.

1.4 Biochemistry: It Works with RNA Polymerase!

It is fortunate that bacterial transcription can be reiterated in an *in vitro* system with purified DNA template and proteins. Landmark studies with bacteriophages DNA as templates clearly demonstrated that *E. coli* RNA polymerase was capable of specific initiation and termination of transcripts as well as elongation. These observations, made in the late 1960s, form the basis of the mechanistic studies described in this book. Of course, we now know of a host of proteins that interact with RNA polymerase to affect transcript initiation, elongation and termination,[38] but the importance of the observation that the "simple" *E. coli* holoenzyme preparation was capable of reproducing the transcription cycle in a test tube cannot be overstated (Figure 1.2). Early experiments showed that the σ subunit was released from RNA polymerase as the nascent RNA chain grew. This led to the establishment of the σ cycle.[39] At the time, it provided a rationale for the observation that *E. coli* cells contained substantially less σ than molar equivalents of core RNA polymerase (a finding that is not supported by a recent quantitative determination[40]). Later, Sankar Adhya, Jack Greenblatt and colleagues showed that NusA was recruited to the core enzyme concomitant with the loss of σ and functioned as an elongation factor.[41]

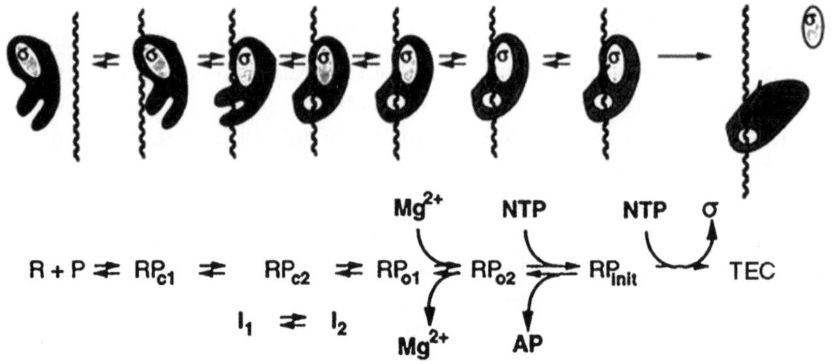

$$R + P \rightleftarrows RP_{c1} \rightleftarrows RP_{c2} \rightleftarrows RP_{o1} \rightleftarrows RP_{o2} \rightleftarrows RP_{init} \longrightarrow TEC$$

$$I_1 \rightleftarrows I_2$$

with cofactors Mg^{2+}, NTP, NTP, σ and Mg^{2+}, AP.

Figure 1.3 Simple, linear scheme displaying several intermediates intervening during the process of promoter recognition and initiation of mRNA synthesis.[7] Besides the free species, R and P, the initial transcribing complex [here RP (init)] and the elongation complex (TEC), already mentioned in Figure 1.2, the closed and the open species are further subdivided according to biochemical and kinetic evidence. The enzyme is depicted as the association of core and σ^{70} subunit. In the closed complexes (where the whole DNA is double-stranded), the "jaws" of the core enzyme can be open (c1 and c2, I_1) or closed on the downstream DNA sequence (c2, I_2). The two open complexes, o1 and o2, differ by the presence or absence of bound magnesium ion, and by the extent of the DNA region that is single-stranded (–12 to –1 and –12 to +2, respectively).

Concerning the study of open complex formation and transcription initiation, the crucial observation was that RNA polymerase holoenzyme could recognize promoter locations with a reasonable degree of specificity, and form stable complexes (Figure 1.3). These "open" complexes could be measured by filter binding assays or visualized by electron microscopy. They were resistant to heparin and were competent for transcript initiation and elongation if nucleoside triphosphates were provided. The fact that open complex formation could easily be separated from transcript formation was exploited extensively from 1975 onwards. The formation of an open complex is generally irreversible, so that the experimental rate with which this entity was formed in the test tube provided a convenient measure of the strength of the corresponding promoter. It became then possible to study, for example, the contribution of the various elements of a promoter sequence to its overall strength. Different footprinting methods were subsequently developed that gave increasingly detailed information about the organization of the open complex in the absence of structural data. For instance, the length of protected DNA and the extent of duplex unwinding could be found from DNase I[42] and permanganate footprinting, respectively,[43] and the wrapping of upstream sequences and their interactions with the C-terminal domains of the RNA polymerase α subunits could be deduced from hydroxyl radical footprints.[44] Increasingly sophisticated chemical crosslinking reagents and fluorescent probes were developed more recently, notably by Richard Ebright and colleagues, to investigate open complexes or to monitor sequential conformational changes at the promoter,[45,46] sometimes through the study of the behavior of complexes formed on single DNA molecules. Later chapters in this book describe some of this work and discuss the kinetics and energetics of open complex formation. A recurring theme is that the final open complex is unaffected by the nature of the different promoter elements used to get there.

Concerning transcript formation, the early literature is dominated by Peter von Hippel's model of an elongation complex, based on the length of DNA duplex unwinding and the RNA-DNA hybrid, and the extent of protection by RNA polymerase of upstream DNA, downstream DNA and nascent RNA (Figure 1.4).[47] However, it rapidly became clear that RNA polymerase did not always progress smoothly into von Hippel's elongation complex, and, at some promoters, large amounts of RNA oligomers were produced before full-length transcript appeared. These products appear to be made by abortive cycling after formation of the first RNA phosphodiester bonds and, if all four nucleoside triphosphates were not provided, RNA polymerase could be trapped in this abortive mode.[24] In the presence of all four RNA precursors, the relative amount of abortive to full length transcripts is dependent on the precise conditions and the base sequence around the transcript startpoint.[48] Later chapters in this book discuss transcript elongation and underscore the impact of single molecule studies on our understanding.

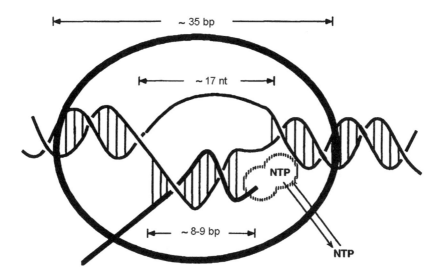

Figure 1.4 Von Hippel's model of an elongation complex.[47] The picture indicates only the extent of base-pairing in the nucleic acid complex (revised according to more recent data), and the length of the DNA sequence protected against nucleolytic agents by the presence of the enzyme. For further information on the NTP sites see Chapters 7 and 8.

1.5 Biochemistry of Promoter Regulation

Many of the first *E. coli* and bacteriophage promoters to be studied *in vitro* are subject to quite sophisticated regulation by transcription factors. Thus, expression from the *lac* operon promoter is dependent on CRP for activation and repressed by the *lac* repressor. Both these factors and their binding sites had been discovered by genetic analysis in the 1960s, and, in the 1970s, both were purified and shown to bind to specific target sequences. CRP binds immediately upstream of the different *lac* promoter elements while *lac* repressor binds to a target, the operator, right over the transcript start site. Once again, biochemistry did not disappoint, and it proved relatively easy to reproduce activation of the *lac* promoter by CRP and repression by the *lac* repressor.[49] Similar pioneering studies were performed with different bacteriophage promoters. By 1980 it was clear that, by combining genetic analysis with biochemistry, it was possible to deduce how transcription repressors and activators worked in the simplest systems studied at the time.

At first sight, the mode of action of bacterial repressors is extremely simple (Figure 1.5). The target for the *lac* repressor is right over the *lac* promoter transcription start, suggesting that the primary mechanism is to prevent RNA polymerase binding. A similar straightforward picture emerged from studies of how the bacteriophage λ repressor repressed the major bacteriophage λ promoters, though repression required cooperative binding between repressors bound at adjacent sites.[50] However, as more examples have been studied, and as *lac* and λ have been revisited, unexpected dimensions have been unearthed. For instance, Benno Muller-Hill, who first purified a transcription factor, the *lac* repressor, noted that, although the *lac* operator can accommodate only two subunits of *lac* repressor, the *lac* repressor is a tetramer.[51] Moreover he noted that upstream and far downstream secondary operators were needed for optimal repression and suggested that these serve to increase the local concentration of the *lac* repressor.[52] The theme of repressors bound at distant sites has reappeared at many regulatory regions and, in some cases, the bound repressors interact and form a zone of exclusion for RNA polymerase.[53] Another intriguing twist to repression is found at promoters where, rather than excluding RNA polymerase, the repressor jams the polymerase and prevents it from escaping the promoter and hence making transcript.[54] This is costly to the cell, since it immobilizes a potentially functional polymerase, but, when repression is lifted, it ensures that RNA synthesis immediately recovers.

Parallel studies with bacterial transcription activators showed that many function by recruiting RNA polymerase to the target promoter. This idea, which emerged from Mark Ptashne's and Will McClure's pioneering studies, supposes that activators, rather than creating new pathways, accelerate the recognition process at pre-existing promoters.[29,55] Hence the *E. coli lac* promoter contains functional −10 and −35 sequences, but they are insufficient to assure promoter activity. Genetic and biochemical analysis showed that upstream-bound CRP makes a direct contact with the C-terminal domain of the RNA polymerase α subunit, which recruits it to an up element and hence the rest of

A. Repression by Steric Hindrance

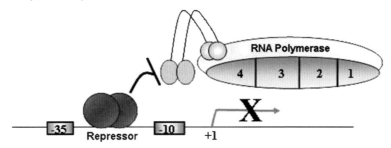

B. Repression by Looping

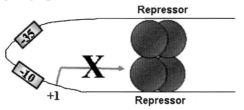

C. Repression by Modulation of an Activator.

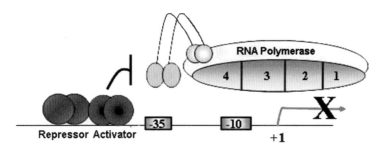

Figure 1.5 Different repression mechanisms. As many regulatory proteins function as dimers, the repressor is shown here as a dimer. RNA polymerase holoenzyme is depicted as a larger multi-subunit enzyme; only the σ subunit (σ^{70} and its four structural domains), and the two α subunits (amino- and carboxy-terminal domain αNTD and αCTD connected by a flexible linker) are depicted in light blue. The –10 and –35 boxes for recognition by the holoenzyme are also shown. (A) Repression by steric hindrance. The repressor binding site overlaps core promoter elements and inhibits recognition of the promoter by RNA polymerase. (B) Repression by looping. Repressors bind to distal sites and interact together by looping the intervening DNA, whereby the intervening promoter is repressed. (C) Repression by modulation of an activator protein. The repressor binds to an activator protein and prevents the activator from functioning. Promoter recognition is inhibited.

RNA polymerase.[56] The CRP-dependence of the *lac* promoter can be circumvented by improving the −10, −35 or UP elements and, in the final open complex, the organization of RNA polymerase is pretty much as at most factor-independent promoters. Subsequent studies with many other *E. coli* gene regulatory regions have shown that other transcription activators bind upstream of promoter elements and function by contacting the C-terminal domain of the RNA polymerase α subunit and recruiting the whole RNA polymerase holoenzyme (Figure 1.6A). Other activators function by making a direct contact with Domain 4 of the RNA polymerase σ subunit (Figure 1.6B). These activators bind to DNA sites that overlap the −35 element at the target promoter and can function either by recruiting RNA polymerase or by accelerating the formation of the open complex from pre-bound RNA polymerase. In principle, an activator that functions by recruitment could contact any part of RNA polymerase and, indeed, as more cases have been studied, it has become clear that many different targets on RNA polymerase are used. Interestingly, activators such as CRP, which can activate transcription at many different promoters, contact different targets in RNA polymerase at different promoters, according to the binding site location relative to the different promoter elements. Although most *E. coli* transcription activators function as recruiters of RNA polymerase, *in vitro* studies revealed other paradigms. For example, the MerR protein – which is a sensor for toxic mercury ions and responds by activating the expression of mercury resistance determinants – binds to a DNA site between its target promoter −10 and −35 elements and twists the DNA (Figure 1.6C). This brings the two elements into register such that they are better recognized by the RNA polymerase σ subunit.[57]

As far as we know, activators that function by recruitment or changing DNA structure can function with RNA polymerase containing different σ factors, with the exception of members of the RpoN (σ^{54}) family.[22] The key point is that most holo RNA polymerases are capable of recognizing promoters and initiating transcription but activator-dependent promoters are handicapped in one way or another. In some cases, the promoter elements are insufficient for RNA polymerase recruitment and the role of the activator is to correct this insufficiency, whilst in others the promoter DNA must be reshaped, and the role of the activator is to induce the necessary conformational changes. The exception is holo RNA polymerase containing σ^{54}, which is not competent to perform factor-independent transcription initiation. We do not have yet a complete structure for any σ^{54} family member. It is clear, however, that it is different to the four-domain structure of most σ factors. The DNA sequence elements recognized by RNA polymerase containing σ^{54} are located at positions −12 and −24 and the key point is that the holo RNA polymerase can form tight binding complexes at target promoters but is unable to unwind the DNA duplex to facilitate transcript initiation.[58] A special class of transcription activators drives transcription initiation at σ^{54}-dependent promoters. These activators all contain a domain (the AAA domain) that can couple ATP hydrolysis to drive conformational changes.[59,60] This domain interacts directly with σ^{54} in the RNA polymerase holoenzyme bound at the target promoter

A. Class I Activation.

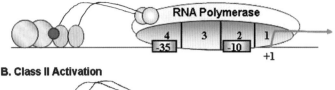

B. Class II Activation

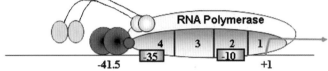

C. Activation by Conformational Change

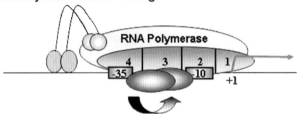

D. Activation by ATP hydrolysis-mediated Conformational Change

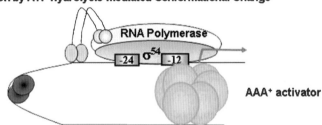

Figure 1.6 Simple activation mechanisms. The activator is depicted in yellow (A, D) or dark blue (B and C). (A) Class I activation. The activator binds upstream of the RNA polymerase binding site and makes specific protein–protein interactions (marked with a red sphere) with αCTD. This allows recruitment of RNA polymerase to the promoter. (B) Class II activation. The activator binds to a target near the −35 element and makes specific protein–protein interactions with σ^{70} domain 4 (red sphere). In most cases, to contact domain 4 of σ^{70}, the activator must bind, at or near to, position −41.5. (C) Activation by conformational change. The activator binds at, or near to, the promoter elements and reorientates domains 2 and 4 of σ^{70}. This allows RNA polymerase to form a productive complex at the promoter. (D) The case of σ^{54}-dependent promoter activation; σ^{54} shown in light cyan associated to RNA polymerase forms a stable closed complex at the promoter that is unable to isomerize to an open complex. The dimeric AAA+ activator protein binds to distal upstream targets on the opposite DNA face to the inactive closed complex and forms a hexamer in the presence of ATP. Upon stimuli and *via* DNA looping, the AAA+ hexamer establishes specific contacts with σ^{54}. ATP hydrolysis remodels the closed complex into a competent open complex. DNA looping is facilitated by binding of a bending protein (shown in dark blue) to the intervening region between the promoter and the distal AAA+ protein target sequences.

and, in an ATP-driven process, powers duplex unwinding to facilitate transcript initiation. This process has been referred to as the second paradigm for transcription activation, since the underlying mechanism of performing work on a preexisting holoenzyme–DNA complex is fundamentally different to mechanisms based on recruitment[61] (Figure 1.6D). Note that RNA polymerase containing σ^{54} still must be recruited to target promoters, and this recruitment can be driven by –12 and –24 elements alone, or in conjunction with an UP element, or with an activator that functions by recruitment.[62]

1.6 A Word about the Intracellular Environment

Although most test-tube studies on RNA polymerase mechanisms use purified protein and purified DNA fragments, this is clearly not a natural environment! To accommodate more than four million base pairs of duplex DNA into a bacterial cell the DNA has to be condensed. Part of the compaction is due to negative supercoiling, and the supercoils appear to be constrained in around 100 domains. Whilst not fixed, the domains lead to a folded structure, often referred to as the bacterial nucleoid. Although still not fully understood, it is clear that many different proteins are needed for the necessary DNA compaction and maintenance of a folded nucleoprotein structure[63] (Chapter 4). Amongst these are the MukA and MukB proteins (related to eukaryotic SMC proteins) and several small DNA-binding proteins such as Fis, IHF, HU and H-NS. These nucleoid-associated proteins are present in large quantities and, because of their relaxed binding specificities, they bind to many targets throughout the bacterial chromosome, inducing bending or wrapping. According to the conditions, H-NS can bring distal targets in DNA together, causing hairpin formation. The effects of nucleoid structure on RNA polymerase activity are still poorly understood but it is now clear that many of the nucleoid-associated proteins play important roles in modulating the activity of specific promoters. Indeed, over 50% of DNA targets for Fis, IHF and H-NS are located in inter-genic regions, suggesting a key role for these proteins in transcriptional regulation.[64]

A big complication in understanding the bacterial nucleoid is that DNA folding is altered with growth conditions. One important player in this is likely to be RNA polymerase itself. A typical *E. coli* cell will contain 3000 to 13 000 core RNA polymerase molecules depending on growth conditions.[40] During exponential growth, many of these will be actively transcribing, with the majority localized to just 90 of the 1800 transcription units. These highly transcribed units, which encode the machinery for protein synthesis and other equipment needed for rapid growth, are spread throughout the genome and are in the same orientation with respect to the replication origin. It is very likely that they play a big role in shaping the nucleoid in growing cells. Possibly they are responsible for setting the boundaries of supercoiled domains. When cells enter stationary phase, or stop growing due to some stress, expression from these highly transcribed regions ceases rapidly and the distribution of RNA polymerases across the chromosome becomes more even, and this may account for some of the growth phase-dependent changes in the shape of the bacterial nucleoid that have been observed.[65] RNA polymerase must have evolved to function in the context of a folded chromosome and it is not improbable that it is an integral part of its structure, whatever that structure may be! For example, mechanisms have evolved to prevent transcribing RNA polymerase from blocking DNA replication forks as they pass through highly transcribed parts of the bacterial chromosome. Similarly, very sophisticated contingency pathways exist to rescue RNA polymerase that is blocked in elongation by backtracking or due to "roadblocks" caused by DNA damage. Thus, backtracking can be resolved by the action of two auxiliary RNAP subunits, GreA and GreB,[66] whilst the Mfd protein removes a core enzyme blocked during elongation and at the same time ensures that recruiting the repair machinery abolishes the DNA damage.[67]

1.7 Coupling Transcription to Changes in a Complex Environment

Most *E. coli* promoters are complicated, and the 1975 view that most promoters would either be factor-independent or regulated by just one or two transcription factors is too simplistic. In hindsight, this is unsurprising since bacterial gene expression is exquisitely sensitive to the environment and promoters have developed to respond to multiple signals, each interpreted by a transcription factor. Although the current picture is far from complete, promoters that are controlled by two repressors are rare. In contrast, most promoters are controlled by activators acting in concert, either with or without a repressor[68] (Figure 1.7).

The *E. coli* genome encodes over 250 transcription regulatory proteins. These include activators and repressors but also σ factors and nucleoid-associated proteins that affect transcription. The activities of many of these proteins simply reflect their cellular concentration. This is set not just by their rate of synthesis, but also by their turnover and by the degree to which they are sequestered and hence unable to interact at target promoters or with RNA polymerase. The activities of many transcription factors may also be specifically modulated either by covalent modification or by interaction with ligands. Such factors often have a separate regulatory domain. For example, many members of the response regulator family of proteins consist of a receptor module that is phosphorylated and a DNA-binding effector module, responsible for transcription activation or repression.[69] Phosphorylation of the receptor module by a sensor-kinase triggers a conformational change that permits the effector module to up- or down-regulate transcription initiation at its target. Sensor-kinases usually span the cytoplasmic membrane and their activity is triggered by signals from outside the cell. In contrast, the activity of other transcription factors is modulated by intracellular ligands that report the availability of specific nutrients or the metabolic state of particular pathways.

A. Independent Contacts

i) Class I/ Class I

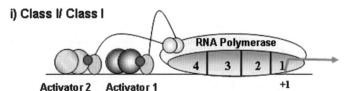

ii) Class I/ Class II

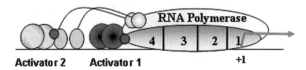

B. Co-operative binding

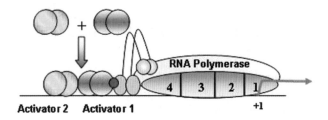

C. Repositioning

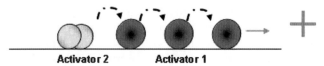

D. Anti-repression

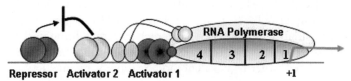

Figure 1.7 Mechanisms of promoter co-dependence on two activator proteins. (A) Independent contacts by both activators at the promoter. (i) Activators 1 and 2 make specific protein–protein contacts (indicated by red spheres) with αCTD. (ii) Activator 1 binds to a site that overlaps the target promoter −35 region and contacts σ^{70} domain 4 whereas activator 2 makes protein–protein contacts with αCTD. (B) Co-operative binding. The binding of activator 1 to the promoter is dependent on activator 2. (C) Activator 2 repositions activator 1 to a position where it can make productive contacts with RNA polymerase (which is omitted from the sketch). (D) Anti-repression. Binding of activator 2 is required to counteract the inhibitory effects of a repressor to allow activator 1 to function.

Most such factors have separate ligand-binding domains that assure ligand-dependent modulation of the activity of a DNA-binding domain. Transcription factors appear to have evolved by fusion of different domains, with each different domain carrying a different function. For instance, many of the activators that drive transcription initiation at σ^{54}-dependent promoters contain a regulatory domain that responds either to ligand binding or covalent modification, an AAA domain responsible for ATP-dependent interactions with σ^{54}, and a DNA binding domain, responsible for targeting activation at specific promoters.[60,68]

Some transcription factors interact at a single locus in the *E. coli* chromosome, whereas others, known as global regulators, interact at hundreds of loci. Promoters have evolved to become sensitive to different transcription factors and several different mechanisms have emerged to couple promoter activity to multiple factors. For example, at many promoters that are dependent on activators to recruit RNA polymerase, one contact, delivered by one transcription factor, is insufficient. Recruitment requires two or more transcription factors each to make independent contacts with RNA polymerase[68] (Figure 1.7A). The net result is that activation is co-dependent on the signals that trigger each transcription factor. The available evidence suggests that this is probably the principal way that bacterial promoters have evolved to be co-dependent on combinations of different signals. Since many transcription factors activate gene expression by recruitment, and since RNA polymerase carries many different target surfaces where such activators make contact, a vast number of combinations of factors are possible, each activating *via* complexes with different architectures. The variety is facilitated by the fact that the two C-terminal domains of the RNA polymerase α subunits can be targets for different activators, and because the flexible linker that connects the α subunit N- and C-terminal domains permits many architectures.

From the point of view of evolution, co-activation by activators making independent contacts with RNA polymerase is an attractive option since no direct contact between the activators is required. The alternative possible mechanism, whereby co-dependence results from the inability of an activator to bind at a promoter without the previous binding of a second factor, requires co-operative interactions between the different factors at the target promoter (Figure 1.7B). To date, very few examples of this type of co-dependence have been found in bacteria, presumably because this mechanism reduces the flexibility of the system by committing one factor to another. In a related mechanism that operates at some promoters, the binding of a second activator shifts the position of the first activator from a location where it is unable to activate transcription to a location where it can activate transcription.[70] Again, this mechanism requires interactions between two factors (Figure 1.7C).

At some promoters, co-dependence of expression on different activators is conferred by nucleoid-associated proteins that create a local structure that stops promoter activity. Here, the activators' role is to reshape the local environment either to allow RNA polymerase to access a factor-independent promoter or to permit activation by recruitment with second activator. To date,

H-NS, IHF and Fis have been found to be involved in silencing, with the precise architecture differing from one case to another.[63]

Once again, at bacterial promoters, variety reigns! In some cases, promoter activity can be scaled up or down with precision according to the bacterial environment. In other cases, promoters have evolved to give all or none responses. Sometimes epigenetic factors, such as methylation, intervene, thereby temporarily locking a promoter into alternative states. Finally, in many bacteria, clonal variation plays a major role in setting promoter activity levels. This is particularly important in pathogens where variation within a bacterial population is essential to combat cell defense mechanisms and to assure bacterial survival.

1.8 A Global View of the RNA Polymerase Economy

The bacterial cytoplasm is a crowded place, crammed full of proteins, metabolites, RNA and DNA. The DNA is folded into the nucleoid and most of the RNA polymerase is associated with this structure. In most bacterial cells, the number of RNA polymerase molecules is comparable to the number of genes. Thus, for *E. coli* K-12, there are up to 10 000 RNA polymerase molecules and 4500 genes organized in ~1800 transcription units. A snapshot of the cell would show that some RNA polymerase molecules are bound non-specifically across the chromosome and are not transcribing, whilst other molecules bind specifically to promoters and enter the transcription cycle. The relative amounts of these two RNA polymerase populations depend on cell growth rate. Remarkably, in rapidly growing cells, nearly 80% of the transcribing RNA polymerase molecules are dedicated to the expression of ~80 transcription units concerned with maintaining the cell's biosynthetic capacity. This raises two key issues. First, how is the bulk of the transcribing RNA polymerase channeled to the 80 highly transcribed regions? Second, how is the distribution of the rest of the RNA polymerase between the thousands of other genes controlled, and how does the cell manage to induce expression of specific genes at the appropriate time in response to environmental signals? Concerning the first question, the key is that the promoters of the ~80 transcription units needed for the cell's biosynthetic capacity are indeed hyper-efficient at recruiting RNA polymerase to initiate transcription (as suggested in the 1980s) and at progressing RNA polymerase from the initiation complex into the elongation mode. Studies from Rick Gourse and colleagues have shown that efficient recruitment is due to the UP elements at these promoters and that efficient promoter escape is due to the instability of the open complex, due to DNA sequence determinants around the transcription start.[65] An important feature of these promoters is that they capture relatively more RNA polymerase as the bacterial growth rate increases and that they can be shut down rapidly if protein synthesis slows down (*e.g.*, due to amino acid limitation). This regulation is mediated by the intracellular levels of two small ligands, ATP and ppGpp, that interact at or near the RNA polymerase active site. An important consequence of the unstable initiation complexes at these promoters is that transcription is especially sensitive to up-regulation by increased ATP levels, associated with rapid growth, or down-regulation by increased ppGpp levels resulting from reduced protein synthesis. More recent experiments by Gourse and colleagues have demonstrated an essential role for DksA protein in ensuring responses to fluctuations in ATP and ppGpp levels.[71] DksA is a small zinc finger protein that binds to RNA polymerase with the finger penetrating the RNA polymerase secondary channel (Chapters 3 and 8): the tip of the finger interacts somehow with the active site and "tunes" RNA polymerase to respond to ATP and to ppGpp.

It has often been noted that the distribution of RNA polymerase between different promoters in rapidly growing *E. coli* is like the distribution of wealth between the people in western democracies! Although some of the promoter-specific mechanisms fixing this unequal distribution are now understood, an intriguing possibility is that these are not the whole story, and that, somehow, the overall structure of the folded bacterial chromosome plays a key role. For example, the chromosome could be folded in such a way that the highly transcribed promoters are positioned in favored locations. Alternatively, the RNA polymerase may be organized in some way, rather than randomly distributed. Whatever the mechanisms, the consequence is that very little RNA polymerase is available for transcription of the remaining 4000 genes and its distribution has to be parsimonious and prudent. Thus, the recruitment of RNA polymerase to the majority of *E. coli* genes is a rare event, and vast tracts of the *E. coli* chromosome rarely see a transcribing RNA polymerase molecule. Our current understanding could be summarized in the following manner: the distribution of RNA polymerases is fixed at several levels, the first being the base sequence of individual promoters and the effects of the local nucleoid structure (*e.g.*, due to nucleoid-associated proteins). The second level is moderated by σ and transcription factors, which assure the coupling of gene expression to environmental conditions, principally responses to nutrient availability or different stresses.[18,19] Thus, activators and repressors interact at different promoters, as described above, in response to different triggers. In some conditions, alternative σ factors "capture" more core RNA polymerase and thus "hi-jack" a sub-population of RNA polymerase molecules that become dedicated to the targets for that specific σ factor.

The most important adaptive change in the RNA polymerase economy takes place when bacteria stop growing and enter stationary phase. The biggest change is that most of the RNA polymerase is no longer channeled to a small number of favored transcription units, probably due to changes in ATP and ppGpp levels. The consequence is that thousands of RNA polymerase molecules become available and, thus, in non-growing cells, RNA polymerase is distributed more evenly between the promoters of the 1800 transcription units. The work of Jay Gralla and colleagues has shown that many of these RNA polymerase molecules are "poised" and unable to escape the promoters, likely due to the supercoiling and ligand regime in the non-growing bacterium[72] but others proceed to make transcripts and are regulated by activators, repressors and σ factors, just as in growing cells.[73] In non-growing *E. coli*, levels of the alternative σ^{38} factor increase and this is essential for the expression of many gene products needed for survival and resistance to different stresses.[74,75] One of the consequences of the massive redistribution of RNA polymerase in non-growing cells is a change in the morphology of the nucleoid and this is accompanied by changes in the expression of different nucleoid-associated proteins.[18] It is not yet clear how these changes affect transcription though we know that the proportion of non-transcribing RNA polymerase molecules increases. Some of these redundant molecules are parked non-specifically around the nucleoid; some are stored in paused complexes whilst others are sequestered by a specific 6S RNA molecule, which mimics the unwound region of a typical *E. coli* promoter.[76]

1.9 The Real World, Emergency Procedures and RNA Polymerase

History shows that Jacques Monod's claim that what is true for *E. coli* is true for elephants should not be taken literally. Of course the molecules are the same, and the basic enzymology transcends species, but the function-driven designs for the two species remind us of chalk and cheese. *E. coli* gene expression has to be fast and flexible and it has to be adaptable, because it is evolving as we perturb its surroundings. Thus, Kevin Struhl and colleagues have argued that bacterial genomes are open and accessible to the transcription machinery whilst eukaryotic genomes are largely dormant.[77] Bacteria mostly exploit promoter-specific factors to modulate the code written in promoter elements to assure balanced growth. In life and death situations, the cell eschews regulation by finely tuning promoter activity and chooses direct intervention at the level of the RNA polymerase. Hence, it is no surprise that many bacteriophages subvert bacterial growth by direct action on host RNA polymerase. The two principal mechanisms are *via* the synthesis of an effector subunit or *via* covalent modification. This is exemplified by bacteriophage T4, which exploits both mechanisms to hi-jack growing *E. coli* cells for its own purpose. Amazingly, though predictably, both mechanisms target parts of RNA polymerase that are directly involved in host promoter recognition. Hence, early in infection, bacteriophage T4 ADP-ribosylates Arg-265 in the C-terminal domain of the RNA polymerase α subunit. Since Arg-265 is the crucial residue in recognition of UP elements, this modification prevents the channeling of RNA polymerase to the promoters controlling the key host products needed for protein synthesis, thereby subverting host growth.[78] A second modification is the early synthesis of the T4 AsiA protein, which binds tightly to Domain 4 of the RNA polymerase σ^{70} subunit. This binding remodels the structure of Domain 4, preventing its interaction with host promoter −35 elements and facilitating interaction with the T4-encoded MotA transcription activator that is specific for activation of T4 middle order genes.[79] Thus, with two lethal blows, the function of the host RNA polymerase is subverted to the benefit of the phage! As more systems are explored, more examples of RNA polymerase being modified by bound effectors or by covalent modification are sure to be found.

1.10 This is just the Beginning!

Our aim in this chapter has been to give the reader a bird's eye view of the field. We wanted to cover some history, introduce the main ideas and also explain the context in which RNA polymerase functions. The following chapters dig deep into different aspects of the topic, and a fascinating read is guaranteed. The subject is barely 50 years old and yet, because of the unique combinations of genetics and genomics with biochemistry and biophysics, we understand RNA polymerase in unprecedented detail. And yet, the fascination is that big fundamental questions still remain, and the field is wide open to the application of emerging technologies and new ideas. New questions are now arising, which are barely touched in this overview or in the following chapters. Let's mention some of them, together with associated technologies, that appear promising:

- How do RNA polymerases, activators and inhibitors reach their target in a bacterial cell? This has been a hotly debated issue for decades. Direct visualization of complexes formed transiently between those proteins and sites of intermediate affinity are now possible, with the use of chip-on-chip methodology.[80] In parallel, several mathematical models refining the earlier theories put forward by von Hippel, Halford and others are under current development.[81,82]
- Where are transcribing RNA polymerases located in a bacterial cell?[83]
- When and how is the resulting message confined in the bacterial space? Here again powerful imaging techniques, relying on the detection of specific mRNA, are being currently developed.[84]
- Not all cells are alike in a given bacterial population. For the expression of specific genes, this variability has been carefully characterized.[85] And, here again, a new type of confrontation is taking place between *in vitro* and *in vivo* results: what kind of models could account for the observed variability in protein content per cell, taking into account the most likely local fluctuations in transcription and translation processes?[86]

So that, despite our great ignorance about the fine details of the control of transcription, *E. coli* becomes the crucible where synthetic biochemists can now probe the efficiency of grafted mini-systems of mutual control, as exemplified by the efficient engineering of elementary molecular clocks.[87]

But, to come back to our main purpose, promoter recognition, this is a research area with a cast of thousands and we make no excuses for deliberately restricting the citations to a few key reviews and landmark papers. When the molecular biology of bacterial RNA polymerases began to be studied, the main motivation was to use it as a general paradigm. Over 50 years, the wheel has turned, and cellular biologists with an interest in eukaryotic transcription no longer have a pressing need for this specific paradigm. However, the need to understand, control and exploit the microbial world has never been more pressing and there can be few more worthy problems for study than their RNA polymerases and associated factors. Furthermore, as pointed out by Peter Geiduschek, for many aspects of transcription, research with the bacterial enzymes continues to point out and light the path. Many chapters of this book will make that plain.

References

1. F. Jacob and J. Monod, *Cold Spring Harbor Symp. Quant. Biol.*, 1961, **26**, 193.
2. R. R. Burgess, A. A. Travers, J. J. Dunn and E. K. Bautz, *Nature*, 1969, **221**, 43–6.
3. D. C. Hinkle and M. J. Chamberlin, *J. Mol. Biol.*, 1972, **70**, 187–95.
4. G. Walter, W. Zillig, P. Palm and E. Fuchs, *Eur. J. Biochem.*, 1967, **3**, 194–201.
5. M. J. Chamberlin, *Annu. Rev. Biochem.*, 1974, **43**, 721–75.
6. P. L. deHaseth and J. D. Helmann, *Mol. Microbiol.*, 1995, **16**, 817–24.
7. P. L. deHaseth, M. L. Zupancic and M. T. Record Jr, *J. Bacteriol.*, 1998, **180**, 3019–25.
8. M. T. Record Jr, W. S. Reznikoff, M.L. Craig, K. Mcquade, P.J. Schlax, in *Escherichia coli and Salmonella: Cellular and Molecular Biology*, ed. F. C. Neidhardt, R. Curtiss III, J.L. Ingraham, E.C.C. Lin, K.B. Low, K. B. Magasani, W.S. Reznikoff, M. Schaechter, H.E. Umbarger, (ASM, Washington, DC), 1996, pp. 792–820.
9. D. C. Hinkle, W. F. Mangel and M. J. Chamberlin, *J. Mol. Biol.*, 1972, **70**, 209–20.
10. D. J. Jin and C. A. Gross, *J. Mol. Biol.*, 1988, **202**, 45–58.
11. A. Ishihama, *Adv. Biophys.*, 1981, **14**, 1–35.
12. K. Igarashi and A. Ishihama, *Cell*, 1991, **65**, 1015–22.
13. G. Zhang, E. A. Campbell, L. Minakhin, C. Richter, K. Severinov and S. A. Darst, *Cell*, 1999, **98**, 811–24.
14. K. S. Murakami, S. Masuda and S. A. Darst, *Science*, 2002, **296**, 1280–4.
15. D. G. Vassylyev, S. Sekine, O. Laptenko, J. Lee, M. N. Vassylyeva, S. Borukhov and S. Yokoyama, *Nature*, 2002, **417**, 712–9.
16. P. Cramer, D. A. Bushnell, J. Fu, A. L. Gnatt, B. Maier-Davis, N. E. Thompson, R. R. Burgess, A. M. Edwards, P. R. David and R. D. Kornberg, *Science*, 2000, **288**, 640–9.
17. R. Losick and J. Pero, *Cell*, 1981, **25**, 582–4.
18. A. Ishihama, *Annu. Rev. Microbiol.*, 2000, **54**, 499–518.
19. T. M. Gruber and C. A. Gross, *Annu. Rev. Microbiol.*, 2003, **57**, 441–66.
20. J. D. Helmann and M. J. Chamberlin, *Annu. Rev. Biochem.*, 1988, **57** 839–72.

21. M. Lonetto, M. Gribskov and C. A. Gross, *J. Bacteriol.*, 1992, **174**, 3843–9.
22. M. J. Merrick, *Mol. Microbiol.*, 1993, **10**, 903–9.
23. K. S. Murakami, S. Masuda, E. A. Campbell, O. Muzzin and S. A. Darst, *Science*, 2002, **296**, 1285–90.
24. W. R. McClure, *Proc. Natl. Acad. Sci. USA*, 1980, **77**, 5634–8.
25. W. S. Reznikoff and J. N. Abelson, in *The Operon*, ed. J. H. Miller, W. S. Reznikoff, New-York, Cold Spring Harbor Laboratory, 1978, pp. 221–243.
26. J. Scaife and J. R. Beckwith, *Cold Spring Harbor Symp. Quant. Biol.*, 1966, **31**, 403–8.
27. W. Gilbert, N. Maizels and A. Maxam, *Cold Spring Harbor Symp. Quant. Biol.*, 1974, **38**, 845–55.
28. D. Pribnow, *J. Mol. Biol.*, 1975, **99**, 419–43.
29. W. R. McClure, *Annu. Rev. Biochem.*, 1985, **54**, 171–204.
30. D. A. Siegele, J. C. Hu, W. A. Walter and C. A. Gross, *J. Mol. Biol.*, 1989, **206**, 591–603.
31. T. Gardella, H. Moyle and M. M. Susskind, *J. Mol. Biol.*, 1989, **206** 579–90.
32. D. Daniels, P. Zuber and R. Losick, *Proc. Natl. Acad. Sci. USA*, 1990, **87**, 8075–9.
33. E. A. Campbell, O. Muzzin, M. Chlenov, J. L. Sun, C. A. Olson, O. Weinman, M. L. Trester-Zedlitz and S. A. Darst, *Mol. Cell*, 2002, **9**, 527–39.
34. K. A. Barne, J. A. Bown, S. J. Busby and S. D. Minchin, *EMBO. J.*, 1997, **16**, 4034–40.
35. W. Ross, K. K. Gosink, J. Salomon, K. Igarashi, C. Zou, A. Ishihama, K. Severinov and R. L. Gourse, *Science*, 1993, **262**, 1407–13.
36. T. Ellinger, D. Behnke, H. Bujard and J. D. Gralla, *J. Mol. Biol.*, 1994, **239**, 455–65.
37. D. F. Browning and S. J. Busby, *Nat. Rev. Microbiol.*, 2004, **2**, 57–65.
38. J. W. Roberts, *Nature*, 1969, **224**, 1168–74.
39. R. A. Mooney, S. A. Darst and R. Landick, *Mol. Cell.*, 2005, **20**, 335–45.
40. I. L. Grigorova, N. J. Phleger, V. K. Mutalik and C. A. Gross, *Proc. Natl. Acad. Sci. USA*, 2006, **103**, 5332–7.
41. J. Greenblatt, J. Li, S. Adhya, D. I. Friedman, L. S. Baron, B. Redfield, H. F. Kung and H. Weissbach, *Proc. Natl. Acad. Sci. USA*, 1980, **77**, 1991–4.
42. A. Schmitz and D. J. Galas, *Nucleic Acids Res.*, 1979, **6**, 111–37.
43. S. Sasse-Dwight and J. D. Gralla, *J. Biol. Chem.*, 1989, **264**, 8074–81.
44. P. Schickor, W. Metzger, W. Werel, H. Lederer and H. Heumann, *EMBO. J.*, 1990, **9**, 2215–20.
45. N. Naryshkin, A. Revyakin, Y. Kim, V. Mekler and R. H. Ebright, *Cell*, 2000, **101**, 601–11.
46. V. Mekler, E. Kortkhonjia, J. Mukhopadhyay, J. Knight, A. Revyakin, A. N. Kapanidis, W. Niu, Y. W. Ebright, R. Levy and R. H. Ebright, *Cell*, 2002, **108**, 599–614.

47. S. C. Gill, T. D. Yager and P. H. von Hippel, *Biophys. Chem.*, 1990, **37**, 239–50.
48. A. J. Carpousis and J. D. Gralla, *Biochemistry*, 1980, **19**, 3245–53.
49. J. Majors, *Proc. Natl. Acad. Sci. USA*, 1975, **72**, 4394–8.
50. A. D. Johnson, A. R. Poteete, G. Lauer, R. T. Sauer, G. K. Ackers and M. Ptashne, *Nature*, 1981, **294**, 217–23.
51. W. Gilbert and B. Muller-Hill, *Proc. Natl. Acad. Sci. USA*, 1966, **56** 1891–1898.
52. S. Oehler, E. R. Eismann, H. Kramer and B. Muller-Hill, *EMBO. J.*, 1990, **9**, 973–9.
53. R. Schleif, *Annu. Rev. Biochem.*, 1992, **61**, 199–223.
54. A. Hochschild and S. L. Dove, *Cell*, 1998, **92**, 597–600.
55. M. Ptashne and A. Gann, *Nature*, 1997, **386**, 569–77.
56. S. Busby and R. H. Ebright, *J. Mol. Biol.*, 1999, **293**, 199–213.
57. A. Z. Ansari, M. L. Chael and T. V. O'Halloran, *Nature*, 1992, **355**, 87–9.
58. M. Buck, M. T. Gallegos, D. J. Studholme, Y. Guo and J. D. Gralla, *J. Bacteriol.*, 2000, **182**, 4129–36.
59. D. J. Studholme and R. Dixon, *J. Bacteriol.*, 2003, **185**, 1757–67.
60. M. Rappas, D. Bose and X. Zhang, *Curr. Opin. Struct. Biol.*, 2007, **17** 110–6.
61. S. R. Wigneshweraraj, P. C. Burrows, P. Bordes, J. Schumacher, M. Rappas, R. D. Finn, W. V. Cannon, X. Zhang and M. Buck, *Prog. Nucleic. Acid. Res. Mol. Biol.*, 2005, **79**, 339–69.
62. G. Bertoni, N. Fujita, A. Ishihama and V. de Lorenzo, *EMBO. J.*, 1998, **17**, 5120–8.
63. C. J. Dorman and P. Deighan, *Curr. Opin. Genet. Dev.*, 2003, **13**, 179–84.
64. D. C. Grainger, D. Hurd, M. D. Goldberg and S. J. Busby, *Nucleic Acids Res.*, 2006, **34**, 4642–52.
65. B. J. Paul, W. Ross, T. Gaal and R. L. Gourse, *Annu. Rev. Genet.*, 2004, **38**, 749–70.
66. A. Perederina, V. Svetlov, M. N. Vassylyeva, T. H. Tahirov, S. Yokoyama, I. Artsimovitch and D. G. Vassylyev, *Cell*, 2004, **118**, 297–309.
67. A. M. Deaconescu, N. Savery and S. A. Darst, *Curr. Opin. Struct. Biol.*, 2007, **17**, 96–102.
68. A. Barnard, A. Wolfe and S. Busby, *Curr. Opin. Microbiol.*, 2004, **7**, 102–8.
69. J. B. Stock, A. J. Ninfa and A. M. Stock, *Microbiol. Rev.*, 1989, **53**, 450–90.
70. E. Richet, D. Vidal-Ingigliardi and O. Raibaud, *Cell*, 1991, **66**, 1185–95.
71. B. J. Paul, M. M. Barker, W. Ross, D. A. Schneider, C. Webb, J. W. Foster and R. L. Gourse, *Cell*, 2004, **118**, 311–22.
72. S. J. Lee and J. D. Gralla, *Mol. Cell*, 2004, **14**, 153–62.
73. R. Hengge-Aronis, *Curr. Opin. Microbiol.*, 1999, **2**, 148–52.
74. H. Weber, T. Polen, J. Heuveling, V. F. Wendisch and R. Hengge, *J. Bacteriol.*, 2005, **187**, 1591–603.
75. E. Klauck, A. Typas and R. Hengge, *Sci. Prog.*, 2007, **90**, 103–27.
76. G. Storz, J. A. Opdyke and K. M. Wassarman, *Cold Spring Harbor Symp. Quant. Biol.*, 2006, **71**, 269–73.

77. J. T. Wade, N. B. Reppas, G. M. Church and K. Struhl, *Genes. Dev.*, 2005, **19**, 2619–30.
78. S. Nechaev and K. Severinov, *Annu. Rev. Microbiol.*, 2003, **57**, 301–22.
79. L. J. Lambert, Y. Wei, V. Schirf, B. Demeler and M. H. Werner, *EMBO. J.*, 2004, **23**, 2952–62.
80. J. T. Wade, K. Struhl, S. J. Busby and D. C. Grainger, *Mol. Microbiol.*, 2007, **65**, 21–6.
81. O. G. Berg and P. H. von Hippel, *Annu. Rev. Biophys. Biophys. Chem.*, 1985, **14**, 131–60.
82. S. E. Halford and J. F. Marko, *Nucleic Acids Res.*, 2004, **32**, 3040–52.
83. D. J. Jin and J. E. Cabrera, *J. Struct. Biol.*, 2006, **156**, 284–91.
84. I. Golding and E. C. Cox, *Proc. Natl. Acad. Sci. USA*, 2004, **101**, 11310–5.
85. M. B. Elowitz, A. J. Levine, E. D. Siggia and P. S. Swain, *Science*, 2002, **297**, 1183–6.
86. H. H. McAdams and A. Arkin, *Trends Genet.*, 1999, **15**, 65–9.
87. C. C. Guet, M. B. Elowitz, W. Hsing and S. Leibler, *Science*, 2002, **296**, 1466–70.

CHAPTER 2
Opening the DNA at the Promoter; The Energetic Challenge

BIANCA SCLAVI

LBPA, UMR 8113 du CNRS, ENS Cachan, 61 Avenue du Président Wilson, 94235, Cachan, France

2.1 Introduction

The holoenzyme structure of *E. coli* RNA polymerase, unlike its archeal and eukaryotic counterparts or any other polynucleotide synthesizing enzymes, integrates *multiple functions in a single machine*. These include the ability to (i) recognize and bind to the specific sequence of the promoter in the absence of accessory factors, (ii) to separate the DNA double strand at the promoter during initiation of transcription, (iii) to maintain separated the DNA double strand during elongation and (iv) to perform the faithful and processive synthesis of RNA during elongation.

As discussed in Chapter 1, the extent of gene expression *in vivo* is highly regulated at the transcriptional level, typically through the use of accessory factors such as transcriptional activators and/or repressors. However, the ability of the holoenzyme to carry out its function independently of such factors allows for a basal level of expression of certain promoters. It also allows for highly efficient expression of essential genes that can be rapidly modulated by changes in global regulators such as nucleoid proteins, cofactors, metabolites or the level of supercoiling.

Characterization of the basic processes carried out by the holoenzyme is necessary if one is to determine the mechanism by which different regulators, both global and promoter specific, affect the level of transcription.

The *pathway* from initial promoter recognition to transcription initiation is populated by several structural intermediates. The kinetic characterization and physical properties of the intermediates were originally determined from the studies of William McClure, Tom Record and their colleagues in the early 1980s. The rate constants of the formation of either an RNAP–promoter open (RPo) complex resistant in the presence of a competitor such as heparin or of a transcriptionally active complex have been measured by steady state kinetics approaches. The rate of formation of these complexes follows Michaelis–Menten kinetics. Thus, the simplest interpretation of this behavior requires a fast equilibrium for initial binding between free and bound enzyme (characterized by the association constant K_B) followed by a rate-determining, isomerization step (k_f), which is generally much faster than the reverse reaction. Further work identified additional intermediates based on their dependencies on temperature and/or salt concentration.[1-4]

A linear reaction pathway representing a common sequence of events that ultimately lead to formation of the elongating complex has since been used for the description and experimental characterization of these intermediates at different promoters:

$$R + P \Leftrightarrow RP_C \Leftrightarrow RP_I \Leftrightarrow RP_O \Leftrightarrow RP_{init} \rightarrow RP_{el}$$

where RP_C represents a heparin-sensitive closed complex, RP_I a heparin-resistant intermediate complex and RP_O the open complex ready to initiate transcription (RP_{init}) and escape into elongation (RP_{el}). The phenomenological constants K_B and k_f that characterize the path from R+P to RPo are thus linear combinations of the microscopic rate constants for the different isomerization steps. Differences in these overall constants have been measured at different promoters in the presence or absence of various modulators. These have been interpreted as being due to changes in the individual, microscopic, rate constants and thus of the relative magnitude of the energetic barriers and the abundance of these species. Different experimental conditions or an intrinsic property of a given promoter sequence might then result in a change in the position of the rate-limiting step within the same linear pathway (Box 2.1). These results were described in detail in a review by deHaseth *et al.*[5]

In this chapter, I concentrate mainly on work since carried out on the characterization of the structures and conformational changes taking place in the pathway to open complex formation by sigma 70 RNA polymerase (RNAP) (for recent reviews on the other sigma subunits see refs 6–8). Recent results have underlined the structural differences in the intermediates formed at different promoters and have put into question the ubiquity of a linear pathway common to all promoters. The multiplicity of potential protein–protein and nucleic acid–protein interactions during promoter binding and open complex formation allows for multiple possible strategies for promoter opening, likely depending on the level of regulation required.

> **Box 2.1 Methods used for the Identification and Characterization of the Intermediates in the Pathway**
>
> **Kinetics of Formation of Intermediates along the Pathway to the Open Complex**
>
> The rate of formation of each postulated species in a linear pathway can be assessed from the kinetic study of RNAP binding to promoter DNA. In general, the process is monitored *via* the properties characteristic of the RNAP–DNA complex, such as stability in the presence of a competitor (*e.g.*, heparin) in filter binding assays or the synthesis of short, abortive, RNAs by a transcriptionally active open complex.[1,125,126] Alternatively, a fluorescence signal has been used to follow the formation of a specific complex, either from modified bases in the DNA or from the intrinsic fluorescence of the protein.[58,117] These approaches have been used to quantitatively measure the dependence of the forward and reverse rates on variables such as temperature, salt concentration or promoter sequence. From these results one can deduce the presence of one or more intermediates in the pathway and the nature of the interactions specific for their formation.[1–4]
>
> **Thermodynamic Properties of the Intermediates**
>
> Quantitative analysis of the kinetics of binding as a function of temperature, salt or concentration of small solutes can provide information on the nature of conformational changes taking place in the protein and the DNA (ref. 30 and references therein). The effect of these parameters on the equilibrium and kinetic constants is used to measure the total changes in entropy, enthalpy, heat capacity and counterion displacement. These, in turn, can be interpreted in terms of burial of polar and nonpolar surfaces and the resulting folding events. Analysis is aided in part by the available structural information obtained from crystallography. The challenge is to infer the most probable conformational changes and their dynamics from the sole knowledge of an equilibrium structure. The approach is rendered difficult at times by the occurrence of coupled conformational changes with opposite dependencies on external factors, causing the magnitude of the effect to be smaller than expected. This can result in discrepancies between the predictions obtained from the analysis of structural data and the values obtained from thermodynamic and kinetic studies. These differences may also be due to structural changes

that cannot be predicted directly from the crystal structure, such as the folding of a disordered domain.[30]

Structural Signatures of the Intermediates

To obtain structural information about the intermediates in the pathway several complementary footprinting approaches have been used. Footprinting is based on the probing of the accessibility of DNA bases to a given reagent, or alternatively in the formation of a covalent base-amino acid crosslink induced by a chemical reagent or UV irradiation. This permits mapping of those bases that are involved in the interaction with the protein, or those sites on the DNA that have been deformed by the formation of the complex. Quantitative analysis of these signals provides affinity and rate constants for the formation of specific interactions.[127] This is carried out either by interpreting the footprints from the relative abundance of each species determined from the known kinetic constants[59,65,66,76,128] or in a time-resolved fashion, by measuring the evolution of each signal as the complex is being formed, allowing for the simultaneous structural and kinetic characterization of the intermediates in the pathway.[49,63,91,114]

In some cases footprints of the intermediates are obtained by trapping the complexes at low temperature. In this case, however, a careful thermodynamic analysis is required to ensure that the complex thus obtained does not reflect a mixture of species or an off-pathway artifact stabilized by the modified conditions (see main text).[53,65,88,90,91]

Single-molecule Studies

The approaches described above measure the properties of a bulk solution sample. The results obtained reflect an average of the population at a given time. In contrast, single-molecule studies can characterize the different steps in the pathway for an individual molecule, and could thus directly determine if a given intermediate is on a linear or branched pathway, or an off-path structure that may be kinetically favored. These studies rely either on the measurement of fluorescence from a single molecule, measurements of the force applied by the protein, or large conformational changes in DNA. The use of these techniques for the study of the intermediates in the pathway to open complex formation is still at its early stages, while it has been already successfully applied to the study of the early steps of promoter search[40] and the later steps of transcription initiation, escape, elongation and pausing (Chapters 5, 6 and 9).

One common property of this process at different promoters is that the enzyme must be able to specifically bind to the transcription initiation site and melt the DNA without forming an overly stable complex that would be unable to escape into elongation within biologically relevant timescales. Recent work has highlighted the negative effect that some of the protein–DNA interactions, formed at different stages of transcription initiation, may have on the stability of the intermediates or the final complex. These negative effects are sometimes attributable to regions of the holoenzyme that are not fully conserved from one species to the next, pointing to a possible source of variability in these mechanisms among different promoters or among the different sigma subunit families, resulting in different sensibilities to regulators depending on the lifestyle demands of the organism.

The process of DNA melting by RNAP is in itself a subject of intense study. The current model proposes a sequential nucleation–propagation pathway. Briefly, interactions of the enzyme with the promoter DNA result in the flipping out of the adenine at position –11, thus nucleating the opening. This is followed by a destabilization of the downstream double helix and subsequent trapping of the single stranded DNA by the RNAP.[9,10] As described below, several protein and DNA conformational changes induced along this pathway may facilitate different steps in this process.

Recent structural data on the RNA polymerase holoenzyme and its complex with DNA can now be used for the interpretation of kinetic data and as a framework for models of the three-dimensional structures of the short-lived intermediates and the conformational changes taking place along the pathway. (See also the Structural Atlas pages 1 to 10.)

2.2 Structural Characterization

2.2.1 Crystal Structure of the Holoenzyme

The resolution of the crystal structures for two different holoenzymes of thermophilic woRNA polymerases by Seth Darst, Dimitry Vassylyev and co-workers and of the structure of one of these in the presence of a partially single-stranded promoter DNA fragment by the group of Seth Darst has constituted a significant contribution to the understanding of the initial steps of the transcription process.[11–13]

The two structures of the holoenzyme allowed for a precise description of the *core–sigma interactions* in relation to other functionally important domains of the core. The different sigma domains connected by flexible linkers sit on the surface of the core, spanning a large area: the downstream DNA binding region, the active site channel and the RNA exit channel (Figure 2.1). The interaction of sigma with core results in a specific orientation of the sigma 4.2 and 2.4 domains necessary for specific simultaneous recognition of, respectively, the –35 and –10 sequences. These observations are in agreement with previous biochemical data showing a large rearrangement in sigma70 upon formation of the holoenzyme.[14] More specifically, these structures confirmed a key role for the beta flap in the orientation of sigma region 4[15] and for the β′

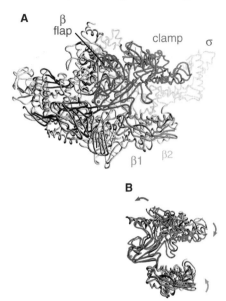

Figure 2.1 (A) A bottom view of *E. coli* RNAP holoenzyme (as in A.1b) where the 5 mobile modules (defined in the legend of A.1) are colored. (B) The 5 modules that move with respect to the core domain are shown superimposed on the core RNAP crystal structure, and their displacements indicated by arrows. (Reproduced from ref. 11 with permission from the American Association for the Advancement of Sciences.)

coiled-coil in the specific interaction with sigma regions 1.2 and 2.2 on the floor of the DNA-binding channel (See Figure 2.1 and Atlas Figure A.1).[16]

The sigma–core interface primarily consists of several rather weak interactions, underlining the flexibility in the orientation of the sigma domains and providing a possible explanation for the temporary existence of this otherwise quite stable complex ($K_d \sim 10^{-9}$ M).[13] From these structures a mechanism for promoter escape was suggested by which the sigma subunit is gradually dissociated from core by a steric clash of the sigma 3.2 linker with the elongating RNA strand as it is extended into the RNA exit channel. In support of this model, it has been shown that the amount of abortive products could be correlated with the modified strength of these interactions in mutants of RNA polymerases.[11,17,18]

2.2.2 Crystal Structure of the Holoenzyme with Fork-junction DNA

The DNA fragment in the *crystal structure of the binary complex* is a "fork junction" fragment that includes nucleotides from −41 to −7 on the non-template (nt) strand and from −41 to −12 on the template (t) strand and contains the consensus −35, −10 and the extended −10 sequences (Figure 2.2). This DNA fragment had been shown to form a stable, heparin resistant

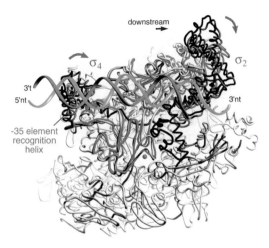

Figure 2.2 DNA–RNAP complex showing the conformational changes taking place upon DNA binding and the bends in the DNA (Reproduced with permission from the American Association for the advancement of Sciences[11]).

complex with RNA polymerase and has been used for biochemical studies of the interactions between the sigma subunit and the –10 sequence.[19–21] In good agreement with previous biochemical data, this structure shows how the sigma region 3.0 can interact with the extended –10 promoter element, keeping region 2.4 in close proximity to –12 and maintaining a conserved tryptophan residue in region 2.3 correctly aligned for its involvement in the nucleation of DNA opening (Figure 2.2). In the same sigma subunit a series of aromatic amino acid residues are responsible for stabilizing the open complex by binding specifically to bases of the single stranded nontemplate strand.[22,23]

In the crystal structure of the binary complex the DNA structure is modified from a canonical B form by its interaction with the protein, resulting in two bends in the double helix: in the spacer region at –25 and just upstream of the extended –10 at –16 (Figure 2.2). An additional bend in the –35 region could result from its interaction with sigma region 4 (ref. 17) but is not observed in this structure, presumably because of the presence of the extended –10 sequence and crystal packing effects. These bends agree with results obtained from DNaseI footprinting experiments (a technique briefly presented in Box 2.1) showing hypersensitivity at these same sites.[12] An additional DNaseI hypersensitive site is also often found at around –45, caused by the wrapping of the upstream DNA by binding of the alpha C-terminal domains (CTDs)[24,25] (the two alpha CTDs are not observed in the crystal structure as they are linked to the enzyme by flexible tethers). The presence of a hypersensitive site in the spacer has often been observed on promoters with a spacer length longer than the consensus 17 base pairs.[12] It has been proposed that this distortion of the DNA is caused by an interaction with the beta′ Zn^{2+} finger and is necessary for the –35 and –10 regions to be positioned on the same face of the double helix. As suggested from the crystal structure and from specific mutations in this region of the polymerase[92] the Zn^{2+} finger is also involved in promoter binding and melting at extended –10 promoters (where a –35 sequence with poor homology to consensus is nevertheless tolerated).

In the presence of DNA, a change in conformation of the holoenzyme is observed resulting in further spatial rearrangement of the sigma domains 4.2 and 2.4. The *flexibility* of the RNAP holoenzyme permits the formation of a stable nucleoprotein complex by adjusting to different spacer lengths and DNA topologies and may play an important role at different steps during promoter binding. In fact, the distance between the beta and beta′ jaws differs when comparing the structures either of the two holoenzymes or of the core and holoenzyme. In the latter case the interjaw distance is 27 Å in the core enzyme and 15 Å in the holoenzyme.[11,13,26,27] Thus, in the presence of downstream double-stranded DNA the jaw of the holoenzyme needs to open to allow the double helix to enter the channel, whereas the subsequent closing of the jaw stabilizes the complex during RNA synthesis.[28,29] Finally, recent evidence points to protein folding in *E. coli* RNA polymerase upon complex formation as stabilizing the final open complex.[30] Additional protein conformational changes take place during RNA synthesis, mediated in part by the bridge helix and the trigger loop (see Chapters 7 and 8).

2.2.3 Structural Model of the Open Complex

The structures of the holoenzyme with fork junction DNA and of sigma region 4 with the –35 fragment were used as the frame from which to build a model of an open complex.[12,17,31] Previous crosslinking and fluorescence resonance energy transfer studies performed on full-length promoter DNA are consistent with the placement of single stranded DNA in the active site proposed by this model and were useful to establish the contacts with the downstream double stranded helix (Figure 2.3B).[29,32–34]

In this model conserved amino acids in sigma region 2.4 and 3 lead the template strand deep into the core enzyme towards the active site through a tunnel lined with positive charges formed by sigma regions 2 and 3, beta region 1, the beta′ lid and the beta′ rudder. Previous work had identified specific interactions between aromatic amino acids in sigma region 2 and bases in the –10 sequence of the nontemplate strand.[20,22,23] These may lead the single stranded DNA into a groove formed by the beta1 and beta2 lobes (Figure 2.3B). Crosslinking and mutational studies have shown that the beta′ clamp and the beta lobe interact with the nucleotides near the catalytic site while the beta′ jaw interacts with the downstream double stranded DNA. These interactions with downstream double stranded DNA are important for stability during open complex formation and elongation (ref. 35 and references therein).

Finally, results from atomic force microscopy and footprinting studies show that the DNA is wrapped around the polymerase by nearly 300° (ref. 127 and references therein). The degree of compaction of the DNA around RNAP is dependent on the promoter sequence under consideration, and more specifically on the nature of the upstream elements (see below).

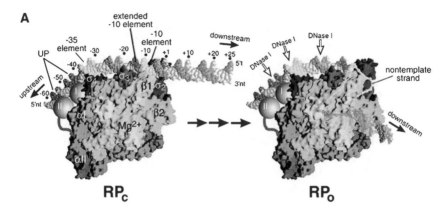

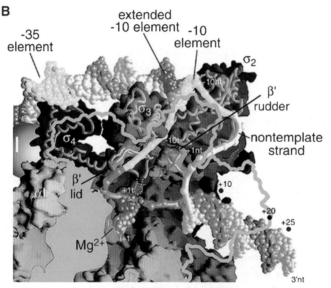

Figure 2.3 (A) Postulated transition from the closed to the open complex.[12] In the final structure, the sigma domain 4 interaction with the −35 sequence and interactions of the alpha CTD with upstream DNA result in bent DNA. (B) Details of the interactions stabilizing the single stranded DNA at the active site in the final complex[12] (reproduced with permission from the American Association for the Advancement of Sciences).

These structural data, combined with earlier footprinting studies of a promoter–RNAP complex trapped at low temperature, led to a proposal for the architecture of the closed complex. ([12] and Figures A.4 and 2.3) In this model, the DNA remains on the surface of the enzyme and does not bend toward the active site. However, the binary complex probed by footprinting reagents at 5 °C might not reflect the conformation of the intermediates that are significantly populated at physiological temperature (see below). Kinetic studies coupled with techniques providing structural signatures of the intermediates in real time may provide a more direct characterization of these short-lived species.

2.3 Physical Characterization and Structure of the Intermediates

As outlined above, the process by which the RNA polymerase holoenzyme specifically forms a transcriptionally active open complex on promoter DNA is populated by several short-lived intermediates. Different approaches have been used in the attempt to characterize the structure of these intermediates and to determine the conformational changes and the mechanism by which the open complex is formed (Box 2.1).

The main objective of these studies is to address some of the unresolved questions about this process, and particularly about the energetics and specificity of DNA melting preceding transcription initiation. Is the double helix actively destabilized by the enzyme after initial promoter binding has taken place, or does the holoenzyme opportunistically trap a structural fluctuation in the DNA such as transient denaturation of the A+T rich −10 element? Does this occur *via* specific contacts with the −10 region or by distortion of the whole promoter region occurring during the initial steps? Is the nucleation–propagation model described in Section 2.1 always valid, or does the enzyme trap single stranded DNA resulting from "breathing" of the double helix? How can the stability of the −10 region be affected by sequences up to 100 base pairs away? Why, when and how do multiple possible open complex structures form at the same promoter? What is the rate-limiting conformational change? How do these steps depend on the different promoter elements? Finally, how are the kinetic properties of these intermediates modified by regulators, such as transcription factors, metabolites and DNA topology?

The promoters for which the process of promoter binding and open complex formation has been best characterized are for the most part naturally strong wild-type promoters, such as early phage (T7A1, λP_R), ribosomal RNA promoters, and variants of natural promoters, such as *lac*UV5, where two mutations in the −10 sequence of the *lac* promoter render its transcription independent of the presence of an activator (Figure 2.4). Comparing the results obtained on these promoters showed very early on that the kinetic constants associated with the linear path, and their dependence on external variables, were not the same from one promoter to the next (*e.g.*, see discussion in ref. 4).

More detailed studies, to be analyzed here, specify the nature of these fundamental differences. From these results it is obvious that the structure of the intermediates can differ from one promoter to another depending on the sequence and architecture of the promoter elements. In other words, the temporal hierarchy of the interactions formed with individual promoter elements during this process can be dictated by the context of the specific promoter.[25] It remains a

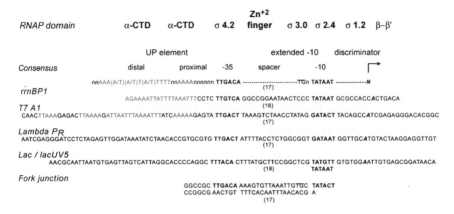

Figure 2.4 Sequences of some of the most studied promoters.

challenge to assess the variation in the conformational changes in both the protein and the DNA that can explain this variability.

2.4 Finding the Promoter. Induced fit and Indirect Sequence Recognition

Protein recognition of a specific site on the DNA takes place in several steps, beginning with the promoter search *via* nonspecific binding and ending with the induction of conformational changes in both the protein and the DNA at the specifically-identified site. This process has been characterized for a series of DNA-binding proteins of different levels of specificity, from nucleoid proteins (IHF and the nucleosome) to more specific transcription factors (CRP, 434 repressor, and Lac repressor) (for recent reviews see refs 36 and 37). The conclusions reached by these studies largely agree with the results of experiments carried out on RNAP and described below.

The *search for the promoter* occurs by a combination of one- and three-dimensional diffusion processes implying sliding, hopping and intersegment transfer.[37–39] From the results obtained by kinetic and single-molecule studies of protein–DNA interactions one can propose a possible scenario for the events leading to promoter binding. During its short forays in one-dimensional diffusion on the DNA, RNA polymerase tracks a groove of the double helix.[39] The interactions formed at this stage are for the most part electrostatic, stabilized by the increased entropy resulting from the displacement of the counterion cloud surrounding the double helix. Indeed the initial binding of RNAP to promoter DNA has been shown to be highly dependent on the salt concentration.[3] The presence of positively charged side chains on the protein screens the negative charges on one face of the DNA, allowing the DNA to adopt a bent structure aided by the repulsion of the unscreened phosphate groups on the opposite face of the DNA.[36]

Thus, it is likely that an initial complex is formed by *indirect readout*, by recognition of such *induced specific DNA structures* that are adopted thanks to the inherent flexibility of the double helix at specific nucleotide sequences.[40,41] For example, the narrow minor groove of the A-tracts in an UP element results in a decreased distance between the phosphate backbone of the two strands. This structural signature is recognized by the small alpha CTD domains,[42] and the rigidity of A-tracts results in increased flexibility at its edges favoring DNA bending.[43,44] Another example is the increased deformability of the TG step, resulting from the poor stacking of these two bases, often conserved in the −35 sequence and in the extended −10 element, or the reduced thermal stability of the TA steps in the −10 region. This flexibility allows the DNA to bend and eventually wrap on the surface of the enzyme, thus increasing the probability that a larger number of interactions are formed at one time, and decreasing the energy barrier for DNA deformation.[41] At the same time the overall energetic cost for DNA wrapping can offset in part the large net negative free energy of binding, destabilizing the final complex.[45]

Evidence that wrapping takes place in the early steps of binding has been obtained from the extensive upstream DNA protection observed in footprinting experiments, and the dependence of binding kinetics on upstream DNA sequences (see below) and on the degree of supercoiling of the DNA.[46-50] DNA wrapping can also favor later steps in the pathway, such as DNA melting, by locally changing the twist of the double helix (see Chapter 3 for a further discussion of this point).

Specific interactions of amino acid side chains with the nucleotide bases are not required for these initial structures to exist. Even though the resulting structure is stabilized mainly by *electrostatic interactions* and the increase in entropy due to the release of counter ions from the DNA, it is a structure that results from the sum of a number of induced changes in various structural elements and thus more stable than the "nonspecific" interactions formed during the initial stages of the search. The difference between specific and nonspecific binding affinities in these early complexes has been estimated to remain relatively small.[4] The formation of these interactions at this stage must be sufficiently rapid, within a biologically-sensible time span, and stable enough for further conformational changes to take place in both the protein and the DNA.[51]

2.5 Formation of the Closed Complex

In general, the closed complex refers to the first significantly populated, competitor-sensitive species that arises prior to melting of the double strand in the −10 region.

2.5.1 On a Unique Structure of the Closed Complex

It was initially believed that, for a given set of conditions, the structure of the closed complex would be the same at all promoters. However, the protein–DNA contacts formed and the resulting structures depend both on the experimental conditions and on the specific promoter under study (see the comparison of closed complex footprints in ref. 52).

For example, a stable, specific complex is formed between RNAP and a short DNA fragment containing just the consensus promoter sequence from −40 to −11. The formation of this complex requires an isomerization step, suggesting that a core promoter sequence could suffice to specifically bind RNAP and induce conformational changes resulting in increased complex stabilization.[53] The presence of an unstable intermediary complex in this context shows that upstream contacts and DNA wrapping are not required for the formation of a closed complex when the core sequence is the consensus one. Studies on such short promoter DNA fragments have helped identify the kinds of interactions that may be important for the formation of a closed complex. They have also shown that deviation from the consensus sequences can result in significant destabilization of the complex, an additional indication of its specificity.[21]

The formation of the early complexes could take place by the combination of at least two parallel pathways, involving interactions of different strength with the consensus sequence and with upstream DNA. One pathway would be favored with respect to the other depending on the relative strength of the interactions formed in each case. Thus upstream interactions would become particularly important in the case of a promoter bearing a non-consensus sequence for example. Some transcription activators (*e.g.*, CRP) act at this step by stabilizing a closed complex structure that would be too short-lived in its absence.[54] The stability of the closed complex also determines the ability of RNAP to compete with nucleoid proteins or specific transcriptional repressors for binding to the promoter.[55]

2.5.2 Role of Upstream Contacts for the Stability of the Closed Complex and in Leading the Complex Towards Subsequent Isomerization

Three different kinds of upstream contacts have been described. The first corresponds to transient nonspecific contacts of the alpha CTDs with upstream DNA, the second to the more specific interactions formed by alpha CTDs with AT-rich UP elements, and the third to complexes stabilized by interactions between the alpha CTD bound to the proximal UP element and sigma region 4 bound to the −35 region.

Transient upstream contacts can increase the rates of initial binding,[56,57] even at those promoters lacking a specific UP element sequence, such as lacUV5 [50] or lambda PR,[49] by stabilizing the early complexes. Differential effects are indeed seen when the upstream DNA or the alpha CTDs are mutated or absent at the lacUV5 (ref. 50) or at the rrn P1 promoters.[56]

Time-resolved footprints and kinetic studies of RNAP mutants support the early formation of the *sigma region 4-alpha CTD interaction* on promoters where the −35 sequence and the proximal UP element are close to the consensus.[59–62] This interaction results in DNA bending. Hypersensitive sites are observed at the –35 and –45 sites in the DNaseI footprints of closed complexes at different promoters.[52,63] The trajectory of the downstream DNA is thus placed in the direction of the sigma subunit domain 3.0 and 2.4, favoring the formation of their interaction with the upstream end of the –10 sequence. The continuation of DNA wrapping by kinking at the upstream end of the –10 sequence and in the spacer[12] then places the downstream DNA within reach of the sigma 2 domain and the beta/beta' jaws, resulting in DNaseI protection down to +25 on the lambda P_R promoter[64] or lacUV5[65] in a closed complex (Sclavi et al. in preparation).

Thus, a full protection in the closed complex appears to be dependent on the successful wrapping of the DNA on the polymerase. This is stabilized either by the presence of a consensus –10 sequence and/or by correct interactions with upstream sequences setting the DNA trajectory. As the temperature is decreased the upstream contacts are favored compared to those with the –10 region, resulting in a shorter footprint at some of the promoters. For example, at the galP1 extended –10 promoter it has been shown that the energetic contribution of the contacts formed in the upstream region becomes preponderant for the stabilization of the complex as the temperature decreases.[61]

A short footprint is also observed at the lambda P_R promoter for the complex formed by a holoenzyme where the aromatic amino acids in sigma region 2.3 have been substituted by alanine residues.[63,66] Not surprisingly these altered proteins are not functional at low temperatures and at 37 °C they still display a decreased isomerization rate compared to wild type.

2.6 The First Isomerization Step. The role of Sigma, Formation of Specific Interactions

Following the formation of the first complex an isomerization takes place that leads to nucleation and opening of the double helix by the sigma subunit and to increased stability of the complex in the presence of heparin.

2.6.1 Upstream Contacts

In some promoters the upstream contacts formed in the early closed complex are only transient, *i.e.*, only observed during the early steps of the pathway.[58,62] They nonetheless set the RNAP on the "right path" to open complex formation by placing the enzyme in the correct register for subsequent steps to occur, as evidenced from their effect on the rate of the subsequent isomerization step.[49,50]

For example, at the lambda P_R promoter in the absence of DNA upstream of –47 the DNaseI protection ends at +2 (on the template strand) and +7 (on the non-template strand), instead of +20, and a change in the pattern of permanganate reactivity is observed in the heparin-resistant intermediate complex,[49] possibly indicating the accumulation of an otherwise short-lived intermediate with an additional distortion of the template strand at the upstream end of the −10 sequence. This result suggests that interactions upstream of −47 are important for proper placement of downstream DNA within the jaw early on in the pathway, an interaction favoring the isomerization to the transcriptionally active complex (see below). However, in the open complex, DNaseI protection in the absence of alpha CTDs extends to +20, as it was also observed at the lacUV5 and rrnBP1 promoters.[24,56,61,67] Finally, the presence of upstream contacts is not necessary to form a transcriptionally active complex; however, it results in an increase in the rate of open complex formation by favoring the formation of the correct contacts either with the upstream end of the −10 region, the downstream jaws, or both.

2.6.2 Nucleation of the Single Stranded Region and its Propagation

As stated above, the current model for DNA opening at the promoter proposes nucleation of opening at −11 followed by propagation of the bubble from upstream to downstream (previously reviewed in refs 9 and 10).

A possible sequence of events can be described from published results, in particular from those performed on simplified DNA structures, such as the fork junction DNA.[19,68] The recognition of a minimal fork junction structure results in a complex that is heparin resistant.[19] The nucleation step involves binding of sigma 2.4 to the double stranded, upstream end of the −10 sequence and the flipping out and trapping of the conserved −11 adenine, a key residue for the melting process (Figure 2.5). Among the amino acid residues in region 2.3, Tyr-430 and Trp-433, and in particular Thr-429, interact with adenine at −11, possibly by the formation of an hydrogen bond with the same side group involved in base pairing.[68–71] The aromatic amino acids of region 2.3 also stabilize the single stranded nontemplate strand by stacking with the DNA bases. The (FYWW) RNAP, where four aromatic amino acids in region 2.3 (Phe-427, Tyr-430, Trp-433 and Trp-434) have been substituted by alanines, is unable to form an open complex and has been used as a model to study the properties of the closed complex.[63] Nucleation of strand separation permits bending of the double helix towards the opening of the jaw, thus facilitating interactions of beta and beta′ with the downstream end of the bubble. These interactions contribute to downstream propagation of DNA melting and the stabilization of the open structure.[29,33,72] Time-resolved footprinting experiments at the T7A1 promoter support this model.[62] Here two intermediates were identified whose protection pattern suggests structures where the upstream end of the −10 region is bent toward the opening of the jaw to different extents, in agreement with the sequence of contacts proposed by Gralla and co-workers.[19]

The identity of the base at −11 becomes less important under conditions destabilizing the −10 region, suggesting that its interaction by RNAP is not an obligate intermediate in DNA opening.[69] The nucleation and propagation steps may occur more or less cooperatively depending on temperature, the presence

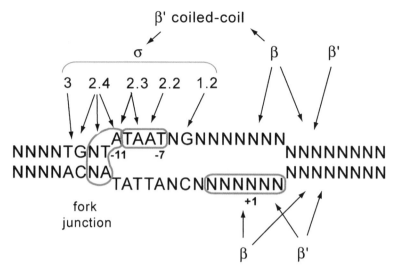

Figure 2.5 Sigma–core, core–DNA and sigma–DNA interactions in the open complex. In the closed complex, stability of the interaction is dependent on the sequence of the −10 element.[21] Subsequently, important interactions are formed with the single stranded nontemplate strand from −7 to −11 by the sigma subunit.[19,22] The single-stranded structure is important for binding stability due to interactions primarily with the backbone of the non-template strand by conserved basic residues in regions 2.2 and 2.3. The nucleotide sequence is important within the single stranded context for inducing a conformational changes in the holoenzyme, resulting in stabilization to competition by heparin.[21,88] Aromatic amino acids of region 2.3 stabilize the single stranded nontemplate strand. Among the amino acid residues in region 2.3, Tyr-430 and Trp-433, and in particular Thr-429, interact with conserved adenine at −11, a key residue for nucleation of the melting process, possibly by the formation of an hydrogen bond with the same side group involved in base pairing.[68–71] The importance of the conserved adenine at −11 is dependent on its context. When downstream DNA is destabilized it becomes less important. A possible source of the favorable energy used for DNA melting comes from the interaction between the −10 region and sigma and core.[69,70] Contacts with the extended −10 are made primarily by sigma region 3 and 2.4 (His-455 and Glu-458).[128,129] Region 2.4 is also involved in interactions with the −10 sequence – Gln-437 and Thr-440 interact with the double-stranded side of the fork junction at −12. The sigma region 1.2 contact with the guanosine base at −5 (−7 in rRNA promoters) contributes to the stability of the open complex. In the unstable open complex formed at rRNA promoters a cytosine is often found at this position.[104]

of magnesium ions at the active site, supercoiling and the sequence of the promoter at the −10 and discriminator regions.[73–75] Alternatively, destabilized −10 sequences are directly bound and trapped by the protein, particularly within negatively supercoiled DNA, aided by a low-melting consensus −10 sequence.[76–78] Thus, a sequential nucleation–propagation pathway is more likely to be observed under conditions unfavorable to DNA melting.

2.6.3 Phasing of −10 and −35 Regions, the Role of the Spacer

An optimal synergy of the various interactions contracted at the upstream and downstream promoter regions requires that they be in register with each other, located on the correct face of the helix. This phasing process may require a significant distortion in the spacer region. The isomerization step is thus strongly dependent on the spacer's length and sequence, the latter due principally to its effects on flexibility of the double helix.[79–82] The distortion of the spacer depends on the nature of the specific interactions contracted with the −10 sequence but not necessarily on DNA melting[63] and can become a rate-limiting step, especially at those promoters that have a longer-than-consensus spacer, such as lacP1 and lacUV5[83] or those that are targets for the MerR family of transcriptional regulators.[84] This distortion results in the appearance of a characteristic DNaseI hypersensitive site at or near −25.[12,25]

The presence of an extended −10 sequence also affects the rate of the isomerization step, probably by facilitating distortion of the double helix and DNA bending during nucleation and open complex formation.[12,63,85]

2.6.4 Probing Possible Sequential Linear Pathways by the use of Temperature

As one of the important roles of RNA polymerase after promoter recognition is strand separation, the overall process was expected to be dominated by enthalpic contributions arising from DNA melting and so it has always been important to consider how it is influenced by temperature.[1,77,86]

In extreme cases one could hope to abort the linear process, described in the introduction, at consecutive steps, RP_I or RP_C, as the temperature of the assay is lowered. These attempts led to contrasting results depending on the promoter considered. Owing to the temperature-dependence of DNA melting, different structures are observed depending on the experimental conditions. As the temperature is decreased, the barrier to DNA opening increases, resulting in the accumulation of species that may not be observed at physiological temperature because of their short lifetime.

At the T7A1 promoter the time-resolved hydroxyl radical footprints of the intermediates observed at 37 °C correspond rather closely to those of the intermediates trapped at lower temperatures.[62,87] The same is observed at the lambda P_R promoter.[64] Thus a kinetic formalism reflecting a linear pathway might be conserved. On such promoters, it is valuable to derive thermodynamic quantities from the temperature-dependence of the relevant rate constants and to try to interpret them (Box 2.1). Indeed at both promoters the −10 sequence strongly deviates from its consensus. Establishing optimal contacts between sigma region 2 and the −10 region might be equally difficult, and equally rate-limiting at the different temperatures considered.

As described above, sigma region 2 can form different types of interactions with the upstream end of the −10 sequence at different steps in the process of open complex formation. In the early steps of the pathway, specific interactions are formed first with the DNA still in its double stranded form, then subsequently with the fork-junction structure and finally with single-stranded DNA (Figure 2.5). At each step the sequence and structure specificity change as does the stability of the interactions.[19,88] Thus at some promoters, such as lambda P_R,[64] sigma region 2 can form stable interactions with the still double-stranded DNA in the closed complex, leading the enzyme to form additional contacts with downstream DNA that may result in an extended DNaseI footprint.

In contrast, at the lacUV5 promoter, temperature more drastically affects the qualitative nature of the positioning of the RNAP subdomains with respect to the DNA template. A linear pathway is no longer acceptable, and the drastic changes of the patterns provide a different type of information than the one obtained in the two preceding cases.

First of all, the DNaseI protection pattern obtained at 0°C ends at –2. It has long been considered as adequately representing the footprint of the first closed complex RP_C at physiological temperature.[89] However, after a temperature upshift the binary complex formed at this lower temperature does not reenter the productive path quickly enough to be assigned to this species.[1] Furthermore, at both 17 and 37 °C the DNaseI footprint of the first observed complex in the pathway reaches downstream to +20, as in all the subsequent species. If a transient species displaying a similar short footprint was indeed present as an early intermediate in the pathway at higher temperatures, it may be very unstable, and too poorly populated to be detected through the current methodologies.[65]

Even though the first complexes at 17 and 37 °C display a similar protection pattern to DNaseI cleavage they do not correspond to the same structure. Indeed the temporal evolution of events occurring at 20 and 37 °C has also been probed *via* UV crosslinking. At 20 °C the –35 sequence becomes crosslinked to the sigma subunit in the first detectable closed complex while the –10 sequence only becomes crosslinked in a subsequent step.[90] At 37 °C the first complex has a full length footprint but no crosslinks. The signals at –35 only appear concomitantly with the –10 signals after the first isomerization step has taken place (Sclavi *et al.* in preparation) (Figure 2.6). The same linear pathway is therefore not followed by the bulk population of binary complexes at these two temperatures.

Increasing the temperature favors the interaction with a DNA structure where melting is nucleated. Hence, at lower temperatures higher energetic barriers result in the accumulation of a complex with a short footprint at 0 °C and a complex with a crosslink at –35 at 20 °C due to the decoupling of –35 –10 interactions. At 20 °C the rate-limiting step is the formation of the interaction with the –10 region while at 37 °C the rate-limiting step may be a protein conformational change (see below) or the distortion in the spacer that takes place when the –35 and –10 regions are specifically bound.[63] (The lacUV5 promoter contains an 18-bp spacer region, longer than the consensus spacer length. This feature decreases the cooperativity of interactions formed at –10 and –35.[80]) Thus, at 37 °C, the formation of the full length protection is faster compared to the formation of contacts at –35, which now appear at the same

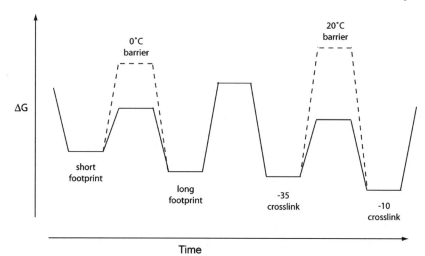

Figure 2.6 Energy barriers in the sequential representation of the pathway to open complex formation on the lacUV5 promoter. Linear representation of the pathway of open complex formation at the lacUV5 promoter. The continuous line corresponds to the landscape at 37 °C, the dashed lines to the increased barriers due to a decrease in temperature. The first barrier is possibly due to the temperature dependence of the conformational change in the holoenzyme stabilizing downstream contacts by the downstream beta lobe. The complex formed at 0 °C results in a short footprint ending at −2 with a regular pattern of DNaseI hypersensitive sites suggesting DNA wrapping.[89] The second barrier is due to the energetic cost of distortion of the spacer that, at this promoter, is one base pair longer than the consensus and results in a specific DNaseI hypersensitive site. The third barrier is due to DNA melting and is thus also temperature-dependent.[86] At 17 or 20 °C the first closed complex is too stable to footprint, instead the stable intermediate that has been observed, RP_I, has a full-length DNaseI footprint and a crosslink at −35, due to of the increased barrier for DNA melting.[1,65,90] At 37 °C the most important energy barrier is due to the distortion of the spacer. Time-resolved DNase I footprinting and UV crosslinking experiments identified a first intermediate characterized by a long DNaseI footprint followed by the distortion of the spacer and the appearance of crosslinks at −35 and −10 (Sclavi *et al.* in preparation).

time as the −10 contacts (Figure 2.6). This change in the rate-limiting step results in a nonlinear temperature dependence of open complex formation at the lacUV5 promoter.[1] Li and McClure also observed important differences in the protection pattern in the −35 region as a function of temperature at the prump-1 Δ265 promoter.[52] In the complex transiently formed at 19 °C the −35 sequence and the bases upstream are less protected than in the complex trapped at 4 °C.

Time-resolved crosslinking experiments on lacUV5 (using a DNA fragment with a chemically modified backbone to increase the efficiency of crosslinking) have shown that at 37 °C specific crosslinks are made at the upstream end of the −10 sequence before the ones with other sites on the promoter, upstream of −35 and at the transcription initiation site (S. Druzhinin and R. Ebright, personal communication). At 37 °C either DNA breathing or the fork junction of the nucleated −10 region may provide the RNAP with stable contacts for an early complex (see below).

2.6.5 Specific Protein Domains Destabilize the Intermediates in the Pathway

A minimal complex made of the beta' coiled coil (aa 1–314) and sigma regions 2 and 3 has been shown to be able to melt an extended −10 consensus sequence on negatively supercoiled DNA.[91] However, due to the exclusive use of negatively supercoiled DNA, it is not absolutely certain that this corresponds to active melting by the minimal complex rather than to opportunistic stabilization of a transiently-denaturing DNA sequence. Nevertheless, this implies that stabilization of the complex and deformation of the double helix by the interaction of elements upstream of the beta' Zn^{2+} finger as well as downstream contacts by the beta subunit are not required for melting to take place under these optimal conditions. In theory, such a small enzyme could suffice to bind and open a consensus −10 sequence. In practice, in the full length enzyme, the presence of additional protein domains not only allows the opening of imperfect promoters but also, as we shall see, destabilizes such simple complexes. Furthermore, the presence of extra domains in the native assembly permits reception of external signals, allowing the system to respond to various levels of regulation.

Sigma domain 1.1: early studies on the process of promoter search had shown that the beta and beta' subunits can be crosslinked to DNA in the first steps of complex formation but that the percentage of crosslinking to the sigma subunit increased as more stable nonspecific complexes were formed.[92] Park et al. proposed that the acidic nature of the sigma subunit decreases the stability of nonspecific binding of beta and beta' to DNA. More recently, a similar proposal was made from the structure of the holoenzyme, where one can see that the negatively charged sigma region 1.1 and linker 3.2 screen positive charges on beta and beta', thus preventing nonspecific interactions with DNA.[11] Decreasing the number of possible orientations in the early complexes may result in faster promoter search.

Sigma domain 1.1 is one of the least conserved domains of sigma and is found primarily in the group 1 ("housekeeping") sigmas, within which a higher level of conservation is found.[6,10,93,94] However, this negatively charged domain is able to screen the other sigma subdomains, preventing both nonspecific and specific sigma–DNA interactions.[95] On the other hand, in the context of the holoenzyme, this autoinhibition is relieved: sigma domain 1.1 binds deep in the beta-beta' cleft; it is subsequently displaced during open complex formation by downstream (+9) DNA entering the jaws, to a positively charged patch on the upstream lobe 1 of the beta pincer.[34,96] The minimal complex described above, missing the whole beta subunit, forms an open complex much less efficiently in the presence of region 1.1.[91]

The presence of region 1.1 within the active site channel has been proposed to widen the entrance of the cleft – thus possibly affecting the structure of the holoenzyme.[12] In fact it has been shown that region 1.1 can directly influence the rate of open complex formation at the isomerization step.[94] Furthermore, the absence of this domain can result in increased transcription levels at promoters with spacer regions shorter than consensus.[97] Its role in the stabilization of sigma–core interactions[98] might result in a more rigid distance between domains 4.2 and 2.4, thus favoring –10 and –35 sequence combinations separated by a consensus spacer over nonconsensus promoters. In addition, promoters more similar to the consensus may more efficiently displace region 1.1 from the cleft due to the increased stability of intermediate complexes (see below).

Beta subunit downstream lobe I: on a linear DNA fragment carrying the full length T7A2 promoter, an RNAP mutant lacking the beta subunit downstream lobe 1 (aa 186–433, part of an evolutionarily variable region) forms a final complex where the bubble is not fully melted. Only the upstream region of the – 10 element is single stranded in the absence of nucleotides. The addition of nucleotides is required to complete the bubble melting process. This half-melted complex is, however, more stable than the one formed by wild-type RNAP and maintains an open bubble even at temperatures as low as $-20\,°C$.[99] This indicates a possible active role of the enzyme in the process of DNA melting in the context of this mutated RNAP. In contrast, in the wild-type enzyme, DNA melting is in part mediated by the trapping of short-lived single stranded bases, a process that is disfavored at lower temperatures, as well as by a temperature-dependent protein conformational change (see below).

In the complex formed by this altered RNAP on the lacUV5 promoter, the sigma subunit can be crosslinked to the single-stranded nontemplate strand at – 11. These contacts, however, are not present in the final open complex once the first nucleotides are added.[100] These results allowed the authors to propose a model where *transient binding* of single-stranded DNA by sigma 2.3 is followed by an interaction of downstream DNA with beta lobe 1. This latter interaction places the double helix in the DNA-binding channel and results in a loss of these sigma contacts to the nontemplate strand and propagation of the bubble. This sequence of events could also occur within the context of wild-type RNAP. The formation of a "normal" open complex upon addition of nucleotides suggests that it could be on the wild-type pathway. It remains to be established whether the presence of this intermediate is dependent on the promoter sequence context.

The downstream beta lobe appears to have two functions, the first is to favor the propagation of the bubble by clamping onto downstream double stranded DNA,[29] and the second is to disfavor uncoupled melting of upstream –10 bases at low temperature by the concerted and temperature-sensitive conformational changes taking place between the upstream and downstream beta lobe. In addition, the heparin resistance of this semi-open intermediate appears to be dependent on the integrity of the upstream lobe.[72]

In other words, the presence of the beta lobe destabilizes the partially opened complex and favors its conversion into RP_O, allowing isomerization

Opening the DNA at the Promoter; The Energetic Challenge 59

to proceed more efficiently. In addition, the loss of the crucial contacts between sigma and the nontemplate strand at –11 also destabilizes the final open complex so that upon transcription initiation the promoter is more easily released as elongation begins.[100] In fact, mutations at positions –11 and –7 likely affecting the nontemplate strand also result in a decreased amount of abortive products.[101]

The presence of a similar, partially opened, stable intermediate during DNA opening was proposed by Gralla and co-workers from the study of fork junction complexes. They showed that two different sets of interactions stabilize the fork junction and the fully opened structures. They also observed that both conserved amino acids in sigma region 2 and single stranded nucleotides may have a negative effect on the stability of the complexes between RNAP and the promoter both before and after strand separation.[19,88,102]

The destabilization of the intermediates could contribute to an increased specificity of binding of RNAP, avoiding the formation of stable complexes at sites outside the promoter region, and allowing for an additional site of regulation of the transcription levels. Decreased stability of the RNAP–promoter complex also results in more efficient escape to transcription elongation.

The sequence of the promoter can also negatively affect the stability of intermediates in the pathway leading to escape. In the case of ribosomal/tRNA promoters the presence of a GC-rich discriminator region does not allow for a propagation of the bubble in the absence of nucleotides.[103] An additional cause for the instability of the open complex at these promoters is the absence of a usual interaction formed between sigma region 1.2 and the non template guanosine located two positions downstream of the –10 sequence (rRNA promoters often have a cytosine at that position).[104] This interaction may mediate the binding of double stranded downstream DNA to the downstream beta lobe. In this case, the combined effect of the beta lobe destabilizing the intermediate and the stability of the double helix in the discriminator disfavoring propagation of strand separation results in a highly unstable half open complex. The low stability of these promoters renders them even more dependent on interactions with the UP element (and under specific condition on interactions with FIS bound upstream), which, as described above, contribute to the correct placement of downstream DNA. Furthermore, the low stability of the open complexes formed at these promoters results in efficient escape in the presence of nucleotides, which renders them sensitive to changes of *in vivo* nucleotide pool concentrations and to regulation by the stringent response *via* the effect of ppGpp, DksA and the omega subunit on the isomerization rate (Chapter 3).[105–107]

2.6.6 Overstabilization is sometimes used as a Regulatory Mechanism

As described above, inefficient isomerization into RPo characterizes some of the most active promoters in the cell. Similarly, an increased stability of the open complex results in a less efficient initiation of transcription that can, however, become a target of regulation. A promoter with consensus features at all of its binding sites does not result in high levels of RNA synthesis due to the excessive stability of the sum of the interactions (Chapter 1).[51]

Recently, it has been shown that a potentially large fraction of RNA polymerases *in vivo* resides most of the time at the promoter.[108] RNAP in this case is poised in a kinetic trap, often mediated by numerous upstream interactions with transcription factors[109–112] or by the stabilization of an incorrect complex geometry as in the case of the malT promoter.[113] A poised enzyme, by avoiding the recruitment steps, can quickly respond to activation signals and, within a short time interval, will perform a first round of transcription. This situation is encountered in particular when initiation of transcription is directly regulated by changes in metabolite concentration, supercoiling and/or the cellular environment, such as salt concentration (Chapter 1).[114,115] This may be important under growth conditions where the amount of free RNAP is low, rendering recruitment of the enzyme to the promoter rate limiting.

2.7 Formation of Transcriptionally Active Open Complex and the Rate-limiting Step: Protein Conformational Changes or DNA Melting?

The nature of the rate-limiting step in the formation of a transcriptionally active open complex is dependent on the promoter sequence. At strong promoters the rate-limiting step at physiological temperature is often either concomitant with or subsequent to open complex formation and may depend on different protein conformational changes.

Non-exponential temperature dependence on the rate of open complex formation was observed at the lacUV5,[1] T7A1,[116] and lambda P_R promoters.[2,4] This indicates a possible change in the rate-limiting step as a function of temperature. One of the possible causes for this change was proposed to be the increased stability of double stranded DNA at lower temperatures, rendering DNA opening rate limiting for open complex formation. As described above, in wild-type RNAP, nucleation and propagation are tightly coupled due to the combined interactions of the bipartite beta lobe 1. However, due to specific mutations in the beta subunit, DNA nucleation and bubble propagation can be decoupled, and partially opened complexes are observed at temperatures as low as $-20\,^\circ\text{C}$. It is possible then that a protein conformational change might become rate limiting as the temperature is decreased, independently of nucleation and fork junction binding. This conformational change could nevertheless be dependent on an interaction either with partially single-stranded DNA within the −10 sequence or with a bent DNA structure resulting from base unpairing.

The nonlinear temperature dependence on the rate of stable complex formation was interpreted by Record, Saecker and co-workers in terms of a change in heat capacity, and thus of the burial of polar and nonpolar side chains.[2,4]

The binding of downstream DNA contributes to the process of placing the template strand at the active site and to the stabilization of the open complex. The presence of double stranded DNA triggers additional conformational changes in RNA polymerase.[64] Studies of the change in fluorescence of RNAP as it is binds to the T7A1 promoter provided evidence for protein conformational changes occurring during both closed complex formation and the following isomerization steps.[116]

Several possible conformational changes have been identified that could be candidates for a rate-limiting step: the displacement of sigma region 1.1 from deep in the cleft to the beta lobe;[34,96] a conformational change in sigma during the isomerization step;[90] the propagation of the bubble and the concomitant conformational changes in the beta lobe 1[72] and/or the conserved G and G' regions of the downstream beta clamp as they bind downstream double stranded DNA.[30]

As discussed by Saecker and co-workers, these folding transitions are driven by binding free energy, and stronger interactions with the DNA can result in an induced conformational change in the protein, stabilizing the final complex. Thus, the sequence and organization of the promoter can favor different conformational changes.[4]

2.8 A Rugged Energy Landscape

As a result of the *structural flexibility* of the macromolecules involved, as well as of the multiple possible interactions that can be formed between RNA polymerase and promoter DNA, the energy landscape for the formation of a transcriptionally active complex may more resemble a rugged mountainside than the usual depiction as a smooth roller coaster (Figure 2.6).

We imagine now that, as the enzyme encounters the promoter, it can contract a large number of interactions, through its various DNA binding domains. Heterogeneity of the early complexes was indeed observed for IHF and RNAP binding to their respective sites on the DNA by time-resolved X-ray footprinting experiments.[62,117] Those largely non-specific encounters are progressively replaced by more specific and stable interactions during the isomerization steps. We have reviewed the evidence indicating that at this stage both induced folding in the protein and phasing of the various initial interactions may occur, thus *reducing* the amount of conformational space available for the two partners.

However, for RNA polymerase this process should not lead to an excessively stable final product. In other words, unlike protein folding, this landscape cannot have the shape of a deep funnel: the formation of the promoter–RNA polymerase complex is not the end of the story, but the beginning of the transcription process. When nucleotides are present, the enzyme needs to overcome the barriers constituting the energy well characteristic of the open complex to begin RNA synthesis and translocation. For this reason, as described above, completely consensus promoters are not very efficient at driving RNA synthesis due the excessive stability of the open complex.[51] The optimal solution, adopted by the most highly transcribed genes coding for ribosomal and transfer RNA, appears to imply a combination of several, non-optimal interactions, both between holoenzyme and DNA and sigma and core.

In such a landscape though, as in protein folding, the presence of a slow, rate-limiting step might allow a *re-equilibration* of the *different species* preceding this higher energy barrier. Thus, the structure of the observed intermediates in bulk experiments is the one that is the most populated after re-equilibration (the most thermodynamically stable) and not the one that has been selected by kinetics and has the fastest rate of formation. Therefore, different structures will accumulate if the barrier to isomerization is increased by changes in environmental variables, such as temperature, or even by the absence of nucleotides.

In some cases this re-equilibration results in the appearance of several branches in the pathway, leading to the formation of multiple populations at the promoter and to more than one possible open complex, as it has been observed at several of the best characterized promoters.[118-121] Ozoline *et al.* have proposed a model of alternative pathways for promoter activation from an analysis of the heterogeneity found in promoter sequences.[122]

Finally, it is unlikely that there is one, linear, sequential pathway shared by all promoters, and it is highly possible that transcriptional regulators work not by changing individual steps within a given pathway but by stabilizing alternative pathways.[123]

2.9 Summary and Conclusions

The study of the conformational changes taking place during promoter recognition has proven to be more challenging than the one encountered for other protein–DNA interactions, due to the complexity of the tasks carried out by this enzyme and of the dynamics of the interactions between the relative partners involved.

A large number of lower affinity interactions can stabilize an early specific complex faster than a small number of highly specific interactions. The structures generated in the very early steps of the pathway (*e.g.*, promoter search) are selected for by their kinetic properties, while the later conformations may result from local equilibria generating multiple possible structures. Initial nonspecific binding-induced fit and DNA wrapping around the enzyme are followed by isomerization into more specific structures where bases contract specific interactions with amino acid side chains and where the DNA structure is further distorted towards the active site.

In some cases upstream DNA can contribute in placing the double helix in the correct conformation to facilitate the formation of the interactions at the upstream end of the –10 region leading to the isomerization step. Active melting can occur by bending and/or untwisting of DNA resulting from the inherent flexibility of the macromolecules and favored by the energy of stable protein–DNA interactions and induced protein folding.[4] Alternatively, destabilized −10 sequences are directly bound and trapped by the protein, particularly within negatively supercoiled DNA, aided by a consensus −10 sequence that melts at low temperature.[76,77] Thus, it is possible, especially for a nearly consensus promoter, that multiple pathways can lead to the formation of an open complex. This strategy leading to a transcriptionally active complex may have the advantage that it results in a process more robust to changes in environmental variables.

Studies of RNAP mutants have shown that the enzyme possesses structural domains that slow down, destabilize or inhibit opening by a direct interaction with the −10 region. Thus, evolution has come up with a few different strategies to improve specificity, destabilize the open complex, increase the kinetics of promoter binding and escape and allow for different levels of regulation, in other words to optimally balance the specificity required earlier on in the pathway with the subsequent destabilization that must result from the transcription activity of the enzyme.

Acknowledgements

I thank Henri Buc for very helpful discussions, Malcolm Buckle, Annie Kolb, Richard Ebright, Anne Olliver and John Herrick for critical reading of the manuscript. My current work is funded by the ANR JCJC grant no. JC05_53151.

References

1. H. Buc and W. R. McClure, *Biochemistry*, 1985, **24**, 2712–2723.
2. J. H. Roe, R. R. Burgess and M. Record Jr, *J. Mol. Biol.*, 1985, **184**, 441–453.
3. J. H. Roe and M. Record Jr, *Biochemistry*, 1985, **24**, 4721–4726.
4. R. M. Saecker, O. V. Tsodikov, K. L. McQuade, P. E. Schlax, M. W. Capp and M. Thomas Record, *J. Mol. Biol.*, 2002, **319**, 649–671.
5. P. L. deHaseth, M. L. Zupancic and M. T. Record Jr, *J. Bacteriol.*, 1998, **180**, 3019–3025.
6. T. Gruber and C. Gross, *Annu. Rev. Microbiol.*, 2003, **57**, 441–466.
7. S. Wigneshweraraj, P. Burrows, P. Bordes, J. Schumacher, M. Rappas, R. Finn, W. Cannon, X. Zhang and M. Buck, *Prog. Nucleic. Acid Res. Mol. Biol.*, 2005, **79**, 339–369.
8. A. Typas, G. Becker and R. Hengge, *Mol. Microbiol.*, 2007, **63**, 1296–1306.
9. J. D. Helmann and P. L. deHaseth, *Biochemistry*, 1999, **38**, 5959–5967.
10. S. Borukhov and K. Severinov, *Res. Microbiol.*, 2002, **153**, 557–562.
11. K. S. Murakami, S. Masuda and S. A. Darst, *Science*, 2002, **296**, 1280–1284.
12. K. S. Murakami, S. Masuda, E. A. Campbell, O. Muzzin and S. A. Darst, *Science*, 2002, **296**, 1285–1290.

13. D. G. Vassylyev, S. Sekine, O. Laptenko, J. Lee, M. N. Vassylyeva, S. Borukhov and S. Yokoyama, *Nature*, 2002, **417**, 712–719.
14. S. Callaci, E. Heyduk and T. Heyduk, *Mol. Cell*, 1999, **3**, 229–238.
15. K. Kuznedelov, L. Minakhin, A. Niedziela-Majka, S. L. Dove, D. Rogulja, B. E. Nickels, A. Hochschild, T. Heyduk and K. Severinov, *Science*, 2002, **295**, 855–857.
16. T. Arthur and R. Burgess, *J. Biol. Chem.*, 1998, **273**, 31381–31387.
17. E. A. Campbell, O. Muzzin, M. Chlenov, J. L. Sun, C. A. Olson, O. Weinman, M. L. Trester-Zedlitz and S. A. Darst, *Mol. Cell*, 2002, **9**, 527–539.
18. B. E. Nickels, S. J. Garrity, V. Mekler, L. Minakhin, K. Severinov, R. H. Ebright and A. Hochschild, *Proc. Natl. Acad. Sci. USA*, 2005, **102**, 4488–4493.
19. Y. Guo and J. D. Gralla, *Proc. Nat. Acad. Sci. USA.*, 1998, **95**, 11655–11660.
20. M. S. Fenton, S. J. Lee and J. D. Gralla, *EMBO J.*, 2000, **19**, 1130–1137.
21. M. S. Fenton and J. D. Gralla, *Proc. Natl. Acad. Sci. USA*, 2001, **98**, 9020–9025.
22. M. Marr and J. Roberts, *Science*, 1997, **276**, 1258–1260.
23. M. Tomsic, L. Tsujikawa, G. Panaghie, Y. Wang, J. Azok and P. L. deHaseth, *J. Biol. Chem.*, 2001, **276**, 31891–31896.
24. A. Kolb, K. Igarashi, A. Ishihama, M. Lavigne, M. Buckle and H. Buc, *Nucleic Acids Res.*, 1993, **21**, 319–326.
25. O. Ozoline and M. Tsyganov, *Nucleic Acids Res*, 1995, **23**, 4533–4541.
26. G. Zhang, E. A. Campbell, L. Minakhin, C. Richter, K. Severinov and S. A. Darst, *Cell*, 1999, **98**, 811–824.
27. S. Darst, N. Opalka, P. Chacon, A. Polyakov, C. Richter, G. Zhang and W. Wriggers, *Proc. Natl. Acad. Sci. USA*, 2002, **99**, 4296–4301.
28. E. Nudler, E. Avetissova, V. Markovtsov and A. Goldfarb, *Science*, 1996, **273**, 211–217.
29. N. Korzheva, A. Mustaev, M. Kozlov, A. Malhotra, V. Nikiforov, A. Goldfarb and S. Darst, *Science*, 2000, **289**, 619–625.
30. W. S. Kontur, R. M. Saecker, C. A. Davis, M. W. Capp and M. T. Record, *Biochemistry*, 2006, **45**, 2161–2177.
31. K. S. Murakami and S. A. Darst, *Biochemistry*, 2003, **13**, 31–39.
32. K. Brodolin, A. Mustaev, K. Severinov and V. Nikiforov, *J. Biol. Chem.*, 2000, **275**, 3661–3666.
33. N. Naryshkin, A. Revyakin, Y. Kim, V. Mekler and R. H. Ebright, *Cell*, 2000, **101**, 601–611.
34. V. Mekler, E. Kortkhonjia, J. Mukhopadhyay, J. Knight, A. Revyakin, A. N. Kapanidis, W. Niu, Y. W. Ebright, R. Levy and R. H. Ebright, *Cell*, 2002, **108**, 599–614.
35. J. Ederth, I. Artsimovitch, L. A. Isaksson and R. Landick, 2002, **277**, 37456–37463.
36. S. Khrapunov, M. Brenowitz, P. Rice and C. Catalano, *Proc. Natl. Acad. Sci. USA*, 2006, **103**, 19217–19218.

37. P. H. von Hippel, *Proc. Natl. Acad. Sci. USA*, 2007, **36**, 79–105.
38. C. S. Park, F. Y. Wu and C. W. Wu, *J. Biol. Chem.*, 1982, **257**, 6950–6956.
39. K. Sakata-Sogawa and N. Shimamoto, *Proc. Natl. Acad. Sci. USA*, 2004, **101**, 14731–14735.
40. O. Ozoline, A. Deev and E. Trifonov, *J. Biomol. Struct. Dyn.*, 1999, **16**, 825–831.
41. A. A. Travers, *Philos. Trans. A Math. Phys. Eng. Sci.*, 2004, **362**, 1423–1438.
42. B. Benoff, H. Yang, C. L. Lawson, G. Parkinson, J. Liu, E. Blatter, Y. W. Ebright, H. M. Berman and R. H. Ebright, *Science*, 2002, **297**, 1562–1566.
43. H. Koo, H. Wu and D. Crothers, *Nature*, 1986, **320**, 501–506.
44. S. Diekmann, *Methods Enzymol.*, 1992, **212**, 30–46.
45. C. Rivetti, M. Guthold and C. Bustamante, *EMBO Journal*, 1999, **18**, 4464–4475.
46. M. Amouyal and H. Buc, *J. Mol. Biol.*, 1987, **195**, 795–808.
47. M. Buckle, H. Buc and A. A. Travers, *EMBO Journal*, 1992, **11**, 2619–2625.
48. I. K. Pemberton, G. Muskhelishvili, A. A. Travers and M. Buckle, *J. Mol. Biol.*, 2002, **318**, 651–663.
49. C. A. Davis, M. W. Capp, M. T. Record and R. M. Saecker, *Proc. Natl. Acad. Sci. USA*, 2005, **102**, 285–290.
50. W. Ross and R. L. Gourse, *Proc. Natl. Acad. Sci. USA*, 2005, **102**, 291–296.
51. R. Knaus and H. Bujard, in *Nucleic acids and Molecular Biology*, ed. D. L. F. Eckstein, Springer-Verlag, Berlin Heidelberg, Editon edn., 1990, vol. **4**, pp. 110–122.
52. X. Y. Li and W. R. McClure, *J. Biol. Chem.*, 1998, **273**, 23549–23557.
53. L. Tsujikawa, O. Tsodikov and P. deHaseth, *Proc. Natl. Acad. Sci. USA*, 2002, **99**, 3493–3498.
54. D. F. Browning and S. J. Busby, *Nat. Rev. Microbiol.*, 2004, **2**, 57–65.
55. D. C. Grainger, D. Hurd, M. D. Goldberg and S. J. W. Busby, *Nucleic Acid Res.*, 2006, **34**, 4642–4652.
56. W. Ross, K. K. Gosink, J. Salomon, K. Igarashi, C. Zou, A. Ishihama, K. Severinov and R. L. Gourse, *Science*, 1993, **262**, 1407–1413.
57. M. G. Strainic Jr, J. J. Sullivan, A. Velevis and P. L. deHaseth, *Biochemistry*, 1998, **37**, 18074–18080.
58. C. A. Davis, C. A. Bingman, R. Landick, M. T. Record and R. M. Saecker, *Proc. Natl. Acad. Sci. USA*, 2007, **104**, 7833–7838.
59. H. Chen, H. Tang and R. H. Ebright, *Mol. Cell*, 2003, **11**, 1621–1633.
60. W. Ross, D. A. Schneider, B. J. Paul, A. Mertens and R. L. Gourse, *Genes Dev.*, 2003, **17**, 1293–1307.
61. L. Minakhin and K. Severinov, *J. Biol. Chem.*, 2003, **278**, 29710–29718.
62. B. Sclavi, E. Zaychikov, A. Rogozina, F. Walther, M. Buckle and H. Heumann, *Proc. Natl. Acad. Sci. USA*, 2005, **102**, 4706–4711.
63. V. Cook and P. deHaseth, *J. Biol. Chem.*, 2007, **282**, 21319–21326.
64. M. L. Craig, O. V. Tsodikov, K. L. McQuade, P. E. Schlax Jr, M. W. Capp, R. M. Saecker and M. T. Record Jr, *J. Mol. Biol.*, 1998, **283**, 741–756.

65. A. Spassky, K. Kirkegaard and H. Buc, *Biochemistry*, 1985, **24**, 2723–2731.
66. G. Panaghie, S. Aiyar, K. Bobb, R. Hayward and P. de Haseth, *J. Mol. Biol.*, 2000, **299**, 1217–1230.
67. S. E. Aiyar, R. L. Gourse and W. Ross, *Proc. Natl. Acad. Sci. USA*, 1998, **95**, 14652–14657.
68. D. Matlock and T. Heyduk, *Biochemistry*, 2000, **39**, 12274–12283.
69. E. Heyduk, K. Kuznedelov, K. Severinov and T. Heyduk, *J. Biol. Chem.*, 2006, **281**, 12362–12369.
70. L. Schroeder, A. Choi and P. DeHaseth, *Nucleic Acids Res.*, 2007, **35**, 4141–4153.
71. L. Schroeder, M. Karpen and P. deHaseth, *J. Mol. Biol.*, 2008, **376**, 153–165.
72. S. Nechaev, M. Chlenov and K. Severinov, *J. Biol. Chem.*, 2000, **275**, 25516–25522.
73. W. C. Suh, W. Ross and M. Record Jr, *Science*, 1993, **259**, 358–361.
74. E. Zaychikov, L. Denissova, T. Meier, M. Gotte and H. Heumann, *J. Biol. Chem.*, 1997, **272**, 2259–2267.
75. O. V. Tsodikov, M. L. Craig, R. M. Saecker and M. T. Record Jr, *J. Mol. Biol.*, 1998, **283**, 757–769.
76. H. R. Drew, J. R. Weeks and A. A. Travers, *EMBO J.*, 1985, **4**, 1025–1032.
77. A. Spassky, S. Rimsky, H. Buc and S. Busby, *EMBO J.*, 1988, **7**, 1871–1879.
78. H. Burns and S. Minchin, *Nucleic Acids Res.*, 1994, **22**, 3840–3845.
79. M. E. Mulligan, J. Brosius and W. R. McClure, *J. Biol. Chem.*, 1985, **260**, 3529–3538.
80. J. E. Stefano and J. D. Gralla, *Proc. Natl. Acad. Sci. USA*, 1982, **79**, 1069–1072.
81. D. G. Ayers, D. T. Auble and P. L. deHaseth, *J. Mol. Biol.*, 1989, **207**, 749–756.
82. S. E. Warne and P. L. deHaseth, *Biochemistry*, 1993, **32**, 6134–6140.
83. M. Liu, M. Tolstorukov, V. Zhurkin, S. Garges and S. Adhya, *Proc. Natl. Acad. Sci. USA*, 2004, **101**, 6911–6916.
84. K. Newberry and R. Brennan, *J. Biol. Chem.*, 2004, **279**, 20356–20362.
85. H. D. Burns, T. A. Belyaeva, S. J. Busby and S. D. Minchin, *Biochem J.*, 1996, **317**, 305–311.
86. K. Kirkegaard, H. Buc, A. Spassky and J. C. Wang, *Proc. Natl. Acad. Sci. USA*, 1983, **80**, 2544–2548.
87. P. Schickor, W. Metzger, W. Werel, H. Lederer and H. Heumann, *EMBO Journal*, 1990, **9**, 2215–2220.
88. M. Fenton and J. Gralla, *Nucleic Acids Res*, 2003, **31**, 2745–2750.
89. R. T. Kovacic, *J. Biol. Chem.*, 1987, **262**, 13654–13661.
90. M. Buckle, I. K. Pemberton, M. A. Jacquet and H. Buc, *J. Mol. Biol.*, 1999, **285**, 955–964.
91. B. A. Young, T. M. Gruber and C. A. Gross, *Science*, 2004, **303**, 1382–1384.
92. C. S. Park, Z. Hillel and C. W. Wu, *J. Biol. Chem.*, 1982, **257**, 6944–6949.

93. M. Lonetto, M. Gribskov and C. A. Gross, *Journal of Bacteriology*, 1992, **174**, 3843–3849.
94. C. Wilson and A. J. Dombroski, *J. Mol. Biol.*, 1997, **267**, 60–74.
95. A. J. Dombroski, W. A. Walter and C. A. Gross, *Genes & Development*, 1993, **7**, 2446–2455.
96. H. Nagai and N. Shimamoto, *Genes to Cells*, 1997, **2**, 725–734.
97. S. Vuthoori, C. W. Bowers, A. McCracken, A. J. Dombroski and D. M. Hinton, *J. Mol. Biol.*, 2001, **309**, 561–572.
98. D. Hinton, S. Vuthoori and R. Mulamba, *J. Bacteriol.*, 2006, **188** 1279–1285.
99. K. Severinov and S. A. Darst, *Proc. Nat. Acad. Sci. USA*, 1997, **94**, 13481–13486.
100. K. Brodolin, N. Zenkin and K. Severinov, *J. Mol. Biol.*, 2005, **350**, 930–937.
101. C. W. Roberts and J. W. Roberts, *Cell*, 1996, **86**, 495–501.
102. M. Fenton and J. Gralla, *J. Biol. Chem.*, 2003, **278**, 39669–39674.
103. K. L. Ohlsen and J. D. Gralla, *J. Biol. Chem.*, 1992, **267**, 19813–19818.
104. S. P. Haugen, M. B. Berkmen, W. Ross, T. Gaal, C. Ward and R. L. Gourse, *Cell*, 2006, **125**, 1069–1082.
105. C. Vrentas, T. Gaal, W. Ross, R. Ebright and R. Gourse, *Genes Dev*, 2005, **19**, 2378–2387.
106. B. Paul, M. Berkmen and R. Gourse, *Proc. Natl. Acad. Sci. USA*, 2005, **102**, 7823–7828.
107. R. Mathew and D. Chatterji, *Trends Microbiol*, 2006, **14**, 450–455.
108. N. Reppas, J. Wade, G. Church and K. Struhl, *Mol. Cell*, 2006, **24**, 747–757.
109. T. Ellinger, D. Behnke, H. Bujard and J. D. Gralla, *J. Mol. Biol.*, 1994, **239**, 455–465.
110. H. Tagami and H. Aiba, *Proc. Nat. Acad Sci USA*, 1999, **96**, 7202–7207.
111. B. Sclavi, C. Beatty, D. Thach, C. Fredericks, M. Buckle and A. Wolfe, *Mol Microbiol*, 2007, **65**, 425–440.
112. L. Paul, P. Mishra, R. Blumenthal and R. Matthews, *BMC Microbiol*, 2007, **7**, 2.
113. P. Eichenberger, S. Dethiollaz, H. Buc and J. Geiselmann, *Proc. Nat. Acad. Sci. USA*, 1997, **94**, 9022–9027.
114. S. Lee and J. Gralla, *Mol. Cell*, 2004, **14**, 153–162.
115. R. Laishram and J. Gowrishankar, *Genes Dev.*, 2007, **21**, 1258–1272.
116. R. S. Johnson and R. E. Chester, *J. Mol. Biol.*, 1998, **283**, 353–370.
117. G. M. Dhavan, D. M. Crothers, M. R. Chance and M. Brenowitz, *J. Mol. Biol.*, 2002, **315**, 1027–1037.
118. D. C. Straney and D. M. Crothers, *Cell*, 1985, **43**, 449–459.
119. D. C. Straney and D. M. Crothers, *J. Mol. Biol.*, 1987, **193**, 279–292.
120. T. Kubori and N. Shimamoto, *J. Mol. Biol.*, 1996, **256**, 449–457.
121. M. Susa, R. Sen and N. Shimamoto, *J. Biol. Chem.*, 2002, **277** 15407–15412.
122. O. Ozoline, T. Uteshev, I. Masulis and S. Kamzolova, *Biochim. Biophys. Acta.*, 1993, **1172**, 251–261.

123. M. Susa, T. Kubori and N. Shimamoto, *Mol. Microbiol.*, 2006, **59** 1807–1817.
124. S. Rosenberg, T. R. Kadesch and M. J. Chamberlin, *J. Mol. Biol.*, 1982, **155**, 31–51.
125. J. H. Roe, R. R. Burgess and M. Record Jr, *J. Mol. Biol.*, 1984, **176**, 495–522.
126. V. Petri and M. Brenowitz, *Curr. Opin. Biotechnol.*, 1997, **8**, 36–44.
127. M. L. Craig, W. C. Suh and M. T. Record Jr, *Biochemistry*, 1995, **34**, 15624–15632.
128. A. Sanderson, J. Mitchell, S. Minchin and S. Busby, *FEBS Lett.*, 2003, **544**, 199–205.
129. K. Barne, J. Bown, S. Busby and S. Minchin, *EMBO J.*, 1997, **16** 4034–4040.

CHAPTER 3
Intrinsic In vivo Modulators: Negative Supercoiling and the Constituents of the Bacterial Nucleoid

GEORGI MUSKHELISHVILI[a] AND ANDREW TRAVERS[b]

[a] Jacobs University, Campus Ring 1, D-28759 Bremen, Germany; [b] MRC Laboratory of Molecular Biology, Hills Road, Cambridge CB2 0QH, UK

3.1 Introduction

The integration of three-dimensional structures induced by DNA superhelicity and transcriptional regulation is a fundamental mechanism facilitating the adaptation of bacteria to different environmental situations. During the bacterial growth cycle changes in DNA superhelicity are paralleled by, and reciprocally dependent on, changes in the composition of RNA polymerase and of the abundant nucleoid-associated DNA-binding proteins stabilizing the three-dimensional structure of the DNA.

3.2 DNA Superhelicity – Structures and Implications

The interaction of RNA polymerase with DNA *in vivo* occurs in the context of bacterial "chromatin" and is thus dependent on both the presence of and also the DNA structures conferred by abundant DNA-binding proteins that organize the bacterial nucleoid. In both the bacterial nucleoid and the eukaryotic nucleus the DNA is maintained largely in a negatively supercoiled state. DNA supercoiling determines both the three-dimensional structures of the DNA polymer and its ability to be untwisted prior to transcription, replication and site-specific recombination. For transcription the question is to what extent the changes in DNA twist and bending conferred by negative supercoiling influence the process as a whole.

In the protein-free DNA molecule, DNA superhelicity is partitioned into a twist component, Tw (which is reflected in a twisting or untwisting of the double helix for positively and negatively supercoiled DNA respectively) and a writhe component, Wr, which is a measure of the three-dimensional path, usually a coil, of the double helical axis. In a closed topological domain these quantities are related to a change in linking number (ΔLk) from the relaxed state such that $\Delta Lk = \Delta Tw + Wr$ (see ref. 1 for a full discussion). The introduction of negative supercoils into DNA requires energy that is then available for driving processes such as transcription and replication initiation. This energy is described by Equation (3.1):

$$E = k_B T (\Delta Lk)^2 / 2 < (\Delta Lk)^2 > \qquad (3.1)$$

where $<(\Delta Lk)^2>$ represents the variance of the Gaussian distribution of DNA topoisomers. Importantly, the available energy is proportional to the square of the difference in linking number conferred by the application of superhelical torsion.

Since the total linking number is a constant for a closed domain, twist and writhe can be partitioned in different ways and hence the DNA can assume different structures. In particular, supercoiled DNA can adopt either toroidal or plectonemic configurations, respectively 1-start or 2-start helical structures. A further important parameter is the surface helical repeat, which is defined as the number of base-pairs between geometrically equivalent points when the path of DNA is constrained on a real, usually protein, or a virtual surface. For example, the surface helical repeat (h) of DNA wrapped on the histone octamer is 10.2 bp but the intrinsic repeat (H) is unchanged.[2] Since $h < H$ the coil is a left-handed toroid assuming a cylindrical geometry.

Another characteristic of supercoiled DNA, the distinction between constrained and unconstrained supercoils is biologically relevant. *In vivo* the wrapping of supercoils by proteins can constrain supercoils that are then less available for utilization in other processes. Similarly a transcribing polymerase constrains the equivalent of approximately one negative supercoil in its transcription bubble. In contrast, in free DNA the superhelicity is available to drive conformational transitions in DNA. It is this latter unconstrained superhelicity or effective superhelical density of DNA that is important for assessing the effect of negative superhelicity on transcription by RNA polymerase. In bacteria the assessment of this quantity *in vivo* is complicated by two factors. First, as measured by plasmid DNA supercoiling, superhelical density is dependent on growth phase[3-5] and, second, the extent to which different topological domains can differ in superhelical density is unknown. Using trimethylpsoralen to crosslink supercoiled DNA Sinden et al.[6] estimated the average superhelical density of chromosomal DNA to be -0.05 ± 0.01. In contrast measurements of the cruciform and Z-DNA formation on plasmid DNA indicated a supercoil density of -0.025 in vivo.[7,8] A similar value was deduced from the characterization of the products of the λ integrase reaction on plasmid DNA in vivo.[9] An important caveat for the lower values is that the transitions observed could depend on the form of superhelical DNA, either toroidal or plectonemic, since the partition between twist and writhe differs depending on the form. Finally, the dynamic nature of constraint – e.g. the supercoil constraining protein HU has fast on/off rates – means that supercoil availability may depend, at least in part, on competition between abundant constraining proteins, such as HU, and utilizing proteins, such as RNA polymerase.

There is abundant evidence that chromosomal DNA is organized into topological domains, which are insulated from their immediate neighbours. The extent of these domains has, until recently, been ill-defined but was initially believed to be ~50–100 kb.[10] One approach to a more precise determination of domain size is to alter local, rather than global topology. Postow et al.[11] accomplished this by the elegant technique of relaxing DNA in the vicinity of supercoiling-sensitive genes using the restriction enzyme SwaI expressed in vivo. By measuring the changes in expression of a number of these genes and taking into account the distance from the cleavage site these authors deduced that the topological domains are, on average, ~10 kb in extent, and importantly their boundaries are dynamic. Very similar conclusions were reached by an entirely independent technique in which the ability of site-specific recombination to take place between sites spaced at different intervals on the Salmonella genomic DNA is determined.[12,13] For the Tn3 resolvase such recombination can only occur if the two res sites are in the same supercoiled domain and consequently this assay provides a measure of the size of the domain. Previous studies using this method suggested that the number of domains in the chromosome was ~150 (~30 kb per domain) but this estimate depends on the assumption that domain boundaries are stable as long as the resolvase itself remains in the cell.[14] With a different assay using an engineered resolvase of reduced half-life the estimated number of domains increased to ~400, also implying that domain boundaries are dynamic. Strikingly the average domain size of ~10 kb corresponds to the average length of the supercoiled loops that are observed in electron micrographs of isolated chromosomes,[11] suggesting, although not proving, that these loops correspond to the functional supercoiled domains.

Discussion of the effects of negative supercoiling on transcription requires an understanding of the structural nature of DNA supercoils in vivo. In vitro negatively supercoiled plasmid DNA adopts the lower energy plectonemic configuration.[15-17] The plectonemic form has also been visualized in crude lysates of E. coli but, since plectonemes and toroids are interconvertible, transitions during preparation cannot be fully excluded. The additional compaction of plectonemes required for packing within the nucleoid can be mediated by the formation of a tightly packed liquid-crystalline mesophase whose structural features are determined by supercoiling.[18] Additionally the chirality of the reaction products of the λ integrase *in vivo* can only be derived by the enzyme acting on the plectonemic form and thus provides direct evidence that DNA can exist in this form, even if only transiently, in bacteria.[19] Supporting this conclusion is the observation that the nucleoid-associated protein H-NS that constrains negative supercoils both *in vivo* and *in vitro* packages DNA as a plectoneme.[20,21]

3.3 Structure of the Bacterial Nucleoid

The bacterial nucleoid is a dynamic entity whose protein composition is strongly dependent on growth phase. The proteins that constrain its structure are generally highly abundant with total intracellular concentrations varying between 20 and 100 μM (Table 3.1).[22] Of these nucleoid associated proteins (NAPs), FIS and the HUα$_2$ dimer are most abundant during early exponential growth while HUαβ heterodimer apparently replaces the HUα dimer and becomes more abundant during late exponential and stationary phases. The paralogues H-NS and StpA show less variation with growth phase, as does CRP, although StpA is more abundant during exponential phase. In contrast, both IHF and Dps, especially the latter, are more abundant in stationary than in exponential phase. In principle, at any one time these proteins can cover most but not all of the *E. coli* genome. Another highly abundant protein, Hfq, commonly included in the NAPs, binds to DNA *in vitro* but likely functions *in vivo* primarily as an RNA chaperone.[23]

NAPs can constrain particular structures of supercoiled DNA and may also be involved in the delimitation of topological domains. The DNA in plectonemes is bent both in the interwindings and at the apical loops. Consistent with this, all abundant NAPs induce and stabilize substantial DNA bends and additionally a subset of these constrains negative supercoils *in vitro*. FIS, H-NS and CRP all bind with high affinity to specific sites but can also bind less specifically to a wide variety of other sequences.[24–26] *In vitro* both HU and H-NS constrain negative supercoils at a superhelical density similar to that observed *in vivo*[27,28] while FIS and CRP also constrain negative superhelicity but at a much lower superhelical density.[21,29] These differences in the level of constrained supercoiling may be related to the type of structure recognized. While H-NS constrains the interwindings of plectonemes by bridging between the two duplexes,[20,21,30] FIS, and probably LRP, constrain DNA loops, which likely correspond to plectonemic apical loops.[31]

In *E. coli* the HU α$_2$ and αβ dimers, but not the β$_2$ dimer, constrain negative supercoils *in vitro*,[32] which is consistent with a loss of negative superhelicity in *hupAhupB* mutants.[33] However, a mutation in the α subunit confers the ability to constrain positive supercoils *in vitro* and increases the normally negligible amount of positive supercoils detected *in vivo*.[34] A crystal structure of HU from *Anabaena* (whose genome encodes only a single HU polypeptide) revealed that

Table 3.1 Summary of the estimated number and concentrations of the most abundant DNA binding proteins in *Escherichia coli*.

Protein[a]	Exponential phase Number/cell	Concentration (µM)	Early stationary phase Number	Concentration (µM)	Structure stabilized
CRP	6000–12000	5–10	Not appreciably modified	Apical loop[b]	
Dps	8000	7	120000	100	
FIS	60000	50	Not detectable		Apical loop[b] (1h toroid)
H-NS (aka H1)	20000	17	15000	13	rh plectoneme
HU	55000	45	25000	8	Lh toroid
IHF	10000	8	50000	41	~175° DNA bend
StpA	25000	28	15000	13	
Totals		155		175	
RNA polymerase	4000	3.5			Apical loop[b]

[a]CRP (aka CAP): Cyclic AMP Receptor Protein (aka Catabolite Activator Protein); Dps protein: DNA Protection during Starvation protein; FIS: Factor for Inversion Stimulation; H-NS (aka H_1) : Histone-like Nucleoid Structuring protein; HU: Stands now for Histone-like Unwinding protein; IHF: Integration Host Factor; StpA: An H-NS paralog; The different subunits of HU and IHF are not distinguished. Some proteins may also bind RNA but Hfq, which is believed to be an RNA chaperone, is excluded. Data mainly taken from.[22]
[b]Only in certain cases.

HU both untwists the DNA and creates two tight bends of $\sim 60°$ each, related by a dihedral angle of $\sim 50°$.[35] This observation implies that HU coils a DNA duplex in a left-handed sense and thus, to constrain negative superhelicity the DNA must assume a toroidal, rather than a plectonemic form. A probable important biological role of HU is to increase the axial and torsional flexibilities of DNA *in vivo*, possibly by creating "flexible hinges" in the DNA. *In vitro* at concentrations equivalent to the *in vivo* concentration of ~ 1 molecule per 100 bp DNA HU substantially decreases the persistence length of DNA[36] in a similar manner to the eukaryotic HMGB proteins.[37–39] This role of HU as a DNA chaperone likely promotes the formation of short DNA loops such as the *lac* O_1-O_3 repression loop[40] and the enhancer loop in the Hin invertasome.[41] This function of HU is antagonized by H-NS,[42] suggesting that HU at low concentrations and H-NS respectively increase and decrease DNA flexibility and may also stabilize DNA coils of different senses (Figure 3.1). However, at higher HU/DNA ratios a more rigid filamentous structure is formed.[36] With the HUαβ dimer this filament crystallizes with the DNA in the form of a left-handed toroid with 64–68 bp for every complete superhelical turn, equivalent to 8 to 8.5 bp per HUαβ monomer.[43] This structure thus agrees well with DNase I footprinting data indicating a periodicity of 8.5 bp[44] and is also probably the same as the one originally visualized in electron micrographs.[27] Similarly, FIS can also wrap DNA toroidally in a left-handed sense.[31]

Like HU the structurally-related protein IHF exists in *E. coli* as a heterodimer that bends DNA by $\sim 170°$,[45,46] but unlike HU does not constrain a significant amount of negative superhelicity. This property is consistent with

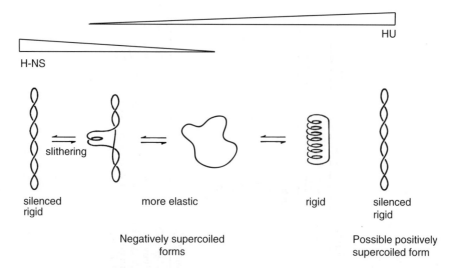

Figure 3.1 Topological forms stabilized by H-NS and HU proteins. The forms containing an isolated DNA coil are toroids; those containing two interwound coils are plectonemes. Toroids and plectonemes can coexist in the same DNA molecule.

the much lower dihedral angle of $\sim 5°$ relating the two induced DNA bends.[35] For LRP the sign of the constraint of superhelicity is not fully established. Whereas the *B. subtilis* protein apparently constrains a positive loop – as visualized by electron microscopy[47] – the *E. coli* protein constrains a negative loop.[48] Not only NAPs but also dedicated transcriptional repressors that constrain repression loops operate in the context of supercoiled DNA. In particular both the LacI and AraC looping proteins constrain loops with a surface repeat of 11.1–11.2 bp.[49–51] This value can be interpreted either in the context of a plectoneme[17] or of a left-handed toroidal coil.

Two NAPs, H-NS and FIS, have been implicated in the establishment of topological domains *in vivo*.[52] The stabilization of plectonemes by H-NS[20,21] would be consistent with such a role. FIS can also delimit apparent domains *in vitro*.[21] The mechanism by which this is accomplished is unclear; possibly FIS may stabilize nodes and/or apical loops.[21,31]

The changes in nucleoid protein composition with growth phase are accompanied by changes in the overall compaction and morphology. During the exponential phase of growth the nucleoid is elongated but in late stationary phase assumes a more spherical, compact shape largely conferred by Dps.[53] Paralleling these changes in morphology are changes in the overall superhelicity, *i.e.* the sum of the constrained and unconstrained supercoils, which in part reflect the ability of the constituent NAPs to constrain negative supercoils. In *E. coli* the average negative superhelical density is highest in early exponential phase and then gradually declines in late exponential phase and the transition to stationary phase.[4] Notably, the NAPs FIS and HUα_2, which are most abundant during the early exponential phase of growth, are among those that potentially stabilize meta-stable toroidal forms of negatively supercoiled DNA.

3.4 Supercoiling Utilization

The process of transcription alters the topology of DNA. The formation of the open initiation complex is accompanied by the unwinding of DNA to create a transcription bubble 10–12 bp in extent.[54,55] This is a measure of the local untwisting. Direct measurements of the total topological change as an open complex is formed yield somewhat higher values of unwinding. [56,57] This additional topological unwinding at some promoters has been interpreted as a wrapping of promoter DNA around the holoenzyme.[57] The extent of wrapping at different promoters has been determined directly by AFM measurements and depends on promoter organization (see below). This measurement of wrapping – essentially a length – is not the same as the constrained writhe deduced from direct measurements of changes in topology. During elongation, if the rotations of both the DNA and the enzyme are impeded, DNA winding will be respectively increased and decreased in front of and behind the transcribing enzyme.[58] However, the net topological change associated with elongation is zero. This implies that if elongation occurs on a supercoiled DNA the increase in DNA winding ahead of the enzyme will depend on the density of the transcribing polymerases within a given transcription unit and, most importantly, will be counterbalanced by a corresponding decrease in winding behind the enzyme (Figure 3.2A). This asymmetry can thus act to maintain promoter DNA in a negatively supercoiled state. If the polymerase density is low – *e.g.* one polymerase/transcription – then the passage of a single enzyme is unlikely to cancel the available negative superhelicity (Figure 3.2B). If, however, the density is high, cancellation is likely. Finally, termination is accompanied by the collapse of the transcription bubble and release of the unwinding contained within the transcription bubble into the body of the adjacent DNA.

These considerations imply that negative supercoiling should promote any stage of the transcription process that requires DNA unwinding. Early studies showed that negative supercoiling could enhance both the initial binding of

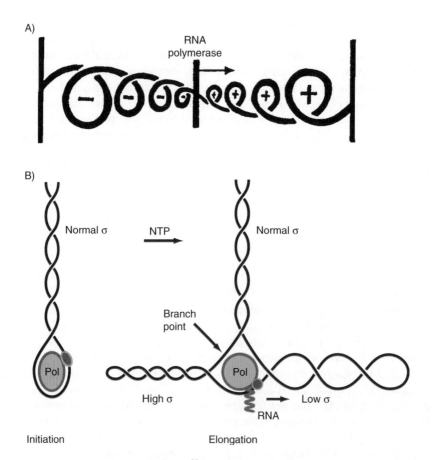

Figure 3.2 (A) The Liu and Wang[58] model for the generation of superhelicity by transcription of a relaxed DNA template. (B) An elongating RNA polymerase within a negatively supercoiled environment. Compaction of DNA is maintained before and after the transcribing enzyme. At high polymerase densities the downstream DNA would be more relaxed or even positively supercoiled.

polymerase, K_B, or increase the rate of open complex formation[59] in a promoter-dependent manner (for the definition of these terms see Chapter 2). Enhancement of K_B implies facilitation of structural recognition while enhancement of k_2 alone suggests a direct effect on DNA unwinding within the promoter.

Evidence that negative supercoiling can promote initial complex formation comes from the nature of the complexes formed on the *lacUV5* and *tyrT* promoters. With a relaxed DNA template the formation of the initial complex at the *lacUV5* promoter constrains ~1 negative supercoil ($\Delta Lk \sim -1$) while subsequent untwisting to form the open complex results in the additional constraint of 0.8–0.9 negative supercoils.[57] The initial constraint of a negative supercoil without untwisting implies that the DNA is writhed and thus that the promoter is likely wrapped by the polymerase holoenzyme.[60] Footprinting of the open complex revealed a polymerase contact ~90 bp upstream of the transcription startpoint.[61] Direct visualization by AFM of an open polymerase/promoter complex at another promoter, λ P_R, showed that polymerase wrapped ~90 bp.[62] This wrapping is mediated by the interaction of an α CTD with an AT-rich sequence resembling a classical UP element.[63] However, the same technique showed that on a 1 kb DNA fragment containing the *tyrT* promoter a single polymerase molecule wraps ~150 bp of DNA (Figure 3.3). In this complex and under these conditions (4 °C) neither the DNA in the –10 region nor that at the transcription start point would be significantly untwisted. This wrapping specifically requires the upstream activating sequence (UAS) proximal to the UP element,[31] suggesting that the enzyme can make direct upstream contacts. Such contacts were initially inferred from UV-photofootprinting that demonstrated that the structure of the DNA 120–130 bp upstream of the transcription start site was distorted by polymerase binding.[64] This sequence again resembles an UP element. Furthermore the rate at which this distortion appeared was more than 100-fold greater on negatively supercoiled DNA than on relaxed DNA.[64] This observation argues that the constrained loop is

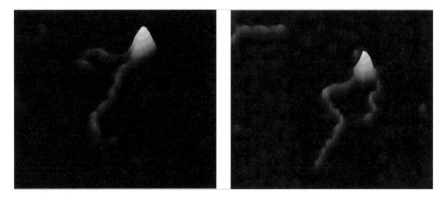

Figure 3.3 AFM image of RNA polymerase σ^{70} holoenzyme bound at the *tyrT* promoter and wrapping ~150 bp of DNA.
(Reproduced from Maurer *et al.*[31] with permission.)

negatively writhed, an inference supported by the finding that the activator FIS, which can bind within the loop constrained by the polymerase, itself wraps the same DNA as a left-handed toroid.[31] In a DNA plectoneme this loop could potentially correspond to an apical loop. The preferential interaction of the upstream region of the *tyrT* promoter with supercoiled DNA implies that this region modulates the dependence of *tyrT* transcription on negative superhelicity, but the core promoter itself is strongly activated by negative superhelicity both *in vivo* and *in vitro*.[65] Without the upstream element the DNA constrained by RNA polymerase is ~60 bp,[31,63] a length that is probably insufficient to induce a significant amount of looping.

The facilitation of untwisting by negative supercoiling *in vitro* can be understood in the context of the structure of the plectonemic supercoiled DNA template. Both nucleation and the formation of an open complex would, in principle, be facilitated by the application of torque. This could be derived from the unwinding of a single node (crossing) in a plectoneme, which would both drive and topologically compensate for the untwisting at the transcription start point (Figure 3.4A).[66] Such transitions would constitute a straightforward mechanism for torsional transmission.[67] A direct consequence of torsional transmission by node unwinding is that if this process is blocked – *e.g.* by a protein such as H-NS bridging between the two duplexes – open complex formation will be impeded (Figure 3.4B). Indeed, blocking has been observed at the *Salmonella typhimurium proU* and the *E. coli rrnB* P1 promoters.[68,69] Further, the release of unwinding associated with transcription bubble collapse at termination should allow the reformation of a node.

In bacteria, transcription elongation also occurs in the context of DNA supercoiling. Functionally there is some evidence that negative superhelicity increases pausing[70] but the mechanism by which this occurs is not established. However, in the context of a supercoiled DNA molecule the size of the enzyme would limit its location. *In vitro* on supercoiled plasmid DNA, which comprises a single closed topological domain, elongating RNA polymerase occupies an apical loop.[71] In this situation any superhelicity generated by transcription, *i.e.* downstream a reduction of negative superhelical density and upstream the converse,[58] would in principle be dissipated by the circularity. However, on domains defined by clamping of a linear stretch of DNA, any generated superhelicity would likely be conserved in adjacent interwound regions of different superhelical density with the polymerase itself residing at a branch point (Figure 3.2B). This feature has the important consequence that the interwindings preceding the polymerase would be loosened (become less negatively supercoiled) by the passage of the enzyme whereas those following the enzyme would become tighter.

3.5 Promoter Structure and DNA Supercoiling

Promoters exhibit widely differing responses to changes in negative superhelicity, an effect that is strongly dependent on their sequence organization. The response to negative superhelicity can depend on multiple elements within the

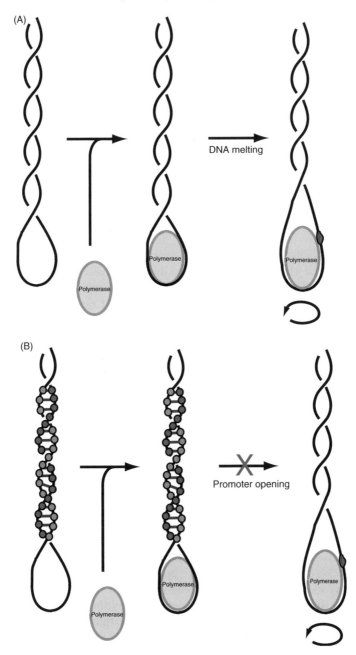

Figure 3.4 (A) A mechanism for negative superhelicity to facilitate untwisting of DNA at or close to the transcription startpoint. Untwisting within the promoter is coupled to the loss of a node in an adjacent plectoneme. (Reproduced from Travers and Muskhelishvili[65] with permission.) (B) Possible mechanism of transcriptional repression by H-NS. H-NS prevents the loss of a node and the polymerase–promoter complex is maintained in the closed state.

core promoter. In general the sequence of promoters whose activity is increased by higher superhelical densities, both *in vivo* and *in vitro*,[65] differs from the optimal sequence for initiation on relaxed DNA templates. Early studies using derivatives of the *lacUV5* promoter showed that the spacing between the −35 and −10 hexamers was important. The less active (on linear DNA) promoters with suboptimal 16 and 18 bp spacers were strongly stimulated by superhelicity.[72] Notably an increase in the spacer from 16 to 17 bp decreases promoter activity at high superhelical densities even though on relaxed DNA templates the converse is observed.[73] The precise response to increasing negative superhelical density depends on the promoter architecture – see for example the *proU* promoter,[74] but in general an enhanced response is apparent in promoters with −10 and −35 hexamers that deviate from the consensus sequence and containing a G/C rich discriminator.[65,72,75]

These early studies also revealed that any stimulatory effects of negative superhelicity on transcription did not increase uniformly with increasing negative superhelical density. Instead, when assayed under equivalent conditions, individual transcription units exhibited different optimal superhelical densities.[65,72,76] This phenomenon has not been investigated in detail but may reflect differing superhelical dependencies of different rate-limiting steps.

Among the natural promoters whose activity is strongly dependent on high superhelical densities both *in vivo* and *in vitro* are some that direct tRNA and rRNA transcription.[65,77–79] Although within this class of promoter there is variability in the dependence on superhelicity certain features of the core promoter region appear critical. Typically this region includes a suboptimal −35 hexamer, a 16 bp spacer and a G/C-rich region (the discriminator) between the −10 hexamer and the transcription startpoint. However, often the −10 region approaches or is a "consensus" sequence. This organization implies that untwisting of the −10 region during nucleation is facile while the subsequent progression of the untwisted region to the transcription startpoint is impeded by the discriminator. Changing the −35 hexamer or the spacer to the consensus concomitantly activates the promoter and abrogates the stimulation by superhelicity both *in vitro* and *in vivo*, while changing the discriminator from a G/C-rich sequence to an A/T-rich sequence both reduces superhelicity dependence and increases activity.[64,65] These effects of changing the base-composition of the discriminator link negative superhelicity to the stringent regulation of stable RNA synthesis. Indeed, in *Salmonella typhimurium* hypernegative superhelicity can override the "stringent regulation" of a tRNA promoter *in vivo*.[80] (The stringent response describes the cessation of stable RNA accumulation on starvation for an amino acid; cells that fail to shut off RNA accumulation under these circumstances are said to be relaxed.)

3.6 Role of RNA Polymerase Composition

To coordinate the gene transcription with physiological demands of the cell the RNA polymerase holoenzyme has to integrate the entire information provided by environmental signals, including various kinds of stress (Chapter 1). Commitment of *E. coli* to growth by subculturing stationary cells in fresh medium and also the growth phase transitions, as well as environmental stimuli eliciting the stress response, affect the global DNA superhelicity[3,69,81–83] and thus necessitate the adaptation of RNA polymerase holoenzyme to altered template topology. How this adjustment of RNA polymerase and template topology is brought about is a central question.

3.6.1 Exchange of σ Factors

The RNA polymerase major σ^{70} initiation subunit (product of the *rpoD* gene) is involved in the transcription of most of the genes during exponential phase of growth, whereas upon exhaustion of nutrients and transition to stationary phase the *rpoS* gene product σ^S directs the transcription of a large set of genes required under poor growth conditions.[84–89] Thus, a possible mechanism to accomplish the adjustment of polymerase to altered template topology would be to shift the balance of competition between alternative σ factors for the RNA polymerase core enzyme,[90] especially since the $E\sigma^{70}$ and $E\sigma^S$ holoenzymes, respectively, confer a preference for highly supercoiled or more relaxed templates for transcription.[4,91] In this model the response to environmental stress or, for example, a mutation leading to relaxation of DNA would entail a shift in RNA polymerase holoenzyme composition by increasing the proportion of $E\sigma^S$ and redirecting transcription to genes required for coping with the challenge under DNA relaxation regime. However, while the intracellular concentration of σ factors varies with growth phase, the balance of σ factor competition can be modulated by anti-sigma factors and also by the "alarmone" ppGpp.[89,92–94] Furthermore, this competition possibly depends on the overall superhelicity of the DNA. Indeed, despite the accumulation of σ^S protein the induction of σ^S-dependent *osmE* promoter is precluded by high negative supercoiling and is only observed with the onset of DNA relaxation.[4]

In addition, recent studies strongly suggest that the ω subunit of RNA polymerase (product of *rpoZ* gene) is intimately involved in both σ factor balance and sensing of superhelicity.[95] This evolutionarily conserved subunit is stably associated with highly purified preparations of RNA polymerase and likely plays a role in stabilizing the RNA polymerase core–σ^{70} interactions.[96–98] The effects of ω on the enzymology of RNA polymerase are discussed in Chapter 8; here we note that structural studies suggest that ω is acting as a molecular "latch" reducing the configurational entropy of the RNA polymerase β' subunit.[97] Furthermore, it was observed that lack of ω affects the ppGpp sensitivity and promoter untwisting by RNA polymerase *in vitro*,[99,100] supporting the hypothesis that ω might be involved in optimization of RNA polymerase function in facilitating the transcription of highly negatively supercoiled DNA.[101] Indeed, the *rpoZ* mutant cells demonstrate an overall relaxation of DNA and a proportional increase of σ^S-programmed transcription during exponential growth.[95,102] In keeping with this observation purified

RNA polymerase holoenzyme preparations isolated from this mutant are severely depleted in σ^{70}.[95] Furthermore, overproduction of σ^{70} in *rpoZ* cells was found to increase the overall superhelicity and also cause early accumulation of σ^S without concomitant induction of σ^S-dependent transcription,[90,95] again suggesting a link between DNA superhelicity and the impact of a σ factor in global transcription. Notably, since in rich medium the *rpoZ* mutant can grow at rates comparable to wild type, this suggests that stable RNA synthesis, which is exquisitely dependent on $E\sigma^{70}$-programmed polymerase, can be sustained in these cells despite global relaxation of DNA. This observation is consistent with a recent study of *E. coli* populations that were allowed to evolve and where it was concluded that reduction of overall superhelicity *per se* does not preclude fast growth.[103]

3.6.2 Auxiliary Subunits

The response to superhelicity is also modulated by auxiliary subunits that engage in the secondary channel of RNA polymerase. These include DksA, GreA and GreB. As discussed in Chapter 8, the latter two subunits are involved in the "backtracking" of RNA polymerase. However, another role of the Gre factors is to facilitate the transition of transcription initiation complexes to elongation,[104–106] although the dependence of this effect on supercoiling has not yet been elucidated. In this context it is noteworthy that the DksA protein structurally related to the Gre factors has been implicated in formation of topological barriers to supercoil diffusion.[52]

3.6.3 Role of ppGpp

The influence of ppGpp on σ factor competition shifts the balance towards the $E\sigma^S$ holoenzyme.[86,93,107] This regulatory nucleotide is thought to function as a general growth arrest signal negatively affecting the stable RNA synthesis by directly interacting with $E\sigma^{70}$ and impairing its ability to initiate transcription at this class of promoters.[108–112] (For a discussion of the enzymology of ppGpp action see Chapter 8.) Cessation of stable RNA synthesis upon binding of ppGpp would release a significant fraction of supercoils constrained by RNA polymerase, making them available to DNA-relaxing topoisomerases, whereas relaxation of DNA would render the chromosomal template preferentially transcribable by $E\sigma^S$ programmed polymerase.[4,101] The production of ppGpp itself is likely linked to DNA superhelicity, as the *gppA* gene converting pppGpp to ppGpp is repressed by DNA relaxation.[113]

3.7 Model and Implications

The sensing of and regulation by DNA supercoiling is a central facet of the integration of environmental signals into the concerted control of gene expression. It appears to be mediated in large part by the bacterial RNA polymerase itself. How is this regulation accomplished at the molecular level? As already mentioned above, sensing of superhelical density depends on both the type of σ factor present in the holoenzyme and also on the presence of the ω subunit[91,95] (Figure 3.5). While, relative to σ^S, σ^{70} can preferentially direct the utilization of templates of high negative superhelical density, this preference is attenuated *in vivo* in the absence of ω.[95] However, since the phenotypic effects of loss of ω can be suppressed by overproduction of σ^{70} the likelihood is that ω determines the relative affinities of σ^S and σ^{70} for the core polymerase and therefore that sensing depends in part on the relative occupancy of core polymerase by the two major σ factors.

The ω subunit not only affects the sensing of superhelicity but also attenuates the response to ppGpp *in vitro*.[114] The effect of ω on ppGpp sensitivity can be compensated for by DksA.[100] In turn, GreA antagonizes the action of DksA on σ^{70}-containing holoenzyme by promoting the initiation of transcription from *rrnB* P1 on supercoiled DNA.[115] Intriguingly, RNA polymerase purified from bacteria lacking ω is enriched in a third auxiliary subunit, GreB (G. Muskhelishvili & M. Geertz, unpublished data), while *in vitro* GreB has an effect similar to DksA on rRNA transcription.[116] However, some phenotypic effects of DksA are independent of, and may even oppose, those of ppGpp.[117] These observations suggest that in addition to determining the relative affinities for

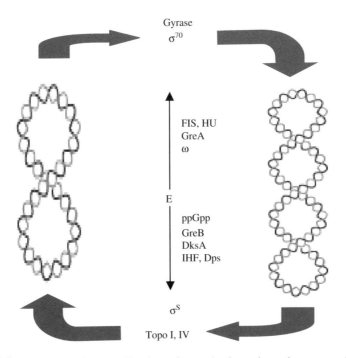

Figure 3.5 Model for the coordination of growth-phase dependent transcription by the integration of the superhelical density of DNA with the availability of different sigma factors, auxiliary polymerase-associated proteins and nucleoid associated proteins.

different sigma factors ω may also modulate the function of the different secondary channel binding auxiliary subunits.

These studies identify at least four components involved in the integrated regulation of RNA polymerase selectivity – the major σ factors, the ω subunit, the auxiliary subunits GreA, GreB and DksA and the regulatory nucleotide ppGpp. These findings can be integrated into a model originally proposed by Travers et al.[118] The ω subunit was initially implicated as an assembly factor for RNA polymerase in vivo.[97] It binds on the surface of the enzyme to β′ distant from the secondary channel. The effects of ω on σ selection and on the requirements for the auxiliary subunits are consistent with a mechanism in which ω subtly alters the structure of the core polymerase, concomitantly facilitating σ^{70} binding at the expense of σ^S and possibly altering the environment of the secondary channel. In addition to altering supercoiling, sensing these changes would be coupled to an increase in the sensitivity to ppGpp. This argument implies that DksA and possibly ppGpp itself could effect similar changes even in the absence of ω.

Evidence from zone sedimentation[119] and from circular dichroism studies[120] suggests that ppGpp elicits substantial conformation changes in the *E. coli* holoenzyme on binding. Yet the crystal structures of *Thermus thermophilus* holoenzyme in the presence and absence of ppGpp revealed no evidence for such changes.[112] However, two essential components of polymerase holoenzyme, the CTDs of the α subunits, are not visible in such crystal structures. Yet these same CTDs are known to be required for binding to the upstream regions of promoters[121–123] that can form structures equivalent to the apical loops of supercoiled DNA plectonemes.[31,63]

A radical interpretation of these observations rests on the finding that the interaction of the α CTDs with the body of the holoenzyme depends, at least in part, on an electrostatic contact between a basic sequence in the σ factors and an acidic one in the α CTDs.[124] In σ^S one residue of the interacting sequence is slightly more basic than in σ^{70} (311-RRLR-314 in σ^S compared with 596-RKLR-599 in σ^{70}) and consequently the interaction with α is likely to be modulated. By favouring the binding of σ^{70}, ω could weaken the interaction with the two α CTDs, releasing them to bind to the apical loops of DNA plectonemes. This weakened interaction is also likely to be apparent in the free holoenzyme. Notably a mutation (R496H) in σ^{70} that further decreases the available positive charge in the sequence both increases the sensitivity to ppGpp and alters the sedimentation coefficient of the holoenzyme.[125]

One effect of ppGpp on transcription initiation at *Escherichia coli rrn* and tRNA promoters *in vitro* is to raise the temperature at which efficient initiation complex formation occurs, suggesting that ppGpp in some way raises the activation energy for productive complex formation. Under less permissive conditions any complexes formed at these promoters in the presence of ppGpp fail to initiate.[126] They are possibly related to moribund complexes described at other promoters.[127] Notably, GreA and GreB mitigate the initiation arrest dependent upon the formation of moribund complexes, but only if they are present prior to initiation complex formation.[104] These observations imply that formally the σ^{70}-holoenzyme (containing ω) can follow one of two distinct initiation pathways at least at *rrn* and stable RNA promoters: a productive pathway enhanced by GreA (and, possibly, GreB) and a "non"-productive pathway dependent on ppGpp and enhanced by DksA.[101] Similar considerations could apply to the holoenzyme lacking ω. In this case GreB is implicated as a major functional auxiliary subunit *in vitro* and *in vivo*[106] (G. Muskhelishvili & M. Geertz, unpublished observations); DksA could act both as an independent effector and as an antagonist to ppGpp[117] in concert with the σ^S form of the holoenzyme which prefers templates of low superhelical density.

3.8 Causality

Available data suggest that while changes of global superhelicity determine genomic transcription, the transcription by RNA polymerase in turn can determine global superhelicity.[113,128–130] This raises the question on causality, but what we observe here is in principle a phenomenon of interdependence, or mutual determination. The RNA polymerase subunits, NAPs and topoisomerases are thought to be connected in a network regulating DNA topology.[101] This explains why mutations in RNA polymerase, similarly to mutations in NAPs and topoisomerase genes, affect global DNA topology. Such mutations in RNA polymerase have been obtained in studies of genetic suppressors over two decades ago.[131–133] It was shown that mutation in the RpoB subunit of RNA polymerase (*rpoB203*) alters the supercoiling of isolated plasmids, suggesting a role for RNA polymerase in determining the overall superhelicity.[131] However, in contrast to *rpoZ* mutation, the *rpoB203* mutation increased the level of plasmid supercoiling, whereas another mutation in the RpoB subunit, having destabilizing effect on initiation complex formation (*rpoB114*;[134]), decreases the level of plasmid supercoiling (G. Muskhelishvili & M. Geertz, unpublished observations).

How can the mutations or compositional changes of polymerase effect alterations of DNA superhelicity? In principle, such alterations could be consequent on reorganization of cellular gene transcription. However, an altered interaction between RNA polymerase and template DNA could also directly affect DNA topology. Indeed, not only the promoter sequence organization but also the composition of RNA polymerase can affect the extent of DNA wrapped in the initial complex.[31,62,63,135] The change in the propensity of RNA polymerase to wrap DNA could in turn modulate the partitioning of constrained and unconstrained DNA supercoils in the chromosome.[136] Another possibility is a change in the interaction of RNA polymerase with DNA topoisomerases.[137,138] In addition to direct binding effects, the strength and extent of the transcriptions utilized by compositionally altered polymerase could modulate the topological barriers to supercoil diffusion and thus affect both the distributions of effective superhelicity and the condensation state of chromosomal DNA.[129,139,140] Physical evidence for the structural role of RNA polymerase in organizing nucleoid DNA is provided by the observation of transcription foci in exponentially growing *E. coli* cells.[141] These foci, thought to be structures analogous to the eukaryotic nucleolus, represent distinct accumulations of RNA polymerase in the nucleoid, which can be observed by fluorescent microscopy *in vivo* using the fusions of RpoC (RNA polymerase β' subunit) with green fluorescent protein.[140] Transcription foci are likely formed by interactions between $E\sigma^{70}$ molecules preferentially engaging in active transcription the strong stable RNA promoters spatially organized on both sides of the origin of replication. In stationary phase the foci are no longer apparent. Notably, the foci also disappear during "stringent response" in wild type cells, but not in the *relA* mutant impaired in production of ppGpp. Furthermore, consistent with the requirement of $E\sigma^{70}$ for stable RNA synthesis the foci do not form in the presence of "stringent" mutants of RNA polymerase, which are both disabled in their interactions with stable RNA promoters and also insensitive to ppGpp.[141] Thus it appears that while negative superhelicity facilitates the selective binding of $E\sigma^{70}$ polymerase at stable RNA promoters, the exceptional strength of the transcription process in turn facilitates the reshaping of DNA. Since the net topological change due to the transcription process itself is zero, the topological changes of DNA during the formation of transcription foci must involve an environment generating topological barriers. The nature of these barriers can be largely determined by polymerase itself. Indeed, it has been proposed that overproduction of σ^{70} alters the intracellular environment to favour σ^{70} transcription and, similarly, induction of σ^S alters the intracellular environment to favour σ^S transcription.[90] Put another way the alteration of supercoiling-dependent transcription is not just a matter of the activities of individual promoters responding to changes of superhelicity but rather the altered composition of the holoenzyme itself determines the spectrum of promoters that are potentially active (including those of NAPs and topoisomerase genes) and, consequently, also the level of global superhelicity.

3.9 Cooperation with Nucleoid Associated Proteins

There is now substantial evidence that NAP expression is coordinated directly or indirectly by both RNA polymerase and the NAP proteins themselves.[20,101] The expression of many NAP genes is strongly affected in both the *rpoB114* and *rpoZ* mutants[95] (M. Geertz and G. Muskhelishvili, unpublished data). The *rpoB114* "stringent RNA polymerase" mutant conferring hyponegative DNA supercoiling also overproduces the DNA architectural protein IHF, which normally accumulates in high amounts on transition to stationary phase.[22,134,142] This coupling of "stringent" polymerase – mimicking the presence of ppGpp with DNA relaxation and overexpression of a chromatin protein abundant in stationary phase – is suggestive of a sensing mechanism adjusting the transcription preferences of RNA polymerase with nucleoid structure.

One clear example of such an adjustment is the interdependence between the nucleoid protein FIS and DNA topology in activating stable RNA transcription during early exponential growth. FIS counteracts the elevation of negative superhelicity by reducing the expression of gyrase, but compensates the stable RNA promoters by constraining local supercoils, buffering transcription from fluctuations of superhelicity and thus antagonizing the inhibitory effect of ppGpp.[5,65,67,79,143-145] As expected for a homeostatic feedback mechanism in the *fis* mutant, both the negative superhelicity and the content of σ^{70} are elevated, as is the level of ppGpp[65,101] (G. Muskhelishvili & A. Travers, unpublished results). In contrast to FIS, the effects of gene mutations of LRP and H-NS that are abundant in stationary phase are coupled with modulation of σ^S content and $E\sigma^S$-dependent transcription.[146,147] In addition, the level of σ^S is regulated by HU.[148] Thus not only the mutations of polymerase affect the expression of NAPs but also, in turn, mutations of these proteins modulate the σ^{70} and σ^S levels, in keeping with the proposed global homeostatic network implied in the regulation of DNA topology[101] (Figure 3.5). Many components of this network serve as sensors for the signals provided by the growth environment and also as modulators of topoisomerase expression and activity, thus directly coupling the environmental information with the nucleoid structure and RNA polymerase composition.[113,128,146,149-154]

3.10 Conversion of Supercoil Energy into Genomic Transcript Patterns

The supercoil energy, E, given in the first equation above, can be regarded as a free energy of unwinding that determines, in part, the extent to which the promoter DNA can be melted or completely untwisted.[66] How is the channelling of DNA torque to bacterial promoters coordinated to produce physiologically meaningful expression of the genome? The growth phase-dependent distributions of genomic transcripts reveal characteristic patterns of supercoiling-sensitivity.[113,128,130] These distributions depend on the presence of abundant NAPs like FIS and H-NS and reveal growth phase-dependent domains of high negative supercoiling and DNA relaxation co-existing in the genome. Genetic screening of *E. coli* mutants with altered topological domains identified the role of *fis* and *hns* genes in the formation of topological barriers in the chromosome.[52] The same function is probably affected by other NAPs and DNA architectural proteins.[34,155,156] Competition between FIS and H-NS for cooperative binding at multiple sites in intergenic regions may serve as a mechanism for formation of dynamic barriers to supercoil diffusion instantly switching the genes and operons on or off depending on the changes of transcriptional environment.[157] This type of regulation employing the spatial proximity of loci strongly depends on the superhelical density of DNA[158] and corresponding structures possibly can spread over large genomic regions comprising entire functional systems.[20]

Thus it appears that binding and reshaping of DNA supercoils by NAPs prevents supercoil diffusion by stabilizing hundreds of topologically isolated domains, the structural dynamics of which will depend on the strength and, perhaps, also length of the transcription(s) delimited by the domain boundaries. Such a "topological differentiation" of nucleoid DNA will determine the genomic distributions of effective superhelicity and thus the spatial organization of supercoiling-sensitive transcription, which specifically differs not only between the mutant cells lacking distinct NAPs but also between the stationary and exponentially growing cells.[101,128]

3.11 Conclusions

The process of transcription is intimately linked to the overall structure and superhelicity of the DNA in the context of bacterial nucleoid. The bacterial RNA polymerase both responds to and organizes the structural environment in which it operates. Apparently, the organization of transcriptional response involves feedback circuits utilizing distinct combinations of NAPs with RNA polymerase holoenzyme and providing redundancy required for flexible adaptation to physiological demands. Nevertheless, since all the regulatory input provided by the NAP network has to be integrated by the transcription machinery, the transcriptional response to environmental signals will be necessarily contingent on altered interactions between polymerase and the chromosomal DNA template. The transcriptional response will therefore depend on the preferred initiation pathway selected in need of the optimization of structural coupling between topological state of DNA sustained by NAPs and the composition of RNA polymerase, regarding both the σ and also the auxiliary subunits. The corollary to this argument is that the physiology-dependent structural coupling between nucleoid DNA topology and the molecular composition of RNA polymerase holoenzyme is an emergent property, which is to be understood in a relation of reciprocal determination.

References

1. N. R. Cozzarelli, T. C. Boles, and J. H. White, in *DNA Topology and its Biological Effects*, N.R. Cozzarelli and J.C. Wang, Editors. 1990, Cold Spring Harbor Press: New York. 139–184.
2. S. C. Satchwell, H. R. Drew and A. A. Travers, *J. Mol. Biol.*, 1986, **191**(4), 659–75.
3. V. L. Balke and J. D. Gralla, *J. Bacteriol.*, 1987, **169**(10), 4499–506.
4. P. Bordes, A. Conter, V. Morales, J. Bouvier, A. Kolb and C. Gutierrez, *Mol. Microbiol.*, 2003, **48**(2), 561–71.
5. R. Schneider, A. Travers and G. Muskhelishvili, *Mol. Microbiol.*, 1997, **26**(3), 519–30.

6. R. R. Sinden, J. O. Carlson and D. E. Pettijohn, *Cell*, 1980, **21**(3), 773–83.
7. J. A. McClellan, P. Boublikova, E. Palecek and D. M. Lilley, *Proc. Natl. Acad. Sci. USA*, 1990, **87**(21), 8373–7.
8. W. Zacharias, A. Jaworski, J. E. Larson and R. D. Wells, *Proc. Natl. Acad. Sci. USA*, 1988, **85**(19), 7069–73.
9. J. B. Bliska and N. R. Cozzarelli, *J. Mol. Biol.*, 1987, **194**(2), 205–18.
10. R. R. Sinden and D. E. Pettijohn, *Proc. Natl. Acad. Sci. USA*, 1981, **78**(1), 224–8.
11. L. Postow, C. D. Hardy, J. Arsuaga and N. R. Cozzarelli, *Genes Dev.*, 2004, **18**(14), 1766–79.
12. N. P. Higgins, X. Yang, Q. Fu and J. R. Roth, *J. Bacteriol.*, 1996, **178**(10), 2825–35.
13. S. Deng, R. A. Stein and N. P. Higgins, *Proc. Natl. Acad. Sci. USA*, 2004, **101**(10), 3398–403.
14. R. A. Stein, S. Deng and N. P. Higgins, *Mol. Microbiol.*, 2005, **56**(4), 1049–61.
15. M. Adrian, B. ten Heggeler-Bordier, W. Wahli, A. Z. Stasiak, A. Stasiak and J. Dubochet, *EMBO J.*, 1990, **9**(13), 4551–4.
16. J. Bednar, P. Furrer, V. Katritch, A. Z. Stasiak, J. Dubochet and A. Stasiak, *J. Mol. Biol.*, 1995, **254**(4), 579–94.
17. T. C. Boles, J. H. White and N. R. Cozzarelli, *J. Mol. Biol.*, 1990, **213**(4), 931–51.
18. Z. Reich, E. J. Wachtel and A. Minsky, *Science*, 1994, **264**(5164), 1460–3.
19. S. J. Spengler, A. Stasiak and N. R. Cozzarelli, *Cell*, 1985, **42**(1), 325–34.
20. B. Lang, N. Blot, E. Bouffartigues, M. Buckle, M. Geertz, C. O. Gualerzi, R. Mavathur, G. Muskhelishvili, C. L. Pon, S. Rimsky, S. Stella, M. Madan Babu and A. Travers, *Nucleic Acids Res.*, 2007, **35**, 6330–6337.
21. R. Schneider, R. Lurz, G. Luder, C. Tolksdorf, A. Travers and G. Muskhelishvili, *Nucleic Acids Res.*, 2001, **29**(24), 5107–14.
22. T. A. Azam, A. Iwata, A. Nishimura, S. Ueda and A. Ishihama, *J. Bacteriol.*, 1999, **181**(20), 6361–70.
23. I. Moll, D. Leitsch, T. Steinhauser and U. Blasi, *EMBO Rep.*, 2003, **4**(3), 284–9.
24. E. Bouffartigues, M. Buckle, C. Badaut, A. Travers and S. Rimsky, *Nat. Struct. Mol. Biol.*, 2007, **14**(5), 441–8.
25. M. Betermier, D. J. Galas and M. Chandler, *Biochimie*, 1994, **76**(10-11), 958–67.
26. C. Q. Pan, R. C. Johnson and D. S. Sigman, *Biochemistry*, 1996, **35**(14), 4326–33.
27. J. Rouviere-Yaniv, M. Yaniv and J. E. Germond, *Cell*, 1979, **17**(2), 265–74.

28. C. S. Hulton, A. Seirafi, J. C. Hinton, J. M. Sidebotham, L. Waddell, G. D. Pavitt, T. Owen-Hughes, A. Spassky, H. Buc and C. F. Higgins, *Cell*, 1990, **63**(3), 631–42.
29. A. Kolb and H. Buc, *Nucleic Acids Res.*, 1982, **10**(2), 473–85.
30. R. T. Dame, C. Wyman and N. Goosen, *Nucleic Acids Res.*, 2000, **28**(18), 3504–10.
31. S. Maurer, J. Fritz, G. Muskhelishvili and A. Travers, *EMBO J.*, 2006, **25**(16), 3784–90.
32. L. Claret and J. Rouviere-Yaniv, *J. Mol. Biol.*, 1997, **273**(1), 93–104.
33. T. T. Paull, M. J. Haykinson and R. C. Johnson, *Biochimie*, 1994 **76**(10-11), 992–1004.
34. S. Kar, E. J. Choi, F. Guo, E. K. Dimitriadis, S. L. Kotova and S. Adhya, *J. Biol. Chem.*, 2006, **281**(52), 40144–53.
35. K. K. Swinger, K. M. Lemberg, Y. Zhang and P. A. Rice, *EMBO J.*, 2003, **22**(14), 3749–60.
36. J. van Noort, S. Verbrugge, N. Goosen, C. Dekker and R. T. Dame, *Proc. Natl. Acad. Sci. USA*, 2004, **101**(18), 6969–74.
37. T. T. Paull and R. C. Johnson, *J. Biol. Chem.*, 1995, **270**(15), 8744–54.
38. D. Payet and A. Travers, *J. Mol. Biol.*, 1997, **266**(1), 66–75.
39. M. McCauley, P. R. Hardwidge, L. J. Maher 3rd and M. C. Williams, *Biophys. J.*, 2005, **89**(1), 353–64.
40. N. A. Becker, J. D. Kahn and L. J. Maher 3rd, *J. Mol. Biol.*, 2005, **349**(4), 716–30.
41. M. J. Haykinson and R. C. Johnson, *EMBO J.*, 1993, **12**(6), 2503–12.
42. N. A. Becker, J. D. Kahn and L. J. Maher 3rd, *Nucleic Acids Res.*, 2007, **35**(12), 3988–4000.
43. F. Guo and S. Adhya, *Proc. Natl. Acad. Sci. USA*, 2007, **104**(11), 4309–14.
44. S. S. Broyles and D. E. Pettijohn, *J. Mol. Biol.*, 1986, **187**(1), 47–60.
45. P. A. Rice, S. Yang, K. Mizuuchi and H. A. Nash, *Cell*, 1996, **87**(7), 1295–306.
46. M. Lorenz, A. Hillisch, S. D. Goodman and S. Diekmann, *Nucleic Acids Res.*, 1999, **27**(23), 4619–25.
47. C. Beloin, J. Jeusset, B. Revet, G. Mirambeau, F. Le Hegarat and E. Le Cam, *J. Biol. Chem.*, 2003, **278**(7), 5333–42.
48. U. Pul, R. Wurm and R. Wagner, *J. Mol. Biol.*, 2007, **366**(3), 900–15.
49. H. Kramer, M. Amouyal, A. Nordheim and B. Muller-Hill, *EMBO J.*, 1988, **7**(2), 547–56.
50. S. M. Law, G. R. Bellomy, P. J. Schlax and M. T. Record Jr, *J. Mol. Biol.*, 1993, **230**(1), 161–73.
51. D. H. Lee and R. F. Schleif, *Proc. Natl. Acad. Sci. USA*, 1989, **86**(2), 476–80.
52. C. D. Hardy and N. R. Cozzarelli, *Mol. Microbiol.*, 2005, **57**(6), 1636–52.
53. S. G. Wolf, D. Frenkiel, T. Arad, S. E. Finkel, R. Kolter and A. Minsky, *Nature*, 1999, **400**(6739), 83–5.

54. J. M. Saucier and J. C. Wang, *Nat. New Biol.*, 1972, **239**(93), 167–70.
55. U. Siebenlist, *Nature*, 1979, **279**(5714), 651–2.
56. H. B. Gamper and J. E. Hearst, *Cell*, 1982, **29**(1), 81–90.
57. M. Amouyal and H. Buc, *J. Mol. Biol.*, 1987, **195**(4), 795–808.
58. L. F. Liu and J. C. Wang, *Proc. Natl. Acad. Sci. USA*, 1987, **84**(20), 7024–7.
59. E. Bertrand-Burggraf, M. Schnarr, J. F. Lefevre and M. Daune, *Nucleic Acids Res.*, 1984, **12**(20), 7741–52.
60. H. Buc, *Biochem. Soc. Trans.*, 1986, **14**(2), 196–9.
61. M. Buckle, H. Buc and A. A. Travers, *EMBO J.*, 1992, **11**(7), 2619–25.
62. C. Rivetti, M. Guthold and C. Bustamante, *EMBO J.*, 1999, **18**(16), 4464–75.
63. S. Cellai, L. Mangiarotti, N. Vannini, N. Naryshkin, E. Kortkhonjia, R. H. Ebright and C. Rivetti, *EMBO Rep.*, 2007, **8**(3), 271–8.
64. I. K. Pemberton, G. Muskhelishvili, A. A. Travers and M. Buckle, *J. Mol. Biol.*, 2002, **318**(3), 651–63.
65. H. Auner, M. Buckle, A. Deufel, T. Kutateladze, L. Lazarus, R. Mavathur, G. Muskhelishvili, I. Pemberton, R. Schneider and A. Travers, *J. Mol. Biol.*, 2003, **331**(2), 331–44.
66. A. Travers and G. Muskhelishvili, *EMBO Rep.*, 2007, **8**(2), 147–51.
67. G. Muskhelishvili and A. Travers, *Nucleic Acids Mol. Biol.*, 1997, **11**, 179–190.
68. R. T. Dame, C. Wyman, R. Wurm, R. Wagner and N. Goosen, *J. Biol. Chem.*, 2002, **277**(3), 2146–50.
69. C. F. Higgins, C. J. Dorman, D. A. Stirling, L. Waddell, I. R. Booth, G. May and E. Bremer, *Cell*, 1988, **52**(4), 569–84.
70. M. Krohn, B. Pardon and R. Wagner, *Mol. Microbiol.*, 1992, **6**(5), 581–9.
71. B. ten Heggeler-Bordier, W. Wahli, M. Adrian, A. Stasiak and J. Dubochet, *EMBO J.*, 1992, **11**(2), 667–72.
72. J. A. Borowiec and J. D. Gralla, *J. Mol. Biol.*, 1987, **195**(1), 89–97.
73. J. E. Stefano and J. D. Gralla, *Proc. Natl. Acad. Sci. USA*, 1982, **79**(4), 1069–72.
74. B. J. Jordi, T. A. Owen-Hughes, C. S. Hulton and C. F. Higgins, *EMBO J.*, 1995, **14**(22), 5690–700.
75. H. Giladi, S. Koby, M. E. Gottesman and A. B. Oppenheim, *J. Mol. Biol.*, 1992, **224**(4), 937–48.
76. S. M. Stirdivant, L. D. Crossland and L. Bogorad, *Proc. Natl. Acad. Sci. USA*, 1985, **82**(15), 4886–90.
77. A. Free and C. J. Dorman, *Mol. Microbiol.*, 1994, **14**(1), 151–61.
78. K. L. Ohlsen and J. D. Gralla, *Mol. Microbiol.*, 1992, **6**(16), 2243–51.
79. M. Rochman, M. Aviv, G. Glaser and G. Muskhelishvili, *EMBO Rep.*, 2002, **3**(4), 355–60.
80. N. Figueroa-Bossi, M. Guerin, R. Rahmouni, M. Leng and L. Bossi, *EMBO J.*, 1998, **17**(8), 2359–67.
81. C. J. Dorman, *Trends Microbiol.*, 1996, **4**(6), 214–6.

82. Y. C. Tse-Dinh, H. Qi and R. Menzel, *Trends Microbiol.*, 1997 **5**(8), 323–6.
83. T. Mizushima, K. Kataoka, Y. Ogata, R. Inoue and K. Sekimizu, *Mol. Microbiol.*, 1997, **23**(2), 381–6.
84. R. Hengge-Aronis, in *Bacterial Stress Responses*, eds. G. Storz and R. Hengge-Aronis, ASM Press, Washington D.C., 2000, pp. 161–178.
85. A. Ishihama, *Annu. Rev. Microbiol.*, 2000, **54**, 499–518.
86. M. Jishage, A. Iwata, S. Ueda and A. Ishihama, *J. Bacteriol.*, 1996, **178**(18), 5447–51.
87. H. Weber, T. Polen, J. Heuveling, V. F. Wendisch and R. Hengge, *J. Bacteriol.*, 2005, **187**(5), 1591–603.
88. A. Ishihama, *Genes Cells*, 1999, **4**(3), 135–43.
89. M. Jishage and A. Ishihama, *J. Bacteriol.*, 1997, **179**(3), 959–63.
90. A. Farewell, K. Kvint and T. Nystrom, *Mol. Microbiol.*, 1998, **29**(4), 1039–51.
91. S. Kusano, Q. Ding, N. Fujita and A. Ishihama, *J. Biol. Chem.*, 1996, **271**(4), 1998–2004.
92. M. Jishage and A. Ishihama, *Proc. Natl. Acad. Sci. USA*, 1998, **95**(9), 4953–8.
93. M. Jishage, K. Kvint, V. Shingler and T. Nystrom, *Genes Dev.*, 2002, **16**(10), 1260–70.
94. G. Becker, E. Klauck and R. Hengge-Aronis, *Mol. Microbiol.*, 2000, **35**(3), 657–66.
95. M. Geertz, A. Travers, N. Shimamoto and G. Muskhelishvili, submitted, 2008.
96. R. R. Burgess, *J. Biol. Chem.*, 1969, **244**(22), 6160–7.
97. L. Minakhin, S. Bhagat, A. Brunning, E. A. Campbell, S. A. Darst, R. H. Ebright and K. Severinov, *Proc. Natl. Acad. Sci. USA*, 2001, **98**(3), 892–7.
98. K. Mukherjee, H. Nagai, N. Shimamoto and D. Chatterji, *Eur. J. Biochem.*, 1999, **266**(1), 228–35.
99. K. Mukherjee and D. Chatterji, *J. Biosci.*, 1999, **24**, 453–559.
100. C. E. Vrentas, T. Gaal, W. Ross, R. H. Ebright and R. L. Gourse, *Genes Dev.*, 2005, **19**(19), 2378–87.
101. A. Travers and G. Muskhelishvili, *Nat. Rev. Microbiol.*, 2005, **3**(2), 157–69.
102. M. Geertz, Elucidation of the functional role of the ω subunit of *Escherichia coli* RNA polymerase, Diploma Thesis. 2004, Phillips Universität Marburg.
103. E. Crozat, N. Philippe, R. E. Lenski, J. Geiselmann and D. Schneider, *Genetics*, 2005, **169**(2), 523–32.
104. R. Sen, H. Nagai and N. Shimamoto, *Genes Cells*, 2001, **6**(5), 389–401.
105. R. N. Fish and C. M. Kane, *Biochim. Biophys. Acta*, 2002, **1577**(2), 287–307.
106. M. Susa, T. Kubori and N. Shimamoto, *Mol. Microbiol.*, 2006, **59**(6), 1807–17.

107. L. U. Magnusson, A. Farewell and T. Nystrom, *Trends Microbiol.*, 2005, **13**(5), 236–42.
108. M. Cashel, *J. Biol. Chem.*, 1969, **244**(12), 3133–41.
109. A. Travers, *Mol. Gen. Genet.*, 1976, **147**(2), 225–32.
110. R. E. Glass, S. T. Jones and A. Ishihama, *Mol. Gen. Genet.*, 1986, **203**(2), 265–8.
111. M. M. Barker, T. Gaal, C. A. Josaitis and R. L. Gourse, *J. Mol. Biol.*, 2001, **305**(4), 673–88.
112. I. Artsimovitch, V. Patlan, S. Sekine, M. N. Vassylyeva, T. Hosaka, K. Ochi, S. Yokoyama and D. G. Vassylyev, *Cell*, 2004, **117**(3), 299–310.
113. B. J. Peter, J. Arsuaga, A. M. Breier, A. B. Khodursky, P. O. Brown and N. R. Cozzarelli, *Genome Biol.*, 2004, **5**(11), R87.
114. K. Igarashi, N. Fujita and A. Ishihama, *Nucleic Acids Res.*, 1989, **17**(21), 8755–65.
115. K. Potrykus, D. Vinella, H. Murphy, A. Szalewska-Palasz, R. D'Ari and M. Cashel, *J. Biol. Chem.*, 2006, **281**(22), 15238–48.
116. S. T. Rutherford, J. J. Lemke, C. E. Vrentas, T. Gaal, W. Ross and R. L. Gourse, *J. Mol. Biol.*, 2007, **366**(4), 1243–57.
117. L. U. Magnusson, B. Gummesson, P. Joksimovic, A. Farewell and T. Nystrom, *J. Bacteriol.*, 2007, **189**(14), 5193–202.
118. A. Travers, C. Kari and H. A. F. Mace. Transcriptional regulation by bacterial RNA polymerase. in Genetics as a Tool in Microbiology. 1981: Cambridge University Press.
119. A. A. Travers, R. Buckland and P. G. Debenham, *Biochemistry*, 1980, **19**(8), 1656–62.
120. A. Y. Woody, R. W. Woody and A. D. Malcolm, *Biochim. Biophys. Acta*, 1987, **909**(2), 115–25.
121. W. Ross and R. L. Gourse, *Proc. Natl. Acad. Sci. USA*, 2005, **102**(2), 291–6.
122. C. A. Davis, C. A. Bingman, R. Landick, M. T. Record Jr and R. M. Saecker, *Proc. Natl. Acad. Sci. USA*, 2007, **104**(19), 7833–8.
123. C. A. Davis, M. W. Capp, M. T. Record Jr and R. M. Saecker, *Proc. Natl. Acad. Sci. USA*, 2005, **102**(2), 285–90.
124. W. Ross, D. A. Schneider, B. J. Paul, A. Mertens and R. L. Gourse, *Genes Dev.*, 2003, **17**(10), 1293–307.
125. A. A. Travers, R. Buckland, M. Goman, S. S. Le Grice and J. G. Scaife, *Nature*, 1978, **273**(5661), 354–8.
126. J. Hamming, G. Ab and M. Gruber, *Nucleic Acids Res.*, 1980, **8**(17), 3947–63.
127. T. Kubori and N. Shimamoto, *J. Mol. Biol.*, 1996, **256**(3), 449–57.
128. N. Blot, R. Mavathur, M. Geertz, A. Travers and G. Muskhelishvili, *EMBO Rep.*, 2006, **7**(7), 710–5.
129. S. Deng, R. A. Stein and N. P. Higgins, *Mol. Microbiol.*, 2005, **57**(6), 1511–21.

130. K. S. Jeong, J. Ahn and A. B. Khodursky, *Genome Biol.*, 2004, **5**(11), R86.
131. G. F. Arnold and I. Tessman, *J. Bacteriol.*, 1988, **170**(9), 4266–71.
132. M. Filutowicz and P. Jonczyk, *Mol. Gen. Genet.*, 1983, **191**(2), 282–7.
133. S. M. Mirkin, E. S. Bogdanova, Z. M. Gorlenko, A. I. Gragerov and O. A. Larionov, *Mol. Gen. Genet.*, 1979, **177**(1), 169–75.
134. Y. N. Zhou and D. J. Jin, *Proc. Natl. Acad. Sci. USA*, 1998, **95**(6), 2908–13.
135. M. Shin, M. Song, J. H. Rhee, Y. Hong, Y. J. Kim, Y. J. Seok, K. S. Ha, S. H. Jung and H. E. Choy, *Genes Dev.*, 2005, **19**(19), 2388–98.
136. K. Drlica, *Mol. Microbiol.*, 1992, **6**(4), 425–33.
137. B. Cheng, C. X. Zhu, C. Ji, A. Ahumada and Y. C. Tse-Dinh, *J. Biol. Chem.*, 2003, **278**(33), 30705–10.
138. R. Gupta, A. China, U. H. Manjunatha, N. M. Ponnanna and V. Nagaraja, *Biochem. Biophys. Res. Commun.*, 2006, **343**(4), 1141–5.
139. A. Travers and G. Muskhelishvili, *Curr. Opin. Genet. Dev.*, 2005, **15**(5), 507–14.
140. D. J. Jin and J. E. Cabrera, *J. Struct. Biol.*, 2006, **156**(2), 284–91.
141. J. E. Cabrera and D. J. Jin, *Mol. Microbiol.*, 2003, **50**(5), 1493–505.
142. M. D. Ditto, D. Roberts and R. A. Weisberg, *J. Bacteriol.*, 1994, **176**(12), 3738–48.
143. G. Muskhelishvili and A. Travers, *Front. Biosci.*, 2003, **8**, d279–85.
144. M. Rochman, N. Blot, M. Dyachenko, G. Glaser, A. Travers and G. Muskhelishvili, *Mol. Microbiol.*, 2004, **53**(1), 143–52.
145. A. Travers, R. Schneider and G. Muskhelishvili, *Biochimie*, 2001, **83**(2), 213–7.
146. M. Barth, C. Marschall, A. Muffler, D. Fischer and R. Hengge-Aronis, *J. Bacteriol.*, 1995, **177**(12), 3455–64.
147. J. Bouvier, S. Gordia, G. Kampmann, R. Lange, R. Hengge-Aronis and C. Gutierrez, *Mol. Microbiol.*, 1998, **28**(5), 971–80.
148. A. Balandina, L. Claret, R. Hengge-Aronis and J. Rouviere-Yaniv, *Mol. Microbiol.*, 2001, **39**(4), 1069–79.
149. Y. Ogata, R. Inoue, T. Mizushima, Y. Kano, T. Miki and K. Sekimizu, *Biochim. Biophys. Acta*, 1997, **1353**(3), 298–306.
150. R. Schneider, A. Travers, T. Kutateladze and G. Muskhelishvili, *Mol. Microbiol.*, 1999, **34**(5), 953–64.
151. D. Weinstein-Fischer, M. Elgrably-Weiss and S. Altuvia, *Mol. Microbiol.*, 2000, **35**(6), 1413–20.
152. M. Malik, A. Bensaid, J. Rouviere-Yaniv and K. Drlica, *J. Mol. Biol.*, 1996, **256**(1), 66–76.
153. A. Bensaid, A. Almeida, K. Drlica and J. Rouviere-Yaniv, *J. Mol. Biol.*, 1996, **256**(2), 292–300.
154. L. Landis, J. Xu and R. C. Johnson, *Genes Dev.*, 1999, **13**(23), 3081–91.
155. S. Kar, R. Edgar and S. Adhya, *Proc. Natl. Acad. Sci. USA*, 2005, **102**(45), 16397–402.

156. F. Leng and R. McMacken, *Proc. Natl. Acad. Sci. USA*, 2002, **99**(14), 9139–44.
157. D. C. Grainger, D. Hurd, M. D. Goldberg and S. J. Busby, *Nucleic. Acids. Res.*, 2006, **34**(16), 4642–52.
158. C. Marr, M. Geertz, M. Huett and G. Muskhelishvili, *BMC Syst. Biol.*, 2008, **2**, 18.

CHAPTER 4
Transcription by RNA Polymerases: From Initiation to Elongation, Translocation and Strand Separation

THOMAS A STEITZ

Departments of Molecular Biophysics & Biochemistry, and Chemistry, Yale University, and Howard Hughes Medical Institute, New Haven, CT 06520-8114, USA

4.1 Introduction

The structures of T7 RNA polymerase (T7 RNAP) captured in the initiation and elongation phases of transcription, as well as an intermediate stage, and those of the multi-subunit RNAPs bound to initiating factors provide insights into how these RNA polymerase proteins can initiate RNA synthesis and synthesize six to ten nucleotides while remaining bound to the DNA promoter site of initiation. Further, the structural basis for the transitions that are responsible for the different functional properties of the initiation and elongation phases of transcription are now becoming well understood. Structural insights into the RNAP's translocation of the product transcript, its separation of the downstream duplex DNA and its removal of the transcript from the heteroduplex are provided by the structures of several states of nucleotide incorporation. A conformational change in the "fingers" domain that is a result of the binding of incoming NTP or PPi appears to be associated with the state of translocation of T7 RNAP. Single-molecule and biochemical studies show a distribution of primer terminus positions that is altered by the binding of NTP and PPi ligands. Movies derived from the structures of the polymerase in different states allow the direct visualization of the conformational changes that the polymerases undergo during the different steps of polymerization as well as the initiation to elongation transition.

Many enzymes not only catalyze chemical reactions, but also act as mechanical machines in the process of executing their functions, and the polynucleotide polymerases utilize the energy of NTP (or dNTP) to also do work. DNA-dependent RNA polymerases can recognize their DNA promoter site at which transcription initiation occurs and remain bound to it throughout the initiation phase of RNA synthesis and then transition to an elongation phase in which the initiation site is released. One question to be addressed is how these enzymes are able to accommodate the elongating duplex product while remaining attached to the promoter site; furthermore, what is the nature of the structural transition from the initiation to the elongation phase and how is it achieved? A second question is the structural basis of promoter clearance upon transition to the elongation phase as well as the source of processivity of this phase. A third property of RNA polymerases to be addressed is how and when the incorporation of nucleotides into a growing primer chain results in translocation of the product duplex and a concomitant separation of the strands of the downstream duplex DNA; additionally, in the case of these RNA polymerases, the RNA transcript must also be peeled off the template by the enzyme as it translocates.

While the most complete description of the structural basis of the transition from the initiation phase of transcription to the elongation phase exists for T7 RNA polymerase,[1-4] significant structural insights are known for both the multi-subunit bacterial and eukaryotic yeast RNA polymerases.[5-9] A common feature of the initiation phase of these polymerases appears to be the accumulation of the duplex product in the active site that in various ways gradually displaces a bound initiation factor or the part of the polymerase that provides the polymerase's specificity for the promoter site.

RNA polymerases translocate along the DNA template as the duplex product is synthesized, and they are able to open the downstream DNA duplex without requiring a separate helicase protein. Structural data on both A family and B family DNA polymerases as well as T7 RNA polymerase are consistent with a mechanism of translocation that is associated with the dissociation of PP_i,[10,11] whereas DNA footprinting as well as single-molecule studies of bacterial RNA polymerase are consistent with a mechanism in which translocation is associated with the binding of the incoming NTP.[12,13] The term "power stroke" has often been used in the literature to describe a mechanism in which pyrophosphate dissociation results in an enzyme conformational change that is in turn associated with or perhaps even "pushes" translocation of the product duplex. The term "Brownian ratchet" has been used to describe a mechanism in which the binding of the incoming NTP captures and stabilizes the translocated state of the product among a number of states that are in equilibrium in its absence. These terms convey two extreme poles of the possibilities. There is always an equilibrium among states of duplex translocation, and generally we can only securely establish that the most stable state is indeed altered by the protein conformational changes associated with the binding or dissociation of the ligands PP_i or NTP, so that the distinction between these terms remains vague. To definitively assess whether the product is being "pushed" by the conformational change or the change is capturing an equilibrium state of translocation generally requires knowing which of these two paths is the fastest, the rate of ligand induced conformational change or the one-dimensional diffusion process of the enzyme on the DNA, data that are generally unknown. Hence, the goal here is to describe what is currently known about the structural, biochemical and kinetic bases for understanding the mechanisms of translocation and strand separation without using the terms "power stroke" or "Brownian ratchet."

4.2 Transition from the Initiation to the Elongation Phase

All DNA-dependent RNA polymerases initiate RNA synthesis after binding to a specific promoter DNA sequence and opening the DNA duplex. With both the multi-subunit RNA polymerases and the single subunit T7 RNA polymerase the enzyme remains bound to the promoter sequence throughout the initiation phase, which involves the synthesis of an 8 to 12 nucleotide transcript. This phase, often referred to as the abortive synthesis phase, results in the synthesis and release of numerous short transcripts. Only after the transition to the elongation phase does transcription proceed with complete processivity until completion of the entire RNA transcript followed by termination. Various models have been previously proposed to account for the enzyme's ability to remain bound to the promoter while synthesizing RNA. These included the inchworm model whereby the enzyme's active site catalyzing nucleotide incorporation progresses down the template while the promoter binding site remains fixed on the DNA, or the scrunching model where the product is accumulated in the polymerase active site.

4.2.1 T7 RNA Polymerase

The RNA polymerase from phage T7 (T7RNAP) is a 98 000 Da molecular weight, single subunit whose polymerase domain is homologous to that of the DNA polymerase I family of polymerases and is completely unrelated to the multi-subunit RNA polymerases. It initiates transcription at a specific site by binding to a 17 base-pair promoter DNA that it recognizes through base specific interactions. These are made by side chains emanating from a β hairpin called the specificity loop that binds in a DNA major groove as well as a loop that interacts within the minor groove.[1,14] The promoter binding site is formed in part by an N-terminal domain of 300 amino acid residues whose structure is unique to this polymerase, as well as the β hairpin specificity loop, which is an insertion into the fingers domain of the polymerase. Comparison of the structure of the T7 RNAP bound to a 17 base-pair promoter DNA (nucleotides –1 to –17) with that of its complex to a promoter duplex containing a five-nucleotide 5′ template overhang (nucleotides 1–5) and a three-nucleotide transcript shows that the promoter recognition is identical, but the template strand has accumulated in the active site after the synthesis of three nucleotides in order to position the +4 template base appropriately for nucleotide insertion; thus, a small amount of "scrunching" of the template has occurred that allowed the polymerase catalytic site to move three nucleotides down the template strand.[1] It was notable, however, that the cleft in which the 3 base-pair heteroduplex resides in the initiation complex is not sufficiently large to accommodate even a fourth base-pair in the post-translocation state. While these data suggested the possibility that the transcript peeled off the DNA after only three base-pairs, biochemical data, such as the crosslinking of template and transcript strands, strongly implied that the heteroduplex would be as long as eight base-pairs.[15] It was suggested at the time that perhaps the co-crystal structure was "... captured in an act of unproductive synthesis" as a consequence of the template sequence used (CCC).[16] Clearly, there was more to this interesting story.

The structure of T7 RNA polymerase captured in an elongation complex, however, was able to accommodate all the biochemical data and show that the transition from initiation to elongation was accompanied by a massive conformational change, particularly in the N-terminal domain of the polymerase.[2,3] This conformational change resulted in the enlargement of the binding cleft for the product, enabling it to accommodate up to eight base-pairs of heteroduplex product, the complete destruction of the promoter binding site and the formation of a tunnel through which the RNA transcript was seen to emerge after being separated from the template DNA.[10] In addition to the eight base-pair heteroduplex, four nucleotides of single stranded RNA transcript at the end are seen to be "peeled" off the template, which is consistent with biochemical data,[15,17] by an α-helix in the thumb domain as well as side chain interactions with a newly refolded H subdomain and the specificity loop (Figure 4.1C and D). Furthermore, a nine-nucleotide bubble is formed, as predicted from biochemical experiments of Huang and Sousa,[17] in the substrate DNA, stabilized on the one hand by the eight base-pair heteroduplex and on the other by extensive interactions of the single-stranded nontemplate strand with the protein (Figure 4.1D).[10] Comparison of the initiation and elongation

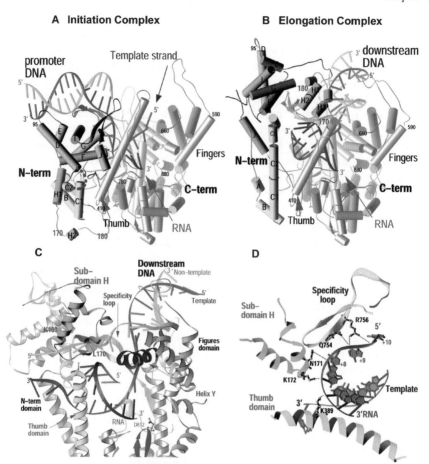

Figure 4.1 Comparison of the structures of the T7 RNAP initiation and elongation complexes (A and B) and views of the transcription bubble (C and D).[2] The initiation complex (A) and elongation complex (B) have been orientated equivalently by superimposing their palm domains. Helices are represented by cylinders and beta strands by arrows. The corresponding residues in the NH_2-terminal domains of the two complexes that undergo major refolding are colored in yellow, green and purple, and the COOH-terminal domain (residues 300–883) is in gray. The template DNA (blue), nontemplate DNA (green) and RNA (red) are represented with ribbon backbones. The proteolysis-susceptible region (residues 170–180) is a part of subdomain H (green) in the elongation complex and has moved more than 70 Å from its location in the initiation complex. The specificity loop (brown) recognizes the promoter during initiation and contacts the 5′ end of RNA during elongation, whereas the intercalating hairpin (purple) opens the upstream end of the bubble in the initiation phase and is not involved in elongation. The large conformational change in the NH_2-terminal region of T7 RNAP facilitates promoter clearance. This figure was made with the program *Ribbons*. (C) Interactions of the transcription bubble and heteroduplex in the elongation complex with domain H (green and red) and specificity loop

structures (Figure 4.1A and B; supplementary Movies 1 and 2[i]) showed that the N-terminal domain could be most conveniently thought of as consisting of three subdomains. In particular, an assembly of six α helices moved as a rigid body through a rotation of 220° and a translation of 30 Å that both provided space for the elongating product and positioned the subdomain in the location formerly occupied by the promoter DNA. Perhaps the most dramatic structural arrangement observed in this transition is a loop of protein that moves from one side of the N-terminal domain to the complete opposite side of the six-helix subdomain and refolds into a two-helix H domain that forms part of the lid of the transcript tunnel through which the RNA product departs.

The structure of the elongation complex immediately explains both promoter clearance and the processivity of the elongation phase. Promoter clearance is achieved because the promoter binding site is completely destroyed (Figure 4.1A and B). Not only is the six-helix domain in the position formerly occupied by the promoter, but the anti-parallel β specificity loop that recognizes the promoter in the initiation complex has now moved to form part of the lid of the transcript exit tunnel and is seen to interact with the departing transcript,[10] which is consistent with biochemical crosslinking data.[15] High processivity is explained by the formation of this transcript exit tunnel that surrounds the transcript and thus functions analogously to the sliding clamp in DNA

---

[i]The movies can be viewed at www.rsc.org/shop/books/2009/9780854041343.asp **Movie 1.** Structure of an initiation complex of T7 RNA polymerase bound to promoter DNA (green and dark blue) with a 5′ extension of the template strand (dark blue). A three nucleotide RNA transcript (red) is bound to the template. The polymerase domain, including the specificity loop interacting in the major groove of the promoter, is light blue. The subdomains of the N-terminal domain include a 6-helix assembly (purple), an extended loop (green) and three helices (yellow). Structure from Cheetham and Steitz[1] and movie by Baocheng Pan. **Movie 2.** Structure of an elongation complex of T7 RNA polymerase bound to DNA (green and dark blue) showing 10 base-pairs of downstream duplex, a melted non-template strand (green) and a melted template strand (blue) that is forming eight base-pairs of heteroduplex with the RNA transcript (red) that peals off and exits through a protein tunnel. The roof the protein exit tunnel is formed by the green domain that forms two helices and the specificity loop (light blue). Protein colors are as in Movie 1. Structure from Yin and Steitz[2] and movie by Baocheng Pan.

Figure 4.1 Continued

(brown). Proteolytic cuts within the red loop in subdomain H reduce elongation synthesis.[43,44] Thumb alpha-helix (yellow) and alpha-helix Y (red) are analogously involved in strand separation. (D) Side chains from subdomain H (green), the specificity loop (brown), and the thumb that interact with the single-stranded 5′ end of the RNA transcript and facilitate its separation from the template. The DNA substrate in the initiation complex consisted of duplex promoter DNA from –1 to –17 and a 5′ nucleotide overhang of the template strand (nucleotides 1–5) at the 5′ end. The DNA used in the elongation complex contained 10 bp of downstream duplex DNA, a 10-nucleotide non-complementary DNA "bubble" and a 10 bp upstream duplex DNA that is disordered in the structure.

replication. Once the transcript is threaded through the tunnel, it can no longer dissociate from the enzyme until transcription is terminated.

The detailed mechanism by which this remarkable conformational change is achieved is at present a matter of speculation though some insights are now emerging from experiments. An interesting proposal has been put forward in which intermediate structural states have been hypothesized[18] (supplementary Movie 3[ii]). Biochemical experiments designed to identify at which stages in the initiation phase particular structural rearrangements occur have been carried out by crosslinking domains whose positions change in the initiation to elongation transition. Formation of a disulfide crosslink between the six-helix domain and the fingers domain that would restrict movement of these domains from the initiation state structure results in synthesis of transcripts of length five or six nucleotides, only one or two more than the initiation state would appear able to accommodate.[19]

Another disulfide crosslink between residue 128 in the helical domain and residue 393 in the fingers domain allows production of transcripts of lengths seven and eight. These and other crosslinking experiments as well as Fe-BABE DNA cleavage studies led Sousa and co-workers[20] to conclude that most of the major conformational changes in the transition from initiation to elongation do not occur during the elongation of the transcript from three to eight nucleotides, but rather occur after the transcript reaches 8–9 nucleotides. Extension to this length also coincides with the release of the promoter.

The emerging structure of the T7 RNA polymerase complexed with a promoter containing primer-template DNA and either a seven or eight nucleotide transcript establishes the nature of the conformational changes that occur in the elongation of the transcript from three to seven nucleotides.[4] Formation of this stable initiation to elongation intermediate complex was achieved using a mutant polymerase (P266L) that makes fewer abortive transcripts.[21] The six-helix bundle domain as well as the bound promoter DNA and the recognition hairpin are observed to rotate, thereby creating a space that is large enough to accommodate the eight nucleotides of heteroduplex. While the C helix and thumb subdomains show modest changes, the rest of the structure remains largely unchanged from that of the initiation complex. Consistent with conclusions from biochemical data, it appears that the other major conformational changes and promoter release occur after synthesis of transcripts larger than eight. Whether there are stable structural intermediates with transcripts between 9 and 12 that can be captured by crystal structures remains to be determined.

[ii] The movie can be viewed at www.rsc.org/shop/books/2009/9780854041343.asp **Movie 3**. A completely hypothetical model of the structural transition undergone by T7 RNA polymerase as the 3-nucleotide RNA transcript (red) is extended until the elongation phase is reached. The colors are as in Movies 1 and 2. The model is based, in part, on the proposal of Theis et al.[18] and employs the program MORPH[42] to model the changes between intermediate states. The starting structure is from Cheetham and Steitz,[1] the final structure is from Yin and Steitz[2] and the movie was made by Baocheng Pan and Yong Xiong.

4.2.2 Multi-subunit RNA Polymerases

As described in the preceding chapters, the multi-subunit RNA polymerase from eubacteria recognizes its promoters using transcription factors that are separate polypeptides. One of a family of sigma factors bound to the bacterial RNA polymerase participates in the initiation phase by recognizing the DNA promoter, and it dissociates from the polymerase either during or following the transition to elongation phase. The structure of the *Thermus thermophilus* holo RNA polymerase including σ^{70} and promoter DNA has been determined, and the synthesis of approximately 12 nucleotides of an RNA transcript has been proposed to displace the 3.2 loop of the σ^{70} factor that is lying in the tunnel through which the transcript is proposed to exit.[5,6] Murakami and Darst[5] further hypothesize that the displacement of the 3.2 loop by the elongating transcript weakens the interactions between the polymerase and σ, which ultimately leads to promoter escape. A more recent structure of a *Thermus thermophilus* RNA polymerase holo enzyme complexed with DNA and a 16 nucleotide transcript[22] shows seven single-stranded RNA nucleotides emanating from an eight base pair heteroduplex and passing through the tunnel that is occupied by the σ 3.2 loop in the initiation complex, which is consistent with the displacement hypothesis.

Eukaryotic RNA polymerase II appears to behave in an analogous but mechanistically different fashion. It is bound to promoters through general transcription factors that are released by the polymerase when it enters into the elongation phase. The binding site on the polymerase for a domain of TFIIB (which together with TATA binding protein recognizes the promoter DNA) is seen in the structure of its complex with the polymerase to overlap with location of the heteroduplex in the elongation complex.[9,23] Thus, once again, the elongating RNA transcript may progressively displace the TFIIB domain, providing a structural basis for the dissociation of the polymerase from the proteins that are binding it to the promoter DNA.

Although the specific mechanisms of initiation site recognition and transitions to the elongation phase differ among these three RNA polymerases, there appear to be some analogous structural principles involved.[2,6,9,24] In each case the elongating duplex or heteroduplex product gradually pushes a part of the initiation site recognition element (the N-terminal domain of T7 RNAP,[2] σ^{70} of bacterial RNAP[6] and TFIIB with yeast Pol II)[9] away from the site of nucleotide incorporation, eventually resulting in the polymerase dissociating from the initiation site. The high processivity of the three RNA polymerases in the elongation phase results from the transcripts exiting through a tunnel in the enzyme.[5–6,10,25]

4.3 Translocation and Strand Separation

4.3.1 T7 RNA Polymerase

The structural basis of heteroduplex product translocation and separation of downstream duplex DNA by the monomeric T7 RNA polymerase has been illuminated by high-resolution crystal structures of the four states of nucleotide incorporation by T7 RNAP in the elongation phase of DNA synthesis.[2,10] These structures include that of the enzyme in a post-translocated product complex,[2] a complex with the post-translocated product with the incoming nucleotide bound in a pre-insertion site,[26] the structure of a ternary complex with the incoming NTP positioned for incorporation and finally the product complex in a pre-translocation state including bound pyrophosphate.[10] Two of these states, the binary complex with primer-template DNA and the ternary complex including the incoming deoxy-NTP, have been determined for several DNA polymerase I enzymes as well,[27–30] and the structures of these states are largely identical to the corresponding complexes with the T7 RNA polymerase. Furthermore, the initiation complex of T7 RNAP with a three-nucleotide transcript, but without PP_i, is in the post-translocated state. The conclusion from these studies is that a conformational change in the "fingers" domain of the pol I family polymerases from an "open" to a "closed" state is driven by the binding of the NTP (or dNTP) to the O helix followed by its positioning in the insertion site, and the change from a "closed" to an "open" state is associated with release of the product pyrophosphate. Further, it is the conformational change resulting from the release of the pyrophosphate product that is associated both with translocation of the heteroduplex product and with strand separation of the downstream DNA.[10]

The structures of the four stages of nucleotide incorporation by T7 RNA polymerase have been determined for the enzyme during its elongation phase of RNA synthesis (Figure 4.2; supplementary Movie 4[iii]). The structure of the complex after nucleotide incorporation and pyrophosphate dissociation is in the post-translocation state and the base that is to template the next incoming nucleotide is located in a binding pocket outside the active site.[2–3,29,31] The substrate NTP binds initially to a distinct "open" conformation of the enzyme at a site located on the O helix. It is there apparently base-paired to the next template base, but still not in a position for insertion.[26] A conformational change that can be described as the rotation of a five-helix assembly about a pivot point then occurs that results in the formation of a closed complex in which the incoming nucleotide is properly positioned for insertion and the templating base moves from a pre-insertion pocket to pair with this incoming

---

[iii] The movie can be viewed at www.rsc.org/shop/books/2009/9780854041343.asp **Movie 4**. The insertion of an incoming NTP into the insertion site, nucleotide incorporation, pyrophosphate dissociation and translocation coupled to a fingers conformational change. NTP initially binds to the O helix (yellow) in the open conformation interacting with Arg627 and a Mg^{2+}. The closing conformational change moves both the incoming NTP and the templating $n+1$ base into base-pairing position while a tyr side chain (Y639, orange) at the end of the O helix moves out of the way. The closed conformation is stabilized by a salt link between the triphosphates and R627 as well as the two Mg ions (blue), M_A and M_B, that are bound to D812 and D537. The strands of the downstream duplex (blue) are separated by a helix (orange) containing F644. The incoming NTP is incorporated at the end of the primer RNA (violet) with the formation of PP_i that is still linked to M_B and R627. The dissociation of PP_i is accompanied by a rotation of the fingers 5-helix bundle, which moves the Y639 to translocate the product heteroduplex. The structures used are the post-translocated product and the ternary complex with NTP as well as the pre-translocated product with PP_i. Movie made by Whitney Yin.

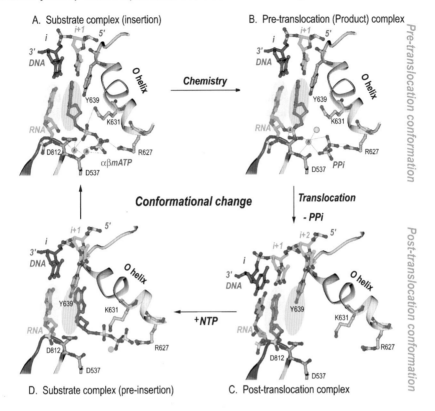

Figure 4.2 Structural changes at the active site of T7 RNAP during a single nucleotide addition cycle.[2] This figure shows the O helix with its phosphate binding K631 and R627, a β turn-β motif bearing the metal binding catalytic D812 and D537, template nucleotides in blue, the RNA primer terminus in green, as well as the P and N sites in green and pink ovals, respectively. (A) The NTP (red) is bound to the N site in position to be inserted with its metal bound triphosphate moiety crosslinking the O helix to the active site aspartic acid residues. Template nucleotide $i+1$ (light blue) forms a base pair with the correct incoming nucleotide. (B) The product of the phosphoryl transfer reaction shows a Mg ion (blue) bound to PPi (red), which crosslinks D537 to R627, thereby maintaining RNAP in an identical conformation as in the substrate complex. The 3′ end of RNA remains in the N site in a pre-translocation state. (C) Release of Mg-PPi results in the loss of the link between the O helix and D537, which promotes the rotation of the O helix and translocation of the 3′ end of the RNA to the P site. The RNAP conformational change also places Y639 into the N site and positions the $i+2$ template nucleotide into the flipped-out, pre-insertion position. (D) A modeled NTP pre-insertion complex with NTP bound to the post-translocated RNAP. Although the base binding site is blocked by the side chain of Y639, the triphosphate binding site on the O helix is accessible.

base.[10] This conformational change is driven by the ionic links that are able to form in the closed conformation between the three phosphates of the NTP and the two magnesium ions bound to the enzyme on one side and an arginine residue of the O helix on the other. The structure of T7 RNAP in the pre-translocation product complex that forms in the presence of pyrophosphate (see again Figure 4.2B) is identical to the structure of the enzyme in the ternary complex (Figure 4.2A). The enzyme continues to be held in the closed conformation after nucleotide incorporation, once again, by the pyrophosphate product forming an ionic crosslink between a magnesium ion bound to the active site carboxylates and an arginine on the O helix. Dissociation of pyrophosphate completes this cycle and results in the formation of an open complex accompanied by translocation of the product heteroduplex.

The conformational change in the enzyme associated with the return to the open state consists largely of a rotation of a five-helix bundle that includes the O helix about a pivot axis. Rotation of the bundle to the "open" conformation results in a 3.4 Å movement of a tyrosine residue (Y639) that becomes stacked on the primer-template bases. This change indeed stabilizes the translocated position of the product duplex (Figure 4.3; supplemental Movie 4). The position of this tyrosine in this open state of the enzyme not only sterically precludes the

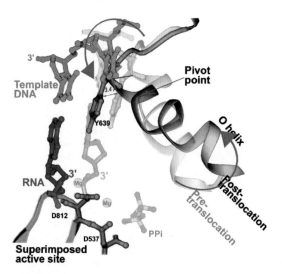

Figure 4.3 A superposition of the pre- and post-translocation structures at the active site showing the pivoted rotation undergone by the O helix that is associated with translocation.[10] In the pre-translocation complex (lighter colors), the O helix (light gray) is positioned in the closed conformation by PPi (light red), which is bound to the catalytic carboxyls through Mg. In this conformation, Y639 allows formation of the new base pair (light red and blue). After PPi release, the O helix rotates around Val634, which results in the positive end of the helix moving away from the active site while the other end of the helix moves Y639 by 3.4 Å into the position of the newly formed primer terminus resulting in translocation.

return of the heteroduplex to its pre-translocation position but also blocks the insertion site for the next incoming NTP. Only after the NTP bound to the pre-insertion site on the O helix in the open state promotes the formation of the closed state does the tyrosine side chain move out of the way, thereby allowing the incoming NTP base and the template base to move into the insertion site and replace it. Since this tyrosine is conserved in all Pol I family polymerases, a similar mechanism of translocation is suggested. The translocated state is a result of the open conformation being more stable than the closed conformation after dissociation of the pyrophosphate and loss of the cognate electrostatic crosslinks. In the open conformation the rotated five-helix bundle buries 130 $Å^2$ of hydrophobic surface area, which would be expected to contribute about 3 kcal mol^{-1} (ref. 32) towards the stabilization of the open conformation. This modest hydrophobic interaction presumably provides the "spring" energy to stabilize formation of the open conformation and the associated translocation upon PP_i dissociation. However, the estimated 3 kcal does not directly yield the equilibrium constant between pre- and post-translocation states of the transcribing complex, since translocation also melts a base pair.

Interestingly, the single-molecule studies of Thomen et al.[33] define the change of the Michaelis constant for substrate incorporation as a load counteracting translocation is applied. They use the Brownian Ratchet formalism, which allows equilibration between the pre- and post-translocation states in the binary complex of the enzyme with DNA (i.e., bound to neither incoming nucleotide nor pyrophosphate) (see Chapters 7 and 9 for discussion). Their estimate for the corresponding equilibrium constant corresponds to a 3-to-1 preference for the post-translocated state at this stage of the reaction, a value compatible with the structural data. In addition, chemical cleavage studies show a distribution of transcript states that favor pre-translocation in the presence of PP_i and post-translocation in the presence of NTP.[20]

While, of course, an equilibrium must exist between the pre-translocated, "closed" product state in the absence of PP_i and the translocated, "open" state it is clear that, in the absence of PP_i, the translocated product state is the most stable in three separate structures of T7 RNAP,[1-3] the structures of three binary substrate complexes with A family DNA polymerases[28-29,31] and the B family φ29 DNA polymerase.[11] Also, the structures of all A and B family apo-DNA polymerases, as well as three structures of T7 RNAP without substrates, are in the open conformation that is associated with the translocated state. Although the DNA substrate can diffuse between the pre- and post-translocation states in the closed conformation of the enzymes, it cannot exist in the pre-translocation state in the open form due to the position of Y639. Since the templating base is likewise blocked from the nucleotide insertion site by Y639, the open to closed state conformational change must occur prior to catalysis, though it is not necessarily the rate-limiting step. While one certainly cannot determine the equilibrium constant between the pre- and post-translocated states by X-ray crystallography, the exclusive observation of the translocated state in all A-family polymerase product complexes in the absence of PP_i suggests that it is significantly energetically favored. These structures appear to describe a plausible pathway for translocation and delivery of the NTP to the incorporation site (supplementary Movie 4). The binding of the incoming nucleotide to the translocated open state and its further conversion into a closed translocation state will stabilize the complex and prevent the primer terminus from entering the pre-translocation position, unless the NTP dissociates. Thus, the binding of NTP can be described as stabilizing the translocated state.

DNA-dependent RNA polymerases and some DNA polymerases function as their own helicases, opening downstream duplex DNA as their primer strand is elongated and the enzyme translocates. The general mechanistic principle appears to be that the energy of nucleotide incorporation and associated translocation "pulls" the downstream template strand through a cleft or a tunnel that is only large enough to accommodate single-stranded DNA.[10,34] In the case of T7 RNAP, and the homologous DNA polymerase I family enzymes that are able to displace the nontemplate strand, the template strand passes through a narrow cleft before entering the active site, and the template and nontemplate strands separate at an α-helix containing a tyrosine that stacks on successive base-pairs to be separated (Supplementary Movie 5[iv]). The RNA transcript is peeled off analogously by a thumb α-helix.[2] Again, it is posited that the energy for melting one base-pair in an incorporation cycle is contributed by PP_i dissociation. It seems likely that the hexameric DnaB helicase works in an analogous manner by "pulling" one strand of DNA through a tunnel that results in the displacement of the other strand. However, the manner in which the hydrolysis of ATP and the associated protein conformational changes are coupled to single strand translocational movement are presently unknown and completely unrelated to those exhibited by polymerases.

4.3.2 Multi-subunit Cellular RNAPs

The structure of the yeast RNAP II complexed with various bound RNA and DNA substrates that represent states of nucleotide incorporation in the elongation phase have been determined[8,25,35,36,45] as well as the corresponding complexes with the *Thermus thermophilus* RNAP.[22,37] While initially it was not transparently obvious what aspect of the complex observed in the absence of incoming NTP determines whether the primer terminus of the heteroduplex resides in the pre- or post-translocation position, a recent structure of a yeast Pol II binary elongation complex shows the presence of both states.[45] The structure of the first complex determined[8] showed electron density for about 20 base-pairs of downstream duplex DNA and an eight base-pair heteroduplex product of RNA transcription synthesized by the enzyme and halted by withholding the next NTP. About 13 nucleotides of a displaced upstream nontemplate DNA strand as well as some upstream template and product were

---

[iv] The movie can be viewed at www.rsc.org/shop/books/2009/9780854041343.asp **Movie 5.** Unwinding of downstream duplex DNA by T7 RNAP upon nucleotide incorporation, PP_i dissociation and product translocation. The dissociation of PP_i results in the "open" conformation change, product translocation, pulling off the template strand through a cleft and separation of the downstream duplex at the orange helix.

not ordered in the complex. The primer terminus of the RNA transcript was observed in the pre-translocation position in this complex (Figure 4.4A).

In contrast, the structure of a complex of yeast RNAP II prepared using chemically synthesized RNA and DNA oligonucleotides including a 10 residue RNA transcript, a 28 residue DNA template spanning -10 to $+18$ and a 14 nucleotide downstream nontemplate DNA showed the primer terminus located in a post-translocated position.[35] Furthermore, the site opposite the next template base $(n+1)$ was empty, but could be occupied by an incoming NTP that was correctly positioned along with two bound metal ions for nucleotide incorporation (Figure 4.4C). Similarly, Kettenberger et al.[36] determined the structure of a Pol II complex with a DNA duplex containing a mismatched "bubble" bound to an RNA transcript. This complex showed density for about nine downstream base-pairs and a template-RNA product whose 3' terminus was also in the post-translocated position. The cognate NTP could bind to what they termed a pre-insertion site, close to but not identical to the positioning of the insertion site.

It is not clear, however, why the earlier complex containing the RNA made by polymerase transcription was in the pre-translocation position while these two complexes containing chemically synthesized transcripts were found in the post-translation position. The structure of the eubacterial RNAP bound to primer template DNA and an RNA transcript also shows the primer terminus in the translocated position.[22] The structural basis of translocated or non-translocated position of the primer terminus in the multi-subunit polymerases may lie in specific experimental differences between these various experiments. For example, the stability of the duplex formed by the specific DNA and RNA sequences used may influence the equilibrium between pre- and post-translocated states. Presumably, the equilibrium constant between the pre- and post-translocation state is near 1, so that any of a number of factors can move the RNA product to pre- or post-translocated positions in the absence of incoming NTP. This is supported by the structure of a yeast Pol II elongation complex in the absence of incoming nucleotides which shows the primer terminus in both pre- and post-translocation positions.[45]

What stabilizes the heteroduplex product in the pre- and post-translocation positions of the multi-subunit RNA polymerases, and what conformational changes in the enzyme are associated with the stabilization of each state have been the subject of ongoing studies. Kornberg and colleagues hypothesized initially that a conformational change in the bridge helix, which is packed against the primer terminal base-pair, from straight to bulged could be responsible for translocation (Figure 4.4B).[8] The bulged form has been observed in the apo-bacterial enzyme and the straight form in all of the published Pol II structures.[7–8,25,35,36,38] Since the translocated state is stable in the absence of a bulged bridge helix, direct evidence for the participation of a bulged helix in translocation has not been established in yeast RNAP II. Furthermore, the recent structure of Thermus thermophilus core RNAP with bound primer-template DNA and a 16 nt RNA transcript also failed to reveal any structural elements that would preclude the backward translocation of the

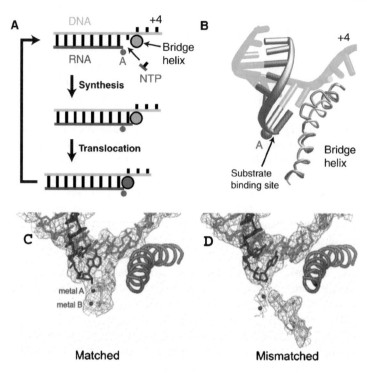

Figure 4.4 An early proposal for the translocation mechanism.[8] (A) Schematic representation of the nucleotide addition cycle. The nucleotide triphosphate (NTP) fills the open substrate site (top) and forms a phosphodiester bond at the active site ("synthesis"). This results in the state of the transcribing complex seen in the crystal structure (middle). Gnatt et al.[8] speculate that "translocation" of the nucleic acids with respect to the active site (marked by a pink dot for metal A) taking place after synthesis involves a change of the bridge helix from a straight (silver circle) to a bent conformation (violet circle, bottom). Relaxation of the bridge helix back to a straight conformation without movement of the nucleic acids would result in an open substrate site one nucleotide downstream and would complete the cycle. (B) Different conformations of the bridge helix in pol II and bacterial RNA polymerase structures. The bacterial RNA polymerase structure[38] was superimposed on the pol II transcribing complex by fitting residues around the active site. The resulting fit of the bridge helices of pol II (silver) and the bacterial polymerase (violet) is shown. The bend in the bridge helix in the bacterial polymerase structure causes a clash of amino acid side chains (extending from the backbone shown here) with the hybrid base pair at position +1. (C) and (D) Downstream end of the DNA–RNA hybrid in transcribing complex structures, showing occupancy of the A and E sites.[35] (C) Transcribing complex with matched NTP (UTP) in the A site. (D) Transcribing complex with mismatched NTP (ATP) in the E site. DNA is blue, RNA is red and NTPs are in yellow. Mg ions are shown as magenta spheres.

DNA from its observed translocated position.[37] The bridge helix is observed to be straight and not bulged in this translocated complex, in contrast to the structure of the apo RNAP from *Thermus aquaticus*[38] whose bulged bridge helix led to the bulge-bridge translocation hypothesis. A recent yeast Pol II structural study shows that binding of the incoming NTP stabilizes the post-translocated state and is associated with changes in the trigger loop conformation.[45] (see Chapter 7).

Another observation that the bridge helix in the bacterial polymerase can change its structure from a bulged to a straight conformation was obtained in the structure of a complex of the bacterial enzyme with the antibiotic streptolydigin.[12] Although the binding site of this antibiotic does not overlap with that of the substrates, it eliminates polymerase activity, leading the authors to propose that the ability of this bridge helix to cycle between straight and bulged is important for activity and perhaps translocation. This is not consistent with the subsequent structure of the translocated DNA complex just described, and another model for a mechanism of allosteric inhibition of polymerase activity by streptolydigin seems more plausible and is further discussed in Chapters 7 and 8. Vassylyev *et al.*[37] observe that in their structure of RNAP with DNA, RNA, incoming NTP and streptolydigin, the NTP is observed in an inactive, pre-insertion position due to the drug-induced stabilization of a trigger loop conformation. Thus, they suggest that streptolydigin inactivates the polymerase by preventing the trigger loop adopting the conformation required for NTP binding to the active, insertion position.

When crystals of the transcribing Pol II complex in the post-translocated state with a chain-terminating residue at the 3' end of the RNA were soaked in a solution containing a cognate NTP, the nucleotide bound in the insertion site, base-paired with the templating base and with its phosphates bound to two Mg^{2+} ions on one side and an Arg residue on the other side,[35] analogously to the case observed with T7 RNA polymerase,[10] and indeed all other polymerases.[39,40] When, however, these crystals were soaked in 15 mM of a mismatched NTP, the nucleotide bound with its three phosphates in an orientation opposite from that of the correct cognate NTP, but interacting through two divalent metal ions and with its sugar-base directed away from the template strand. Kornberg and colleagues proposed that the NTP bound in this opposite orientation site corresponds to an entrance site, or E site.[35] A far more plausible alternative explanation of this reversed binding would be that at this high concentration of NTP the mismatched nucleotide binds adventitiously largely through the three phosphates that are interacting with the magnesiums and arginines but in an opposite direction, since the base is unable to base-pair with the template base. It would be of some interest to know the binding constants for nucleotide binding to the E site as well as the insertion site. It has been proposed by Kornberg and co-workers that a correct nucleotide binds first to the E site, in the wrong orientation, and then rotates to bind to the insertion site.[35] Since the interactions of the phosphates and Mg^{2+} ions differ significantly in the two orientations, the NTP would have to first dissociate and then finally rebind in the productive insertion orientation.

A very similar pre-insertion site for NTP has been observed in both the structures of Pol II ternary complexes[36] and the bacterial substrate complex.[37] The NTP binding refolds the trigger loop with which it interacts strongly but does not have its phosphates positioned close enough for catalysis. Also the base is not paired with the template base, but is in the vicinity.

Single-molecule studies as well as biochemical results on transcribing bacterial RNA polymerase are consistent with a model of translocation in which the binding of the incoming NTP is responsible for stabilizing the translocated state.[12,13] Using an ultra-stable optical trapping system Abbondanzieri et al.[13] showed discreet steps of 3.7 Å and concluded that RNAP translocates along DNA by a single base-pair per nucleotide addition. More significantly, they determined force–velocity relationships for transcription at both saturating and sub-saturating nucleotide concentrations. Global fits of models to the data were inconsistent with a model for movement that is tightly coupled to pyrophosphate release, but consistent with a model that included a secondary NTP binding site in addition to the insertion site, full translocation being associated with NTP binding.

4.4 Additional Similarities between Single and Multi-subunit Polymerases

Although T7 RNA polymerase and the multi-subunit polymerases differ in several fundamental aspects, most significantly with the complete lack of a structural relatedness, there are aspects that are remarkably similar. Westover et al.[35] opined that the involvement of two Mg ions in the mechanism of catalysis by the single subunit and multi-subunit polymerases is " . . . only incidental . . . " and "a consequence of the association of Mg^{2+} with nucleotides in their various forms, and especially with NTPs . . . ". On the contrary, all DNA and RNA polymerases utilize the same two metal ion mechanism of catalysis.[39,40] The orientation of the two metal ions relative to the attacking 3′ OH, which metal ion A activates, and to the phosphates of NTP, which they contact identically, is the same in all polymerases; likewise, metal ion B plays the same initial catalytic role. Indeed, one might surmise that the presumed RNA based enzyme predecessors of the present protein polymerases also utilized the same two metal ion mechanism. As recently demonstrated for the group 1 intron, orientation of the Mg^{2+} ions was achieved using backbone phosphates rather than carboxylates.[41] All polymerases have converged on the same chemical basis of catalysis promoted by two metal ions.

Both families of RNA polymerases achieve nearly complete processivity in the elongation phase by having the transcript pass through an exit tunnel. Perhaps surprisingly, the length of the heteroduplex product appears to be the same in T7 RNAP[2] in yeast Pol II[8,35,36] and in *Thermus thermophilus* RNAP.[22] The multi-subunit polymerases have a pore or tunnel through which the incoming NTPs must presumably pass (see Chapter 8) to reach the active site as do all of the A family monomeric polymerases, though it is less deep and approaches the heteroduplex along its axis rather than perpendicular to it, as occurs with the multi-subunit polymerases.[35] Finally, the product heteroduplex and the downstream duplex are at right angles to each other, but with differing relative rotational orientation, in both T7 RNAP and yeast Pol II.[2,35] The bend between the upstream duplex DNA and the downstream duplex, along with a change in relative twist, probably assists formation of the open "bubble" in the duplex, as suggested for T7 RNA polymerase by Yin and Steitz.[2]

Acknowledgements

I thank Yong Xiong and Baocheng Pan for providing unpublished supplementary Movies 1, 2 and 3. Funding for research on polymerases in the laboratory was provided by NIH grant GM57510.

References

1. G. M. T. Cheetham and T. A. Steitz, *Science*, 1999, **286**, 2305.
2. Y. W. Yin and T. A. Steitz, *Science*, 2002, **298**, 1387.
3. T. H. Tahirov, D. Temiakov, M. Anikin, V. Patlan, W. T. McAllister, D. G. Vassylyev and S. Yokoyoma, *Nature*, 2002, **420**, 43.
4. K. J. Durniak, S. Bailey and T. A. Steitz, *Science*, 2008, **322**, 553.
5. K. S. Murakami and S. A. Darst, *Curr. Opin. Struct. Biol.*, 2003, **13**, 31.
6. K. S. Murakami, S. Masuda, E. A. Campbell, O. Muzzin and S. A. Darst, *Science*, 2003, **296**, 1285.
7. D. G. Vassylyev, S. Sekine, O. Laptenko, J. Lee, M. N. Vassylyeva, S. Borukhov and S. Yokoyama, *Nature*, 2002, **417**, 712.
8. A. L. Gnatt, P. Cramer, J. Fu, D. A. Bushnell and R. D. Kornberg, *Science*, 2001, **292**, 1876.
9. D. A. Bushnell, K. D. Westover, R. E. Davis and R. D. Kornberg, *Science*, 2004, **303**, 983.
10. Y. W. Yin and T. A. Steitz, *Cell*, 2004, **116**, 393.
11. A. J. Berman, S. Kamtekar, S. Goodman, J. M. Lázaro, M. de Vega, L. Blanco, M. Salas and T. A. Steitz, *EMBO J.*, 2007, **26**, 3493.
12. S. Tuske, S. G. Sarafianos, X. Wang, B. Hudson, E. Sineva, J. Mukhopadhyay, J. J. Birktoft, O. Leroy, S. Ismail, A. D. Clark Jr., C. Dharia, A. Napoli, O. Laptenko, J. Lee, S. Borukhov, R. H. Ebright and E. Arnold, *Cell*, 2005, **122**, 541.
13. E. A. Abbondanzieri, W. J. Greenleaf, J. W. Shaevitz, R. Landick and S. M. Block, *Nature*, 2005, **438**, 460.
14. G. M. T. Cheetham, D. Jeruzalmi and T. A. Steitz, *Nature*, 1999, **399**, 80.
15. D. Temiakov, P. E. Mentesana, K. Ma, A. Mustaev, S. Borukhov and W. T. McAllister, *Proc. Natl. Acad. Sci. USA*, 2000, **97**, 14109.

16. K. Severinov K, *Proc. Natl. Acad. Sci. USA*, 2001, **98**, 5.
17. J. Huang and R. Sousa, *J. Mol. Biol.*, 2000, **303**, 347.
18. K. Theis, P. Gong and C. T. Martin, *Biochemistry*, 2004, **43**, 12709.
19. K. Ma, D. Temiakov, M. Anikin and W. T. McAllister, *Proc. Natl. Acad. Sci. USA*, 2005, **102**, 17612.
20. Q. Guo and R. Sousa, *J. Mol. Biol.*, 2006, **358**, 241.
21. J. Guillerez, P. J. Lopez, F. Proux, H. Launay and M. Dreyfus, *Proc. Natl. Acad. Sci. USA*, 2005, **102**, 5958.
22. D. G. Vassylyev, M. N. Vassylyeva, A. Perederina, T. H. Tahirov and I. Artsimovitch, *Nature*, 2007, **448**, 157.
23. H. T. Chen and S. Hahn, *Cell*, 2004, **119**, 169.
24. S. Kamtekar, A. J. Berman, J. Wang, J. M. Lázaro, M. de Vega, L. Blanco, M. Salas and T. A. Steitz, *EMBO J.*, 2006, **25**, 1335.
25. K. D. Westover, D. A. Bushnell and R. D. Kornberg, *Science*, 2004, **303**, 1014.
26. D. Temiakov, V. Patlan, M. Anikin, M. T. McAllister, S. Yokoyama and D. G. Vassylyev, *Cell*, 2004, **116**, 381.
27. S. D. Doublié and T. Ellenberger, *Curr. Opin. Struct. Biol.*, 1998, **8**, 704.
28. S. Doublié, S. Tabor, A. J. Long, C. C. Richardson and T. Ellenberger, *Nature*, 1998, **391**, 251.
29. Y. Li, S. Korolev and G. Waksman, *EMBO J.*, 1998, **17**, 7514.
30. S. J. Johnson, J. S. Taylor and L. S. Beese, *Proc. Natl. Acad. Sci. USA*, 2003, **100**, 3895.
31. J. R. Kiefer, C. Mao, J. C. Braman and L. S. Beese, *Nature*, 1998, **391**, 304.
32. C. Chothia, *Nature*, 1974, **248**, 338.
33. P. Thomen, P. J. Lopez and F. Heslot, *Phys. Rev. Lett.*, 2005, **94**, 128102.
34. S. Kamtekar, A. J. Berman, J. Wang, J. M. Lázaro, M. de Vega, L. Blanco, M. Salas and T. A. Steitz, *Mol. Cell*, 2004, **16**, 609.
35. K. D. Westover, D. A. Bushnell and R. D. Kornberg, *Cell*, 2004, **119**, 481.
36. H. Kettenberger, K. Armache and P. Cramer, *Mol. Cell*, 2004, **16**, 955.
37. D. G. Vassylyev, M. N. Vassylyeva, J. Zhang, M. Palangat, I. Artisimovitch and R. Landick, *Nature*, 2007, **448**, 163.
38. G. Zhang, E. A. Campbell, L. Minakhin, C. Richter, K. Severinov and S. A. Darst, *Cell*, 1999, **98**, 811.
39. T. A. Steitz, *Nature*, 1998, **391**, 231.
40. T. A. Steitz, *J. Biol. Chem.*, 1999, **274**, 17395.
41. M. R. Stahley and S. A. Strobel, *Science*, 2005, **390**, 1587.
42. W. Krebs and M. Gerstein, *Nucl. Acids Res.*, 2000, **28**, 1165.
43. R. A. Ikeda and C. C. Richardson, *J. Biol. Chem.*, 1987, **262**, 3800.
44. D. K. Muller, C. T. Martin and J. E. Coleman, *Biochemistry*, 1988, **27**, 5763.
45. F. Brueckner and P. Cramer, *Nature Struct. and Mol. Biol.*, 2008, **15**, 811.

CHAPTER 5
Single-molecule FRET Analysis of the Path from Transcription Initiation to Elongation

ACHILLEFS N. KAPANIDIS[a] AND SHIMON WEISS[b]

[a] Department of Physics and IRC in Bionanotechnology, Clarendon Laboratory, University of Oxford, Parks Road, Oxford, OX1 3PU, United Kingdom; [b] Department of Chemistry and Biochemistry, Department of Physiology, and the California NanoSystems Institute, University of California at Los Angeles, CA 90095, USA

5.1 Introduction

RNA polymerase (RNAP), the protein that orchestrates transcription, is a sophisticated and dynamic molecular machine that uses the energy of nucleotide hydrolysis to swiftly operate on DNA and transfer genetic information to RNA.[1–3] Previous chapters have focused on genetic, biochemical and structural studies of transcription mechanisms; arguably, the most important recent contribution comes from high-resolution X-ray crystallographic structures (see Structural Atlas, refs 4–8 and Chapters 2, 4 and 7). The structures represent a quantum leap in our understanding of transcription; they led to proposals for the mechanism of many multi-step transitions along the pathway, and inspired experiments that test these proposals and further examine RNAP function and its interactions with nucleic acids and transcription factors.

However, it is difficult for X-ray crystallography to capture *directly* the dynamic nature of the RNAP machinery. Crystal structures are static

RSC Biomolecular Sciences
RNA Polymerases as Molecular Motors
Edited by Henri Buc and Terence Strick
© Royal Society of Chemistry 2009
Published by the Royal Society of Chemistry, www.rsc.org

snapshots of transcriptional states, and are often affected by inherent limitations of X-ray crystallography (*e.g.* effect of crystal packing forces, and absence or distortion of disordered protein and nucleic acid structures). For example, the mechanistically important flexibility of many RNAP modules results in absence of these modules from the structural data. Moreover, the transient nature, dynamics and heterogeneity of some intermediates (such as the short-lived RNAP–promoter DNA closed complex in σ^{70}-dependent transcription, or the series of initial transcribing complexes formed during abortive initiation) hinders their trapping and examination. Finally, it is often unclear whether a structure corresponds to a kinetically observed ("on-pathway") intermediate. Nonetheless, the structures set the stage for addressing long-standing questions through a combination of biochemical and biophysical studies; for instance, the structures are instrumental to the experimental design of such studies.

Analysis of intermediates is vital for understanding transcription, since it can reveal the exact order of steps during transcription and help identify steps wherein transcriptional regulators and sequence elements (such as activators, repressors, DNA bending proteins and promoter sequences) exert their function. Analysis of intermediates also reports on conformational changes in DNA, RNAP core and σ factors, and can help address important questions such as the mechanochemical coupling achieved in ATP-dependent transcription initiation, a feature of σ^{54}-dependent transcription.

Consider, for example, the transition from RNAP-σ^{70} holoenzyme to the RNAP–promoter DNA open complex (RP$_o$), a critical regulatory checkpoint targeted by transcriptional activators. A simplified kinetic scheme[9–14,4–5] includes the reactions:

$$\text{RNAP} + \text{P} \rightarrow \text{RP}_c \rightarrow \text{RP}_i \rightarrow \text{RP}_o$$

where P, RP$_c$ and RP$_i$ stand for promoter DNA, RNAP–promoter DNA closed complex, and RNAP–promoter DNA intermediate complex, respectively. New methods such as time-resolved X-ray-generated hydroxyl radical footprinting (Chapter 2) have provided insight into the unfolding of the above process in the context of T7 A1 promoter.[15] However, the overall transition entails additional, less-characterized transient intermediates that involve conformational changes in RNAP or DNA, occurring either sequentially or concurrently. Such changes include RNAP-claw closing, ejection of σ^{70} region 1.1 (σR1.1, an N-terminal RNAP region that acts as molecular placeholder for downstream DNA in the RP$_o$; see also Section 5.2 and Structural Atlas Figure A.4) from the main channel, and multi-step DNA strand separation. The order of steps, the timescale of steps and the protein and DNA determinants of each step are not fully elucidated yet. This precludes any definitive conclusions about the exact mechanism of the transition from holoenzyme to RP$_o$.

Further down the pathway, major questions remain regarding abortive initiation, promoter escape and early elongation. Despite recent advances,[16,17] the detailed mechanism, kinetics and protein and DNA determinants of these processes are unclear: how and when are the RNAP contacts with the promoter broken? How does RNAP translocate from its DNA register in RP_o to its DNA register in the first elongation complex? Regarding σ factor release (*i.e.* the dissociation of the σ factor from holoenzyme upon entry of RNAP into processive elongation), what are the release determinants on RNAP, on σ factor, on DNA and on RNA? Does the release occur in single or multiple steps, and how stochastic is it? What is the structure of σ-containing elongation complexes? Equally important questions exist regarding transcriptional pausing, arrest and termination.[18–20]

Single-molecule methods (as opposed to ensemble methods that report on the mean properties of large populations of molecules) offer much promise for directly observing transient intermediates and complex reaction kinetics in real-time and free from ensemble averaging. Single-molecule methods allow (i) the study of asynchronous reactions, (ii) the discovery of short-lived transient intermediates, (iii) the observation of full time-trajectories of pathways and (iv) the combination of optical and mechanical measurements on a single molecule. This chapter reviews single-molecule fluorescence approaches – especially single-molecule FRET and the related method of alternating-laser excitation of single molecules – that, along with single-molecule DNA nano-manipulation (Chapter 6), have improved our understanding of transcription and created opportunities for real-time, single-molecule kinetic views of the process *in vitro*.

5.2 Methodology: FRET and ALEX Spectroscopy

FRET stands for fluorescence resonance energy transfer (or Förster resonance energy transfer, in honour of Theodore Förster who first correctly described the FRET process[21]). Excellent reviews and book chapters exist for FRET;[22–24] only a brief introduction is presented here. FRET is a photophysical interaction (a non-radiative transfer of excited-state energy between two fluorescent probes) that reports on the proximity of two sites within a biomolecule (or within a complex of biomolecules). The two sites are usually modified by two different probes that act as the FRET donor and the FRET acceptor. For example, to measure a distance between a protein site and a DNA site within a protein–DNA complex, fluorophores need to be incorporated site-specifically in the DNA and the protein; in the latter case, labelling with reactive forms of fluorophores is preceded by genetic modification of the protein to introduce a single reactive site on the protein surface; this site is often a single, surface-exposed cysteine residue.[23,25]

For a typical donor–acceptor FRET pair, the probes are selected in such a way that the emission spectrum of the donor overlaps significantly with the absorption spectrum of the acceptor. When the fluorophores are in close proximity, the fluorescence of the donor decreases and the fluorescence of the acceptor increases (Figure 5.1A). The FRET efficiency E is a sensitive function of donor–acceptor distance R, since $E = [1 + (R/R_o)^6]^{-1}$ (Figure 5.1B) – thus FRET can be used as a nanoscale "spectroscopic ruler". The characteristic distance R_o, known as the Förster radius, is a fluorophore- and sample-dependent constant that equals the donor–acceptor distance at $E=50\%$; typical R_o values fall in the 3–7 nm range, allowing distance measurements in the 1–10 nm scale. The presence of intermolecular FRET also allows detection and analysis of molecular interactions. If there is a conformational change that alters the interprobe distance, it can be detected by changes in the fluorescence intensity of the donor and the acceptor, provided that the interprobe distances before and after the conformation change are within the 1–10 nm dynamic range of FRET (Figure 5.1C) and provided that the conformational change does not result in an equidistant configuration of the FRET pair.

Since FRET is a dipole–dipole interaction, its efficiency is not only a function of distance but also a function of the relative orientation of donor and acceptor dipoles; this dependence is summarized in the orientation factor κ^2, the value of which cannot be measured accurately and can significantly affect the value of R_o. It is therefore important to ensure that the orientation factor does not change significantly when two structural states are being compared. Some of the

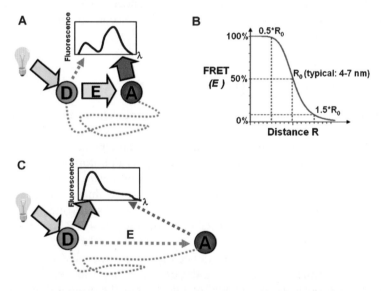

Figure 5.1 Principle of FRET and related observables. (A) A biomolecule is labelled by a FRET donor–acceptor pair, and the donor and acceptor are in close proximity. Upon excitation of the donor, D, with light, part of the excited-state energy is transferred to the acceptor, A, this decreases the emission from the donor and increases the emission from the acceptor. (B) Relation of FRET efficiency and interprobe distance; this relation defines the FRET dynamic range for distance measurements. (C) If a conformational change moves the acceptor away from the donor, the FRET efficiency decreases; this increases the emission from the donor and decreases the emission from the acceptor.

uncertainty can be lifted by showing that the fluorophores enjoy significant rotational freedom; this can be done by performing steady-state and time-resolved fluorescence anisotropy measurements. Substantial rotational freedom for the probes allows approximations of orientational averaging (where $\kappa^2 = 2/3$), which is useful if FRET measurements are used to measure distances.[26] Notably, high rotational freedom of one of two fluorophores is sufficient to minimize the error in the average donor–acceptor distance.[27] If, on the other hand, both dipoles show restricted rotation one can exploit this feature to monitor conformational changes using single-molecule fluorescence anisotropy or defocusing optical imaging.[28,29]

Ensemble FRET, typically performed by measuring the *mean* FRET efficiency for a macroscopic population of molecules, has been used extensively for structure and function analysis of *E. coli* RNAP, mainly due to the efforts of the groups of Tomasz Heyduk and Richard Ebright. Such studies identified conformational changes in σ^{70} resulting from formation of holoenzyme and open complex,[30–32] and showed that σ^{70} is retained after promoter escape at the *lac* promoter.[33] Ensemble FRET studies (along with earlier ensemble kinetic studies) also identified a large number of RNAP sites where incorporation of a fluorophore does not significantly affect the enzymatic properties of RNAP. Recently, ensemble-FRET measurements were reported for σ^{54}-dependent transcription.[34]

Arguably the most significant ensemble-FRET study was the generation of solution structures for the RNAP-σ^{70} holoenzyme and the RNAP–DNA open complex;[35] these structures resulted from docking of existing and modelled structures based on more than 100 FRET-based distance restraints between σ^{70} and core RNAP (for the holoenzyme), as well as σ^{70}, core RNAP and DNA (for RP$_o$). These structures are consistent with published X-ray structures. The most important contribution of the FRET-based structural work was the observation that σR1.1 (Section 5.1) occupies the downstream DNA channel in the holoenzyme, but moves out of this channel in RP$_o$: thus σR1.1 essentially acts as a "molecular placeholder" for downstream dsDNA. The same conclusion was reached by the Darst group, albeit in a more indirect fashion, since σR1.1 was absent from the X-ray structures of the holoenzyme and the RP$_o$ due to its inherent flexibility.[4,5] This work demonstrated that FRET measurements may help piece together "puzzles" of large and transient transcription complexes, provided that the individual "pieces" of the puzzle are available from conventional high-resolution structural studies.

Not surprisingly, ensemble FRET comes with its own set of shortcomings. Since ensemble methods report on the mean properties of populations of billions of molecules (*e.g.* a 16 µL solution of 10 nM DNA contains 1 billion DNA molecules), sample heterogeneity can skew the mean properties to a degree that complicates the interpretation of experiments. This is especially true in cases of large, unstable or dynamic biomolecules such as transcription complexes.

Consider, for example, the formation of RP$_o$; if the reaction mixture prepared for RP$_o$ formation is used without any separation (*e.g.* native gel electrophoresis), a sample will contain RP_o but also free σ^{70}, free RNAP holoenzyme, free DNA and σ^{70} aggregates. This "static" heterogeneity (defined as the presence of molecular subpopulations that do not interconvert during the timescale of the observation) often sets a requirement for purification of the species of interest (*e.g.* RP_o) for further analysis; otherwise, one needs to account for the heterogeneity, a difficult task exacerbated by variable heterogeneity between experiments and between different protein or DNA derivatives. Moreover, labelling reactions often do not produce 100% labelled biomolecules; thus, calculations of fluorescence indices, such as FRET efficiency, have to account for incomplete labelling. To complicate matters further, static heterogeneity is not always apparent. Even if electrophoretically pure RP_o complexes can be obtained, there is no guarantee that they will be 100% functional, since subtle protein modifications can abolish transcription activity without necessarily altering electrophoretic mobility. A functional assay is therefore needed to establish the fraction of active complexes; although such corrections are feasible, they add uncertainty and tedium to the overall measurement. One also has to consider that transcription complexes have finite lifetimes due to trivial dissociation, inactivation or aggregation.

Ensemble methods are also limited by their inability to monitor stochastic (and thus, unsynchronizable) dynamic motions; such motions produce "dynamic heterogeneity" (the presence of molecular conformations wherein a single molecule dwells for a measurable time during the timescale of observation). Consider, for example, the dynamics of σR1.1 that result in two conformational states of the σ^{70} holoenzyme: a long-lived state, where σR1.1 occupies the downstream dsDNA channel, and a transient state where σR1.1 is out of this channel. The ability to observe the conformational status of a *single* molecule can directly report the extent of the conformational change and its kinetics, provided that the timescale of the change is faster than the timescale of the entire observation. In contrast, the unsynchronized dynamics of one billion molecules may easily mask the presence, the kinetics and the exact nature of a transient state within a multi-step reaction pathway, such as the path from initiation to early elongation. We should note, however, that information obtained from ensemble and single-molecule data is complementary and that reliable ensemble measurements lay a strong and necessary foundation for single-molecule measurements. Moreover, both in the case of static and dynamic heterogeneity, one should check carefully for convergence of the ensemble results with the population-averaged, time-averaged mean of the single-molecule results.

Single-molecule methods circumvent the strict requirement for 100% pure preparations, and allow analysis of difficult-to-purify, unstable complexes. Therefore, single-molecule FRET (the observation of FRET occurring within a FRET pair placed on a single molecule) emerges as a powerful method for studying transcription. Single-molecule FRET (aka single-pair FRET) was first described in 1996[36] and has already revealed important information about biological machines and mechanisms.[37,38] In a typical smFRET experiment, a single molecule or complex labelled with a donor–acceptor FRET pair is exposed to light that excites the donor; after identification of the signals arising from single molecules, FRET-related ratios (derived from detected donor and acceptor fluorescence) report on donor–acceptor distance. For example, for experiments on single diffusing molecules (Figure 5.2A), the emissions for each fluorophore plotted as a function of time appear as "bursts" of fluorescence (green and red curve, Figure 5.2B); this information is summarized in one-dimensional FRET histograms (Figure 5.2C), which report on biomolecular structure, on the presence of static and dynamic heterogeneity, and on the presence and kinetics of conformational changes in biomolecules. Single-molecule FRET has been used to study dynamics of proteins, nucleic acids and their complexes.[39–44]

However, single-molecule FRET is not a general method for quantitative analysis of structure; it has mainly been used to identify distance changes and their kinetics. This limitation is due to the large number of corrections needed to accurately measure FRET efficiencies within single molecules, and to the presence of chemically or photophysically induced species that obscure FRET measurements when $R > 6-8$ nm. Single-molecule FRET is also not a general platform for quantitative analysis of molecular interactions. For example, consider the equilibrium $M^A + L^D \rightleftarrows M^A L^D$, where M^A is an acceptor-labelled macromolecule, and L^D is a donor-labelled ligand. First, smFRET yields a measurable signal only when donor–acceptor distances in the $M^A L^D$ complex are sufficiently short (typically $R_{D-A} < 6-8$ nm) to distinguish complexes from free L^D species. This proximity constraint often cannot be satisfied, especially for large complexes or interacting proteins of unknown structure. Second, inactive states of FRET acceptors (e.g. due to photoswitching[45–47] or photobleaching) result in ML^D species that exhibit donor-only characteristics,[48] leading to apparent increases in the free L^D species. Third, no M^A species are detected, since direct acceptor-excitation at the wavelength of donor-excitation is minimized to reduce crosstalk. Fourth, complexes with stoichiometries other than 1 : 1 (e.g. $M^A[L^D]_2$) cannot be identified by smFRET. The cumulative effect of such limitations, combined with complications arising from fluorophore photophysics, photobleaching and aggregation have prevented the widespread use of smFRET, at least in experiments involving diffusing molecules. Some of these limitations have been addressed in smFRET experiments on surface-immobilized molecules; such experiments are easier to interpret, especially with new buffer conditions that minimize fluorophore photophysics.[49]

To address the above shortcomings, single-molecule FRET has been recently extended through the use of alternating-laser excitation (ALEX) spectroscopy.[26,50–54] The ALEX-based methods provide additional, direct information on the presence and the state of both donor and acceptor fluorophores in the molecules of interest; the additional information leads to multidimensional histograms of FRET and fluorophore stoichiometry, which, in turn, report on biomolecular structure and stoichiometry, respectively.

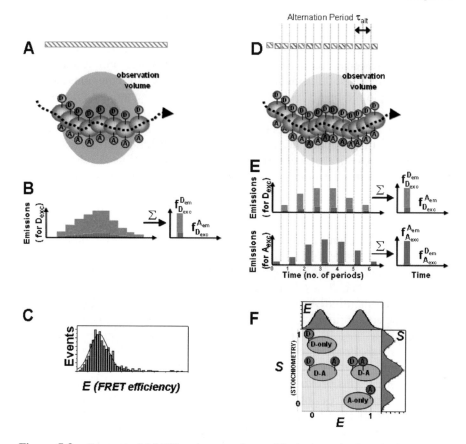

Figure 5.2 Concept of ALEX and comparison with single-excitation single-molecule FRET. (A–C) Single-molecule FRET using single-laser excitation; diffusing-molecule example. A fluorescent molecule traverses a focused green laser beam and emits photons in the donor- and acceptor-emission wavelengths. The photon counts at these two wavelengths are used to generate one-dimensional histograms of FRET efficiency E. (D and E) Single-molecule FRET using alternating-laser excitation. A fluorescent molecule traverses an observation volume illuminated in an alternating fashion using focused green and red laser beams. Using the photons emitted in the donor- and acceptor-emission wavelengths for each laser excitation, one can generate a two-dimensional histograms of FRET efficiency, E, and relative fluorophore stoichiometry, S, enabling molecular sorting (see text for details).

Initial applications of ALEX focused on studies of gene transcription[16,47,55] and protein folding.[53]

As with conventional smFRET, ALEX is compatible with studies of both diffusing and immobilized molecules. Diffusing molecules can be studied in solutions, gels, or even porous materials; immobilized molecules can be studied using confocal microscopy and scanning, or total-internal-reflection wide-field microscopy.[47]

ALEX is achieved by obtaining donor-excitation and acceptor-excitation-based observables (Figure 5.2D and E) for each single molecule; this is achieved by implementing an excitation scheme wherein the sample is illuminated, in an alternating fashion, by a laser that primarily excites the donor and a laser that primarily excites the acceptor (Figure 5.2D; Figure 5.3).

This scheme recovers distinct emission signatures for all diffusing species (Figure 5.2E) by calculating two fluorescence ratios: FRET efficiency, E, which reports on donor–acceptor distance in the $M^A L^D$ complex, and relative probe stoichiometry ratio S, which reports primarily on the donor–acceptor stoichiometry of all species. The stoichiometry ratio, S, provides information even in the absence of close proximity between fluorophores; it allows

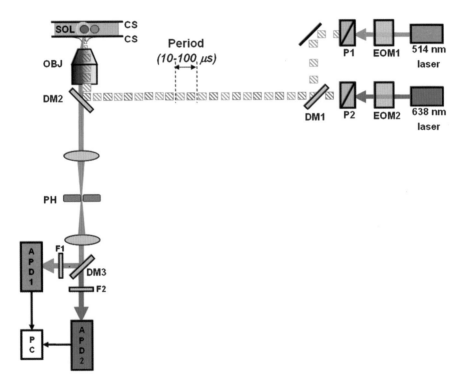

Figure 5.3 Alternating-laser excitation (ALEX) spectroscopy. When EOM-polarizer combinations are used, lasers yield an excitation alternating between 514 nm (D_{exc}) and 638 nm (A_{exc}) light (green and red hatched boxes). The alternating laser light excites the sample, which emits fluorescence (orange line) collected in donor-emission (D_{em}) and acceptor-emission (A_{em}) channels. Abbreviations: EOM, electrooptical modulator; P, polarizer; DM, dichroic mirror; OBJ, objective; PH, pinhole; F, filter; APD, avalanche photodiode.

thermodynamic and kinetic analysis of interactions, identification of interaction stoichiometry and study of local environment.

We define S as:

$$S = \frac{F_{D_{exc}}}{F_{D_{exc}} + F_{A_{exc}}} \tag{5.1}$$

where $F_{D_{exc}}$ is the sum of donor and acceptor photon counts for each molecule upon excitation with the donor-excitation laser, and $F_{A_{exc}}$ is the sum of donor and acceptor photon counts for each molecule upon excitation with the acceptor-excitation laser. Use of S allows stoichiometry observations that are independent of the diffusion path. Notably, S ($0 \leq S \leq 1$) assumes distinct values for all species in mixtures of interacting components (Figure 5.2F). After adjusting the excitation to obtain $F_{D_{exc}} \sim F_{A_{exc}}$ for a donor–acceptor species, S for donor-only species is high, ~ 1 (due to $F_{A_{exc}} = 0$), and S for acceptor-only species is low, in the 0–0.2 range (due to low $F_{D_{exc}}$); S for donor–acceptor species assumes intermediate values, in the 0.3–0.8 range. Provided that the local environment of one of the probes does not vary (preferably the acceptor), the S ratio can also report on quantum yield changes during a conformational transition (note that any quantum yield changes in the donor or acceptor fluorophores can be extracted from the single-molecule data directly by inspecting the $F_{D_{exc}}$ and $F_{A_{exc}}$ photon-counting histograms for all molecular transits; a shift of the histogram to lower photon count values indicates a decrease in the quantum yield). Combination of E and S in two-dimensional histograms enables virtual molecular sorting and quantification of sorted species (Figure 5.2F), while maintaining donor–acceptor distance information. The principle of ALEX has been implemented with millisecond-timescale temporal resolution on single immobilized molecules (millisecond-ALEX[47] see also Figure 5.4), and with nanosecond-timescale temporal resolution for distributions of single diffusing molecules (nanosecond-ALEX;[53] or pulsed interleaved excitation[56]).

5.3 Transcription Mechanisms Addressed using Single-molecule FRET and ALEX

A wide range of questions pertaining to transcription can be addressed with FRET. One set of questions involves detection of the presence and the extent of conformational changes: in essence, do two molecular parts move relative to each other at a certain stage of transcription, and, if they do, what is the nature and direction of motion? Collecting multiple distance constraints between the moving parts may also be used to localize the position of a mobile module during the course of the transcription pathway. In transcription initiation, three main sets of distances can be probed: protein–protein distances, DNA–DNA distances and protein–DNA distances. Studies of transcription elongation can also include labelled RNA, thus increasing the sets of molecular partners that can be used for distance measurements. In all cases, the high-resolution

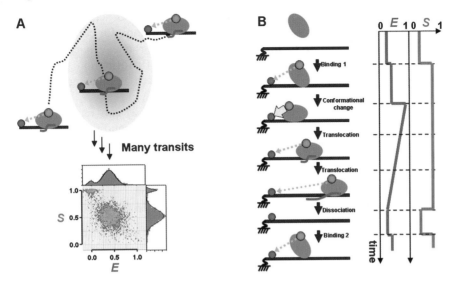

Figure 5.4 Use of alternating-laser excitation to study transcription complexes formed using diffusing and immobilized molecules. (A) Many molecular transits generate one- or multi-dimensional histograms that reflect an equilibrium view or steady-state view of the systems. Only slow kinetic changes (occurring at the timescale of minutes) can be observed by monitoring changes in the relative abundance of the subpopulations present on the histograms. (B) Schematic of ms-ALEX-based observation of the entire transcription pathway through changes in structure (as ratio E) and stoichiometry (as ratio S). Initial binding of RNAP on immobilized DNA (sensed as an increase in S) is followed by a conformational change (detected as an E increase). Subsequent translocation decreases E, until RNAP dissociates (decrease in S). Binding of a new RNAP molecule resumes the transcription cycle.

structures serve as an irreplacable guide for selecting labelling sites that will maximise the desired observable.

Equally important information is present in the dynamics of the complexes, as well as in the kinetics of interconversion between states during a non-equilibrium transition. Although slow kinetics can be monitored by studying diffusing molecules, dynamics/kinetics are best studied using imaging of immobilized single molecules; the timescale for kinetics should be slower than 10 ms to ensure that the current temporal resolution of single-molecule fluorescence imaging is adequate for monitoring the transitions between states (otherwise, time-averaging of the FRET efficiencies will occur). Depending on the position and the number of the probes one or multiple transcription steps can be monitored in real-time, as was done previously using probe-free methods (magnetic tweezers[17,57,58] or optical tweezers[59,60]).

Measurement of protein–protein distances (Figure 5.5A) can report on conformational changes occurring in the RNAP holoenzyme during the formation of transcription complexes and during initial transcription. For example, one may use donor–acceptor derivatives of σ^{70} (prepared using double surface-exposed-cysteine derivatives of σ^{70}) to look for conformational changes occurring upon binding to core RNAP, or one can examine the position of σR1.1 in the intermediates leading to RP_o formation. Moreover, one can examine possible conformational changes such as opening/closure of the RNAP β' clamp during transcription (Figure 5.4A).

Measurement of DNA–DNA distances (Figure 5.5B) can report on conformational changes occurring in DNA during the formation of RP_o and during initial transcription. For example, one may use donor–acceptor labelled DNA with the template strand labelled with a donor and the non-template strand labelled with an acceptor (prepared using reactive fluorophore conjugates and promoter DNA modified with reactive groups such as amines and thiols) to look for DNA opening during RP_o formation, or compaction of DNA during initial transcription.[16]

Measurement of protein–DNA distances (Figure 5.5C and D) can report on conformational changes occurring in both protein and DNA, and is particularly suited for monitoring general relative movement and, more specifically, translocation of RNAP on DNA (or pulling of DNA within RNAP, depending on the preferred frame of reference) during all phases of transcription. Two assays that can monitor translocation of RNAP *versus* DNA are known as the leading-edge-FRET assay and the trailing-edge-FRET assay. In leading-edge FRET, the donor is introduced at the leading edge of RNAP and the acceptor at the downstream end of DNA; thus, downstream translocation of RNAP is monitored as a FRET increase. In contrast, in trailing-edge-FRET, the donor is introduced at the trailing edge of RNAP and the acceptor at the upstream end of DNA; thus, downstream translocation of RNAP is monitored as a FRET decrease. Such assays have been used to monitor relative movement of DNA *versus* RNAP in abortive initiation and upon promoter escape.[16,55,61]

Combination of these methods can also be used to monitor the timing and coupling of conformational changes. This can be achieved by analysing multiple complexes (each with a specific donor–acceptor pair) and either "trapping" the states of interest (for diffusing molecules) or correlating their dwell times (for immobilized molecules). A more elegant but technically demanding way to study the timing and coupling of conformational changes involves simultaneous monitoring of three distances within a single transcription complex,[62] thereby providing a direct report on which distances change during a certain step.

5.4 Fate of Initiation Factor σ^{70} in Elongation

The first tests of the ability of ALEX spectroscopy to address questions in transcription came during collaborative studies of the group of one of us (S.W.) with the group of Richard Ebright (Rutgers/HHMI); the first study examined the controversy regarding the fate of σ^{70} in transcription elongation. Textbook

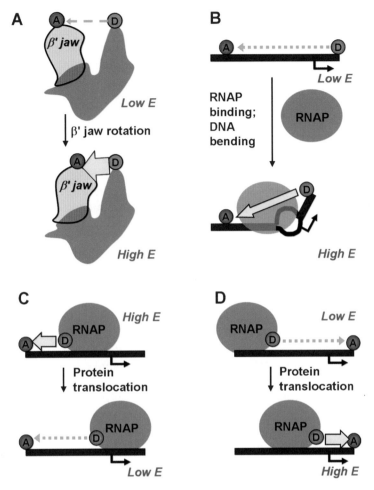

Figure 5.5 Strategies for monitoring movement within transcription complexes. In all cases, we assume that the fluorophores have adequate rotational freedom (see main text for details). (A) Monitoring conformational changes in RNAP by monitoring distances between two probes introduced to RNAP. For example, if one wishes to monitor closing of the β′ jaw, labels can be introduced on parts that move relative to each other during jaw closure, increasing FRET efficiency. (B) Monitoring conformational changes in DNA by monitoring distances between two probes introduced to DNA. For example, to monitor DNA bending during open complex formation, labels can be introduced on either side of the transcription bubble; DNA bending and bubble formation will bring the labelled sites closer, increasing FRET efficiency. (C) Monitoring RNAP translocation on DNA: trailing-edge FRET. A fluorophore is incorporated at the trailing-edge of the protein, and a complementary fluorophore is incorporated at the upstream end of the DNA. RNAP translocation towards a downstream direction increases the interprobe distance, and decreases FRET efficiency. (D) Monitoring RNAP translocation on DNA: leading-edge FRET. A fluorophore is incorporated at the leading-edge of the RNAP, and a complementary fluorophore is incorporated at the downstream end of the DNA. Protein translocation towards a downstream direction decreases the interprobe distance, and increases FRET efficiency.

descriptions of transcription routinely refer to obligatory release of σ^{70} upon formation of the first stable elongation complex (RD$_e$), formed upon synthesis of a 9–11 nucleotide RNA. This conclusion had been reached from several studies that determined the σ^{70} content of early elongation complexes.[63–67] However, one should note that early methods for quantifying σ^{70} release in elongation included separation steps that may have translated decreased affinity of σ^{70} for elongation complexes into release of σ^{70} from elongation complexes. This possibility became more real in 2001, with reports of analysis of elongation complexes using ensemble-FRET,[33] or purification *via* gentle separation steps,[68] combined with observations of σ^{70}-dependent promoter-proximal pausing at phage promoters[69] and the presence of σ^{70} in elongation complexes[70,71] identifying a fraction of elongation complexes that retain σ^{70}. Combined, these results suggest that σ^{70} release upon promoter escape is non-obligatory. The growing evidence for the presence of σ^{70} in elongation complexes has led to new proposals for the regulatory roles of σ in elongation.[68,72,73] However, the extent of σ^{70} retention in early and mature elongation complexes, the point(s) where release occurs, and the determinants of the extent and point of release are still unclear, in part due to shortcomings of the available methodology for measuring σ^{70} release.

The ensemble-FRET assay for σ^{70} retention[33,61] used the complementary assays of trailing-edge-FRET and leading-edge-FRET (see previous section and Figure 5.5C and D) to show that σ^{70} translocates with RNAP and enters transcription elongation. However, the assay was complicated by the extensive heterogeneity of transcription complexes (Section 5.2). For example, most ensemble-FRET measurements in ref. 33 were performed in polyacrylamide gels, raising the possibility that gels may stabilize a complex due to the "caging" effect of the gel, potentially changing the extent of σ^{70} retention and the affinity of σ^{70} for the elongation complex. Moreover, the ensemble-FRET work required many separate control measurements, such as parallel in-gel FRET measurements of reference complexes, measurement of transcriptional activities by quantifying RNA, or parallel in-gel FRET measurements on complementary complexes. Finally, the ensemble-FRET work assumes that all complexes in question are structurally homogeneous. However, if a conformational transition or translocation event changes the donor–acceptor distance in a way that cannot be accounted for with reference complexes, the σ^{70} content cannot be recovered. This was evident in the study of mature elongation complexes using leading-edge-FRET, where RNAP backtracking increased the donor–acceptor distance, and necessitated the use of an alternative, more indirect assay for evaluating σ^{70} retention.[33] Given such complications, it was difficult to explain the large differences in σ^{70} retention in mature elongation complexes measured using ensemble-FRET (60–70%)[33] or using a gentle purification method (6–30%, albeit on T7A1 promoter[68]).

To address these concerns, we used ALEX spectroscopy to measure σ^{70} retention at various points along elongation. Use of ALEX eliminates most of the above complications, since it is a direct assay that evaluates interprobe distances and interaction stoichiometry *independently*. We used ALEX in combination with the trailing-edge-FRET and leading-edge-FRET approaches (Figure 5.5C and D); both can be used to evaluate σ factor release and distances within open and elongation complexes. We used σ^{70} derivatives labelled with a tetramethylrhodamine (TMR) donor fluorophore in σR4 (the σ^{70} domain responsible for recognition of the promoter −10 element; probes were placed on position 569 or 596) for trailing-edge-FRET, and in σR2 (the σ^{70} domain responsible for recognition of the promoter −10 element; probes were placed on position 369 or 396) for leading-edge-FRET. We also used labelled DNA fragments based on promoter *lacUV5* (*placUV5*); DNA was labelled by a Cy5 acceptor fluorophore at the upstream end (position −40) for trailing-edge-FRET experiments, or the downstream end (position +25) for leading-edge-FRET experiments.

We prepared open complexes using unlabelled RNAP core, labelled σ^{70}, and labelled DNA, removed non-specific complexes, diluted the complexes to picomolar concentrations, and used ALEX to sort and identify open complexes, and, upon NTP addition, elongation complexes. From a single sample without additional controls, ALEX determines: (i) the fraction of active open complexes, (ii) the fraction of stalled elongation complexes that can resume transcription ("chaseable complexes") and (iii) the fraction of transcription complexes that retain σ^{70}. Significantly, the σ^{70} content of an elongation complex can be measured by simple molecular coincidence of σ^{70} and DNA, where donor and acceptor are beyond FRET range, with donor–acceptor distances longer than 100 Å; this ability obviates labelling at sites that guarantee close proximity of the probes.

The principle of the ALEX-based σ release assay (Figure 5.6) is based on single-molecule observations of changes in intermolecular distances and binding stoichiometries upon promoter escape and formation of the elongation complex. Owing to this capability for analytical sorting, the ALEX assay can distinguish between σ^{70} retention and σ^{70} release in elongation: for a leading-edge FRET assay, forward translocation and formation of a σ^{70}-containing elongation complex results in conversion of a donor–acceptor ($S \sim 0.5$) species with low E into a donor–acceptor ($S \sim 0.5$) species with high E (Figure 5.6A, right-hand panel). On the other hand, formation of elongation complex lacking σ^{70} results in conversion of a donor–acceptor ($S \sim 0.5$) species with low E to donor-only ($S > 0.8$) and acceptor-only ($S < 0.3$) species (Figure 5.6A, middle panel).

Figure 5.6A presents leading-edge FRET ALEX results for the open complex and RD$_{e,11}$ at *lacUV5*. In these experiments, fluorescence ratio E^*, a simplified form of FRET efficiency was used.[55] In samples of open complex, two species are observed (Figure 5.6B, left). One species exhibits $S \sim 0.55$ and $E^* \sim 0.23$ (corresponding to an apparent donor–acceptor distance of 77 Å). This species is the open complex, since it displays stoichiometry corresponding to a donor–acceptor species and displays a donor–acceptor distance consistent with that observed in previous ensemble and single-molecule experiments[26,33,35,61] and with the donor–acceptor distance predicted based on structural models of open complex.[35,74] We note that the distance measurement's accuracy within the

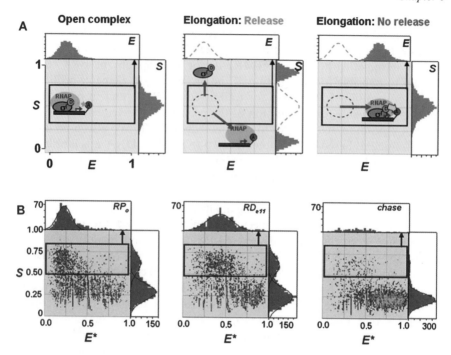

Figure 5.6 Single-molecule analysis of σ release: using leading-edge FRET to analyze σ release. (A) Release of σ^{70} from the transcription apparatus results in dissociation of a low-E donor–acceptor RNAP-DNA complex (left) to free σ^{70} (donor-only species) and a σ^{70}-free RNAP-DNA elongation complex (acceptor-only species) (middle). Retention of σ^{70} converts a low-E, donor–acceptor species into a high-E, donor–acceptor species where σ^{70} and RNAP–DNA complex remain co-localized (right). (B) σ^{70} is retained in the first stable elongation complex: E^*–S histograms for RP_o, $RD_{e,11}$ and chased $RD_{e,11}$ complexes studied by leading-edge-FRET. The donor–acceptor species and acceptor-only species are separated on the S histograms; the E^* histogram is shown only for the bound species (molecules $0.5 < S < 0.8$, in solid rectangle).

10–90% FRET range is ~ 5 Å. The remaining species exhibits $S < 0.3$; this species is free promoter DNA, the by-product of disruption of nonspecific RNAP–promoter complexes and removal of free RNAP holoenzyme by heparin challenge. No species with $S > 0.8$ are shown in Figure 5.6(B), since the molecule-search criteria employed solely identify acceptor-containing molecules; information about donor-only molecules is available, however, and can be used where necessary.[52]

Upon addition of the NTP subset permitting formation of $RD_{e,11}$, $\sim 80\%$ of open complexes are converted into a species that exhibits the same stoichiometry ($S \sim 0.55$) but higher FRET efficiency ($E^* \sim 0.44$; corresponding to a donor–acceptor distance of 58 Å) and that can be "chased" upon adding all four NTPs. This species is a σ^{70}-containing $RD_{e,11}$ and displays donor–acceptor distance consistent with the donor–acceptor distance observed in ensemble experiments;[33] the remaining ~20% of open complexes are converted into species with $S<0.3$ (i.e. σ^{70}-free $RD_{e,11}$). This observation led to the conclusion that most or all complexes are functional and competent to undergo transition from initiation to elongation, and that ~80% retain σ^{70} upon the transition from initiation to elongation. The σ^{70} retention value of ~80% is a lower bound, mainly because the data-acquisition time (~30 min) is significant relative to the half-life of σ^{70} retention. The pattern of σ^{70} retention was similar when measured using trailing-edge FRET.

The kinetics of dissociation of σ^{70} can be accessed directly using ALEX to monitor the fraction of σ^{70}-containing transcription complexes as a function of time (essentially by monitoring the decrease of the relative abundance of donor–acceptor species as a function of data acquisition time). Figure 5.7(A) presents results of leading-edge-FRET kinetic experiments assessing the initial extent and half-life of σ^{70} retention in $RD_{e,11}$. These experiments were performed by monitoring molecule-count ratios (involving counts of donor–acceptor and acceptor-only molecules) as a function of time after NTP addition, followed by comparisons with identical ratios for the open complex (which sets the ratio for 100% σ^{70} retention) and the "chased" complex (which sets the ratio for 0% σ^{70} retention). The results indicate that the initial extent of σ^{70} retention in $RD_{e,11}$ is ~90%, and the half-life of σ^{70} retention in $RD_{e,11}$ is ~90 min. Control experiments on RP_o showed that σ^{70} release or complex dissociation in RP_o occurs significantly more slowly (half-life of σ^{70} retention >2 h).

We also tested how σ^{70} content and dissociation is affected by a determinant for sequence-specific σ^{70}-DNA interaction (hereafter referred to as a "downstream −10 element") found in the initial-transcribed-region sequence of

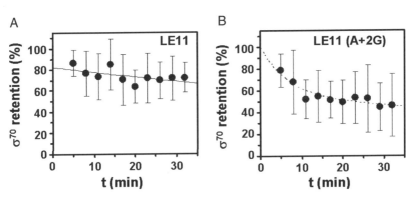

Figure 5.7 Retention and stability of σ^{70} in the first stable elongation complex ($RD_{e,11}$). Leading-edge-FRET experiments on p*lacUV5* (pause-containing promoter) (A) and p*lacUV5(A+2G)* DNA (pause-free promoter) (B). In the absence of the pause, the dissociation of σ^{70} from the elongation complex is significantly faster (see main text for details).

lacUV5; this downstream −10 element was shown to increase the initial extent and half-life of σ^{70} retention and mediate σ^{70}-dependent transcriptional pausing.[72,73] To assess effects of the downstream −10 element on the initial extent and half-life of σ^{70} retention, we performed leading-edge FRET ALEX experiments assessing open complex and $RD_{e,11}$ at *lacUV5(A+2G)*,[73] a substituted *lacUV5* derivative that lacks the downstream −10 element (Figure 5.7B). Here the initial extent of σ^{70} retention is $\sim$80%, and the half-life of σ^{70} retention is $\sim$30 min. These values are lower than the corresponding values for *lacUV5* ($\sim$90% and $\sim$90 min; Figure 5.7A). Parallel control experiments on RP_o showed that σ^{70} release or complex dissociation in RP_o occurs significantly more slowly (half-life of σ^{70} retention >2 h). These results are consistent with conclusions from ensemble FRET experiments[73] that the presence of a downstream −10 element increases the initial extent of σ^{70} retention and half-life of σ^{70} retention (albeit the initial σ^{70} retention remains high, >80%, even in the absence of the downstream −10 element).

Using the same methods on DNA fragments that permit formation of early elongation complexes containing longer RNA transcripts, it was shown that most of the elongation complexes retain σ^{70} after synthesizing 14 nt of RNA (a point where the RNA exit channel is expected to be filled). The translocation activity was >80%. The ability of the elongation complexes to resume elongation upon addition of the full set of nucleotides was $\sim$80%. More intriguingly, a large fraction ($\sim$50%) of mature elongation complexes (containing 50 nt of RNA) retained σ^{70}. These results support the proposal for non-obligatory σ^{70} release and are consistent with immunodetection of σ^{70} in elongation complexes *in vitro* containing RNAP prepared from stationary-phase cultures,[68] in elongation complexes *in vivo* in stationary-phase cultures[73,75] and in elongation complexes *in vivo* in stationary-phase and exponential-phase cultures.[76,77]

This study also served as a benchmark for ALEX spectroscopy and defined the range of questions that can be addressed by this new method. To allow for proper comparison between the ensemble-FRET study[33] and the ALEX study of σ^{70} release, identical DNA fragments (both in sequence and fluorophores used), labelled protein derivatives, reaction conditions and incubations were used (*e.g.* transcription complexes were kept at 37 °C during data acquisition on the microscope). Moreover, all labelled protein derivatives were fully characterized and fully functional.[33] This way, it was ensured that the genetic and chemical modification of proteins did not perturb function, that the Förster radius R_o for the FRET pairs was identical, and, in general, that the studies could be directly compared. Although this need for consistency precluded use of brighter and more photostable fluorophores, this comparative study provided a crucial and reassuring link from solution-based ensemble-FRET experiments to solution-based ALEX experiments, and set a new benchmark for comparison with ALEX measurements on immobilized complexes; in the latter measurements, one should ensure that the immobilization chemistry, the presence of the nearby surface, and the lengthy exposure to excitation lasers does not perturb transcription or the FRET measurements to a significant degree.

5.5 Mechanism of Initial Transcription

The validation of the ALEX method and the high translocational activity of the transcription complexes has paved the way for mechanistic analysis of initial transcription, and, in particular, of its abortive initiation phase. Specifically, after RP_o formation, conformational changes lead to the formation of an RNAP–promoter initial transcribing complex (RP_{itc}) that engages in abortive cycles of synthesis and release of short RNA products. Upon synthesis of an RNA product of ~9–11 nt, RNAP breaks its interactions with promoter DNA, breaks or weakens its interactions with initiation factors, escapes from the promoter, and enters into processive synthesis of RNA as an RNAP–DNA elongation complex.[2–3,78,79] Although the process of abortive initiation was described in 1980,[80] the mechanism by which the RNAP active centre translocates relative to DNA in initial transcription has remained controversial.[78] DNA footprinting studies had shown that the upstream boundary of the promoter DNA segment protected by RNAP is unchanged in RP_{itc} vs. in the open complex.[80–82] To reconcile this observation with the documented ability of RP_{itc} to synthesize RNA products up to ~9–11 nt long, three models have been proposed (Figure 5.8).

The first model, known as the transient-excursions model,[81] treats the footprinting results as a population- and time-averaged picture wherein RNAP is rigid and spends most of the time at the state of the open complex, but translocates transiently downstream during abortive initiation, and returns to the open complex state after abortive-RNA release (Figure 5.8, top). In this model, RNAP translocates as a unit, translocating by a single bp along DNA for each nucleotide added to the nascent RNA chain (just as in elongation, see ref. 59); upon release of abortive RNA, RNAP reverse translocates as a unit, regenerating the initial state. According to this model, the cycles of forward and reverse translocation are so short in duration and so infrequent in occurrence that, although they occur, they are not detected by a time-averaged, population-averaged method such as DNA footprinting.

The second model, known as the RNAP-inchworming model,[64,83] invokes a flexible, spring-like element in RNAP (Figure 5.8, middle); this element connects a DNA-recognition domain (assumed to remain fixed the process) and a "polymerizing" domain (assumed to contain the RNAP active centre and to move during the process). According to this model, abortive cycling extends the flexible element due to downstream translocation of the polymerizing domain relative to DNA (translocating 1 bp per phosphodiester bond formed) and relative to the DNA-recognition domain. Strain accumulated in the extended spring provides the driving force for disruption of σ^{70}–DNA and σ^{70}–RNAP contacts that bind RNAP to the open promoter, thus allowing for promoter escape. Otherwise, upon release of the short abortive RNA the accumulated strain provides the driving force for returning the complex to its initial state, RP_o.

A third model, known as the DNA-scrunching model,[78,81,84] invokes a flexible element in DNA; specifically, this element is formed by the single-stranded

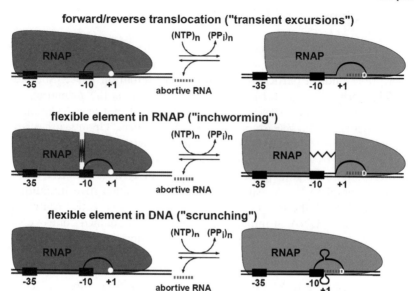

Figure 5.8 Models of abortive initiation Three models have been proposed for RNAP-active-centre translocation during initial transcription: transient excursions, inchworming and scrunching. White circles: RNAP active centre; red dashed lines: RNA; black rectangles: promoter −10 and −35 elements. See main text for details.

DNA of the transcription bubble (Figure 5.8, bottom). According to this model, abortive cycling contracts the flexible element due to "reeling-in" of downstream DNA in the cleft of the active site of RNAP (reeling in 1 bp per phosphodiester bond formed), with the accumulated DNA contracting by forming single-stranded DNA bulges ("scrunched" DNA) within the transcription bubble. Here, strain accumulated in the scrunched DNA–RNAP complex provides the driving force for disruption of σ^{70}–DNA and σ^{70}–RNAP contacts that bind RNAP to the open promoter, thus allowing for promoter escape. If escape fails to occur, then upon release of the short abortive RNA the accumulated strain provides the driving force for returning the complex to its initial state, RP_o.

Hybrid mechanisms that combine features of the three models are also possible, since the models are not mutually exclusive; in principle, combinations of mechanisms might be at work, or different mechanisms might be used at different stages of initial synthesis (*e.g.* one for synthesis of short RNA, and another for synthesis of longer RNA).

Since the ALEX work on σ^{70} release in elongation showed that, at the *lacUV5* promoter, the large majority of open complexes was capable of escaping to elongation, we reasoned that it should be possible to monitor distance changes occurring during abortive initiation, which constitutes an intermediate phase between the formation of open complex and the formation of the first stable elongation complex. Thus, we designed experiments that directly test the predictions of the three models in Figure 5.8 by monitoring distances and changes in distances within single open complexes and initial transcribing complexes formed on promoter *lac*UV5-CONS (a derivative of promoter *lacUV5*), and comparing the results with the predictions of the proposed models. We directly monitored distances between specific sites of σ^{70} and DNA, or between two sites within DNA (as in Figure 5.5). The ALEX-based distance measurements were performed in solution, taking advantage of the fact that the assay can analyze complexes with considerable heterogeneity in terms of composition and translocation position. Four sets of experiments were performed, assessing: movement of the RNAP leading edge relative to DNA; movement of the RNAP trailing edge relative to DNA; expansion/contraction of RNAP; and expansion/contraction of DNA. In each case, we performed FRET measurements within the open complex stabilized by the initiating dinucleotide ApA (referred to hereafter as RP_o) and within the initially-transcribing complex obtained by adding UTP and GTP to RP_o, referred to hereafter as $RP_{itc, \leq 7}$ as it engages in abortive synthesis of RNA products up to 7 nt in length.

All three models predict that, during abortive initiation, the RNAP leading-edge translocates downstream *relative* to the downstream end of DNA in fragments extending from positions around −40 to around +20 relative to transcription start site. To assess possible movement of the RNAP leading edge relative to downstream DNA in initial transcription, we monitored a leading-edge FRET pair similar to the one used in the σ^{70} release studies: we used a donor incorporated at the RNAP leading edge (σ^{70} residue 366, located in σR2) and an acceptor incorporated at a site in downstream DNA (position +20) (Figure 5.9A). A positive result also ensures that our assay has the temporal resolution to detect initially-transcribing complexes with decreased distances between the RNAP leading-edge and the downstream end of DNA. This RP_o has a donor : acceptor stoichiometry of 1 : 1, manifested as a species with $S \sim 0.6$, easily distinguished from free species (such as free DNA, with $S<0.3$; as in Figure 5.6B). If the RNAP leading-edge translocates towards DNA downstream of the transcription start site, leading-edge FRET complexes maintain their component stoichiometry (no S changes) and show a FRET increase since translocation *decreases* interprobe distance R. The results indicate that during abortive initiation the mean FRET efficiency E^* significantly increases, implying that the mean donor–acceptor distance R significantly decreases (Figure 5.9A, right). The increase in mean E^* corresponds to a decrease in mean R of ~ 7 Å (Figure 5.9A, right). Upon addition of NTPs to form $RP_{itc, \leq 7}$ (Figure 5.9A), the FRET distribution widens and shifts to higher E^* compared to RP_o, showing that a large fraction of complexes spends most of its time in a state where the interprobe distance is shorter than that observed in RP_o. This observation also implies that the timescale for abortive initiation is similar or slower than that of diffusion (1–5 ms). Indeed if abortive–initiation-related interconversions were faster than diffusion, they would have averaged out during diffusion within the confocal spot, resulting in no changes in the

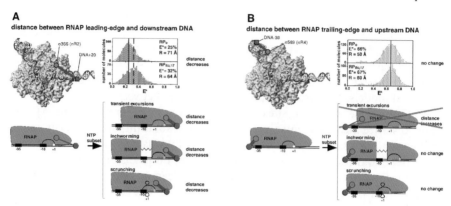

Figure 5.9 Initial transcription does not involve transient excursions. (A) Experiment documenting movement of the RNAP leading edge relative to downstream DNA [TMR as donor at σ^{70} residue 366 (located in σR2); Cy5 as acceptor at DNA position +20]. Top left: structural model of RP_o (ref. 74) showing positions of donor (green circle) and acceptor (red square). RNAP core is in grey; σ^{70} is in yellow; the DNA template and non-template strands are in red and pink, respectively. Top right: E^* histograms for RP_o and $RP_{itc, \leq 7}$. The vertical line and vertical dashed line mark mean E^* values for RP_o and $RP_{itc, \leq 7}$, respectively. Bottom: predictions of the three models. (B) Experiment documenting absence of movement of the RNAP trailing edge relative to downstream DNA [TMR as donor at σ^{70} residue 569 (located in σR4, the σ^{70} domain responsible for recognition of the promoter −35 element); Cy5 as acceptor at DNA position −39]. Subpanels as in part (A).

mean of the FRET distribution. This observation also underscored the feasibility of performing kinetic analysis of abortive initiation using surface-immobilized initially-transcribing complexes.

To confirm that this FRET change indeed reflects distance changes due to abortive initiation and not due to changes in the local environment of the probes or other trivial sources, we performed parallel experiments using a donor incorporated at a different site at the RNAP leading edge (σ^{70} residue 396, also located in σR2), or using an acceptor incorporated at a different site in downstream DNA (position +15 or position +25); in both cases, decreases in mean donor–acceptor distance (in the range of $\sim$4 to $\sim$15 Å) were observed. Control experiments performed in the presence of rifampicin,[85] an antibiotic that blocks synthesis of RNA products longer than 2 nt (and therefore blocks abortive initiation and initial transcription), show that the observed decreases in mean donor–acceptor distance require synthesis of RNA products longer than 2 nt. These results strongly suggested that the RNAP leading edge translocates relative to downstream DNA in initial transcription.

Moreover, the results led to the conclusion that, during initial transcription at the *lacUV5* promoter, complexes predominantly occupy states in which the

RNAP leading edge is translocated relative to downstream DNA, and not states in which the RNAP leading edge is positioned as in RP_o; therefore, the rate-limiting step in initial transcription is abortive-product release and RNAP-active-centre reverse translocation (see also ref. 47). The finding that the RNAP leading-edge translocates relative to downstream DNA is consistent with the predictions of all three models for initial transcription (Figure 5.8). However, the finding that RNAP predominantly occupies states in which the RNAP leading edge is translocated is inconsistent with the transient-excursion model since that model invokes infrequent and short-lived translocated states.

To assess possible movement of the RNAP trailing edge relative to upstream DNA in initial transcription (a crucial test for the transient excursion model), we monitored FRET between a donor incorporated at the RNAP trailing edge (σ^{70} residue 569, located in σR4) and an acceptor incorporated at a site in upstream DNA (position –35) (Figure 5.9B). In this case, during abortive initiation mean E^* remains unchanged (Figure 5.9B top right), implying that the mean donor–acceptor distance remains unchanged. Moreover, control experiments clearly showed that the probe sites used were well-positioned to detect a translocation-dependent change in mean donor–acceptor distance and would have detected a change if it had occurred.

Indeed even when the interprobe distance is within the FRET-accessible range it is possible that molecular movement leads to a state with a similar FRET efficiency, for instance if the net donor–acceptor distance is unchanged by the molecular movement. To check whether the absence of a FRET change is due to an equidistant position or to absence of relative motion, it is necessary to evaluate two or more distances that can report on the putative motion. Parallel experiments performed using a donor incorporated at a different site at the RNAP trailing edge (σ^{70} residue 596, also located in σR4) also suggested that the mean donor–acceptor distance remains unchanged.

From these experiments, we concluded that at the *lacUV5* promoter the RNAP trailing edge does not translocate relative to upstream DNA in initial transcription and that σR4 does not alter its interactions with DNA in initial transcription. This is true even for reaction conditions where the RNAP leading edge translocates relative to downstream DNA (Figure 5.9A). These findings contradict the fundamental prediction of the transient-excursions model, *i.e.* any molecule having the RNAP leading edge translocated relative to DNA also must have the RNAP trailing edge translocated relative to DNA (Figures 5.8 and 5.9). Thus initial transcription does not involve transient excursions.

To assess possible expansion/contraction of RNAP in initial transcription (a major prediction of models involving RNAP inchworming), we first monitored FRET between a donor incorporated at the RNAP leading edge (σ^{70} residue 366, located in σR2) and an acceptor incorporated at a site in –10/–35 spacer DNA (position –20) (summarized in Figure 5.10B). In this case, upon transition from RP_o to $RP_{itc,\leq 7}$, the mean E^* remained unchanged, implying that the mean donor–acceptor distance remained unchanged. Parallel experiments performed using a donor incorporated at a different site at the RNAP leading edge (σ^{70} residue 396, shown to translocate relative to downstream DNA) also imply

138   Chapter 5

that mean donor–acceptor distance was unchanged. Similar results were obtained after monitoring FRET between a donor incorporated at the RNAP trailing edge (σ^{70} residue 569 or σ^{70} residue 596, located in σR4) and an acceptor incorporated at a site in the –10/–35 spacer DNA (position –20) (summarized in Figure 5.10B). These results strongly suggested that RNAP does not expand or contract in initial transcription, and that the leading edge of RNAP does not translocate relative to DNA upstream of the unwound

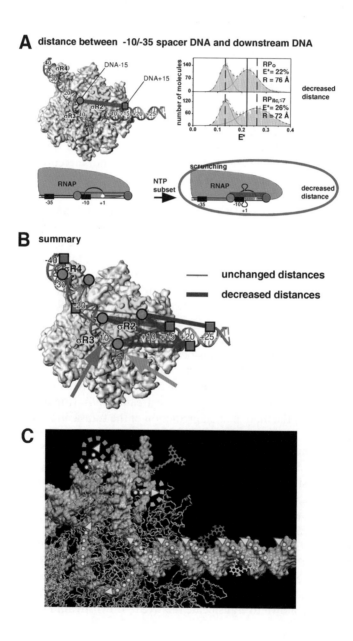

region – even under reaction conditions where we have shown that the RNAP leading edge translocates relative to downstream DNA. These findings are inconsistent with the fundamental prediction of the inchworming model, *i.e.* any molecule having the RNAP leading edge translocated relative to downstream DNA also must have the RNAP leading edge translocated relative to DNA upstream of the unwound region. Therefore, it was concluded that initial transcription does not involve inchworming.

To assess possible expansion/contraction of DNA in initial transcription (the main prediction of models involving DNA scrunching), we monitored FRET between a donor incorporated at a site in the –10/–35 spacer DNA (position –15) and an acceptor incorporated at a site in downstream DNA (position +15) (Figure 5.10A). The results indicate that, during abortive initiation, the mean E^* significantly increases (Figure 5.10A, top right), implying that the mean donor–acceptor distance significantly decreases. The quantitative increase in mean E^* corresponds to a decrease in mean R of $\sim 4\,\text{Å}$ (Figure 5.10A, top right). The FRET distribution for $RP_{itc,\leq 7}$ also becomes much wider than in RP_o, reflecting the fact that this sample represents a set of complexes with a variable degree of DNA contraction; high-resolution FRET measurements (L.C. Hwang and A.N. Kapanidis, unpublished) may be able to separate the various subpopulations present within the broad E^* distribution for $RP_{itc,\leq 7}$.

Additional experiments performed using an acceptor incorporated at a different site in downstream DNA (position +20) also show a $\sim 6\,\text{Å}$ decrease in mean donor–acceptor distance. Control experiments performed in the presence of rifampicin again showed that the observed decreases in mean

Figure 5.10 Initial transcription involves scrunching. (A) Experiment documenting contraction of DNA between positions –15 and +15 (Cy3B as donor at DNA position –15; Alexa647 as acceptor at DNA position +15). Subpanels as in Figure 5.9(A). The two donor–acceptor species in the E^* histograms comprise free DNA (lower-E^* species) and RP_o or $RP_{itc,\leq 7}$ (higher-E^* species; higher FRET attributable to RNAP-induced DNA bending). (B) Summary of results. Structural model of RP_o,[74] showing all donor–acceptor distances monitored in ref. 16. Distances that remain unchanged upon transition from RP_o to $RP_{itc,\leq 7}$ are indicated with thin blue lines. Distances that decrease upon transition from RP_o to $RP_{itc,\leq 7}$ are indicated with thick blue lines. The red and pink arrows show the proposed positions at which scrunched template-strand DNA and scrunched nontemplate-strand DNA, respectively, emerge from RNAP (*i.e.* near template-strand positions –9 to –10 and near nontemplate-strand positions –5 to –6). (C) A model showing an open promoter complex during transcription initiation, with arrows designating motions of DNA segments occurring during abortive initiation. Downstream DNA (on the right) rotates inward and separates in the RNAP active site channel. Template (orange) and nontemplate (pink) DNA strands follow the indicated paths, moving as the RNA chain (not shown) is polymerized on the template DNA. Sites of extrusion of scrunched single-stranded DNA are shown by outlined arrows at the top. σ^{70} (blue); RNAP core chains (grey); FRET donor (green) on σ^{70} region 2 (in cyan); FRET acceptor on downstream DNA (initial position in yellow; scrunched position in red).

donor–acceptor distance require synthesis of an RNA product longer than 2 nt. We infer that the DNA segment between –10/–35 spacer DNA and downstream DNA contracts in initial transcription. These findings document the fundamental prediction of the simplest version of the scrunching model, *i.e.* any molecule having the RNAP leading edge translocated relative to downstream DNA also must have contraction – scrunching – of the DNA segment between –10/–35 spacer DNA and downstream DNA (Figure 5.8). We conclude that initial transcription involves scrunching. Interestingly, FRET distributions for DNA→DNA FRET become narrower upon addition of initiating dinucleotide, pointing to the heterogeneous nature of the transcription bubble in the absence of the initiating NTP, and to the possibility that initiating nucleotides act as "anchors" that limit the conformational search of bubble DNA. The flexibility of the bubble might be the key factor in the ability of RNAP to reprogram the transcription start site when necessary (L.C. Hwang, R.H. Ebright, S. Weiss and A.N. Kapanidis, unpublished).

Mapping the distance changes onto the structure of the open complex clearly shows that they fall into two distinct categories: all measured distances between RNAP and upstream DNA or –10/–35 spacer DNA remain unchanged during abortive initiation (Figure 5.10B, thin blue lines). In contrast, all measured distances between RNAP and downstream DNA, or between the –10/–35 spacer and downstream DNA, decrease during abortive initiation (Figure 5.10B, thick blue lines). We infer that DNA scrunching occurs exclusively within the DNA segment consisting of positions –15 to +15. This DNA segment contains the transcription bubble that, in structural models of the open complex,[4,6,35,74,86] is located within and immediately upstream of the RNAP active-centre cleft. Inspection of structural models of the open complex and elongation complex[4,6,35,74,86,87] indicates that there is insufficient space within the RNAP active-centre cleft to accommodate scrunched DNA without conformational changes in the complexes. Therefore, it is likely that scrunched DNA emerges from RNAP immediately upstream of the RNAP active-centre cleft (Figure 5.10C); although the locations at which scrunched DNA emerges are not known, we propose that the scrunched template DNA strand and nontemplate DNA strand emerge at or near the points where the respective DNA strands normally emerge from RNAP immediately upstream of the RNAP active-centre cleft, *i.e.* at or near positions –9 to –10 of the template strand and positions –5 to –6 of the nontemplate strand (Figure 5.10B, red and pink arrows; Figure 5.10C). An alternative possibility is that the cleft of RNAP widens to accommodate the scrunched single-stranded DNA, perhaps in preparation for the process of promoter escape.

These results, along with the magnetic tweezer studies of abortive initiation (Chapter 6 and ref. 17), clearly establish that initial transcription involves DNA scrunching. In contrast, processive transcription elongation involves simple translocation, not DNA scrunching.[59] Therefore, there is a fundamental difference in the mechanisms of RNAP-active-centre translocation in initial transcription and processive transcription elongation; this difference may reflect the presence or the absence of the sigma factor from the initiation and late elongation processes, respectively. Moreover, the two studies provide strong support for existence of a "stressed intermediate" in initial transcription,[72,83] specifically a stressed intermediate with accumulated DNA-scrunching energetic stress. This stress arises from many factors, including electrostatic repulsions between negative phosphate charges in each of the two scrunched DNA segments, and formation of weaker RNA–DNA interactions (in the DNA–RNA hybrid) upon breaking stronger DNA–DNA interactions (in the downstream DNA). As proposed earlier by Crothers *et al.*, the energy accumulated in the stressed intermediate provides the driving force for abortive initiation and also provides the driving force for promoter escape and productive initiation. The stress can be resolved in two ways: (i) by releasing abortive RNA, retaining interactions with promoter DNA, retaining interactions with initiation factors, retaining an unchanged position of the RNAP trailing edge, extruding scrunched DNA, and re-forming the open complex (abortive initiation); or (ii) by retaining the RNA transcript, breaking interactions with promoter DNA, breaking interactions with initiation factors, translocating the RNAP trailing edge, and forming the first stable elongation complex (promoter escape and productive initiation). Abortive-RNA synthesis can be thought of as the process that charges a spring element that can return to its relaxed state by expanding either in the direction opposite to the charging force (returning to the initial uncharged state) or in the same direction as the charging force (generating a translocated uncharged state).

The existence of a DNA-scrunching mechanism for abortive initiation by multi-subunit RNAP strongly suggests that DNA scrunching is a universal feature of all RNA polymerases, since the available information on single-subunit RNAP also points to DNA scrunching[88–90] (see also Chapter 4). In addition to its importance in abortive initiation and promoter escape, the flexibility of the single-stranded DNA in the transcription bubble might explain the variable spacing between the downstream end of −10 element and the transcription start sites,[91,92] as well as the start-site reprogramming by dinucleotide triphosphates.[93,94] Our results show also that the "mobile" element in abortive initiation is DNA, and therefore are consistent with proposals that transcription *in vivo* might occur by immobile RNA polymerases residing in transcription factories.[95]

5.6 Kinetic Analysis of Initial Transcription and Promoter Escape

ALEX studies of diffusing single molecules cannot report on the milisecond-to-second kinetics of transcription because of the short transit time of molecules through the laser focus (∼1 ms) and the need to accumulate significant statistics to extract kinetic information (such as in Figure 5.7). The long lifetime of initial transcribing complexes (timescale of seconds[17]) pointed to the possibility of real-time kinetic analysis of the conformational changes in initial transcription, promoter escape, and early transcription elongation using active surface-immobilized complexes. This analysis also requires that changes in

FRET arise indeed from conformational changes and not from light-induced fluorophore photophysics, such as the formation of transient non-absorptive, non-emissive states[96] or photoswitching.[45-47]

Using total-internal-reflection fluorescence (TIRF) microscopy equipped with alternating-laser excitation implemented at the millisecond timescale (ms-ALEX[47]), we were able to detect abortive initiation and promoter escape within single immobilized transcription complexes. Complexes prepared were identical to the ones used for the analysis of σ^{70} release and abortive initiation but for the addition of a biotin group incorporated at either end of the promoter DNA fragment (Figure 5.11A). This format allows immobilization of DNA fragments or stable transcription complexes on glass or quartz surfaces modified by hydrophilic poly(ethylene glycol) (PEG) groups and coated with streptavidin; the immobilization strategy has been developed for the study of other DNA-processing enzymes and has been used extensively.[39,41,97]

The ALEX-based translocation assays used to study σ^{70} retention upon promoter escape (Figures 5.6 and 5.7) were used to measure the fraction of immobilized open complexes able to escape to elongation. An example is given in Figure 5.11(B). Initially, open complexes for leading-edge FRET were surface-immobilized and ms-ALEX was used to identify donor–acceptor species and generate a one-dimensional histogram of E^*. Upon addition of the NTP subset that allows formation of the first stable elongation complex, ~70% of the molecules move to higher E^*, which essentially means that they escape to elongation and retain σ^{70}. Similar results were obtained using the trailing-edge FRET assay. These results also confirm that σ^{70} is retained during promoter escape and that it does not dissociate from RNAP prior to escape only to rebind to it after escape. Indeed, as there is virtually no free σ^{70} in the solution surrounding the immobilized complexes, the probability of σ^{70} rebinding is negligible. Therefore, the σ^{70} molecule present in transcription initiation is carried into elongation after promoter escape.

Using ms-ALEX, we also observed small but reproducible and abortive-product-length-dependent, decreases in distance between the RNAP leading edge and DNA downstream of RNAP upon abortive initiation. Inspection of population distributions and single-molecule time traces for abortive initiation indicates that, at a consensus promoter under saturating NTP concentrations, abortive-product release is rate-limiting, *i.e.* abortive-product synthesis and RNAP-active-centre forward translocation are fast, whereas abortive-product dissociation and/or RNAP-active-centre reverse translocation are slow.

5.7 Comparison of FRET Approaches with Magnetic-trap Approaches

Studies of abortive initiation using single-molecule FRET and magnetic traps (Chapter 6) demonstrate how the two techniques can be used to provide complementary information on transcription mechanisms. But how do these

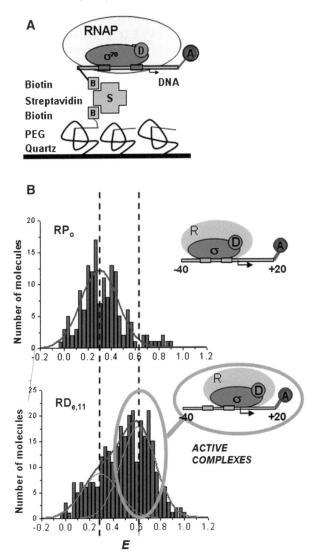

Figure 5.11 Millisecond-ALEX on immobilized transcription complexes. (A) Schematic of a surface-*immobilized* biotinylated transcription complex for leading-edge FRET experiments. (B) Functional assays for immobilized open complexes: measurement of translocational activity using leading-edge FRET. More than 70% of the open complexes are competent to perform transcription and enter elongation.

two methods compare in terms of their spatial and temporal information, requirements, instrumentation and reagents?

Single-molecule FRET is preferable over magnetic traps if one needs to monitor distance changes occurring between sites separated by 3–10 nm along any direction within biomolecules. In contrast, magnetic traps rely on events that change the length of the tethered DNA (*e.g.* due to unwinding/rewinding and DNA wrapping), and cannot easily report on conformational transitions for all biomolecules, especially conformational changes in the RNAP core or holoenzyme. Moreover, processes that do not substantially change the topology of tethered DNA (*e.g.* processive elongation) are unlikely to generate a signal for magnetic traps.

At present, single-molecule FRET imaging of multiple isolated fluorescent molecules excited in a TIRF field can reach ~ 10 ms time resolution,[97,98] and confocal imaging extends this limit to the 1–10 ms timescale.[99,100] Time-resolved approaches such as nanosecond ALEX[53] (see also pulsed-interleaved excitation[56]) can probe events down to nanosecond timescales for subpopulations of single molecules. As a comparison, the temporal resolution achieved with rapid mechanical measurements on short (50–100 bp) DNA fragments is in the 1–10 ms range.[101,102]

Single-molecule FRET, especially as implemented using ALEX, allow multi-point monitoring of conformational changes within single biomolecules. At this point, using the method of three-color ALEX, multiple-stoichiometries and up to three distances can be observed within a transcription complex (see next section).

Single-molecule FRET experiments have a higher throughput than magnetic traps: experiments with diffusing molecules can record thousands of molecules within 10 min, and TIRF-based imaging can record time-trajectories of hundreds of molecules for several minutes. Magnetic traps usually record observables from a single molecule over several hours, although methods for increasing throughput have been described.[103]

Finally, single-molecule FRET measurements do not require placing the template under any tension (which may complicate some processes, *e.g.* processes that rely on transient DNA looping), and are compatible with observation within living cells.

However, there are several limitations of single-molecule FRET *vs.* magnetic traps and other single-molecule nanomanipulation methods; most are related to the incorporated fluorophores. FRET observations of immobilized molecules are usually limited in duration due to photobleaching of either the donor and acceptor probe. Until recently, typical observations of a single FRET pair rarely exceeded a couple of minutes, even with the use of oxygen-scavenging additives (enzymatic systems that remove oxygen to minimize photoinduced oxidation reactions that destroy the fluorescent molecule). The limited photostability of the fluorophores thus limits the number of turnovers that can be observed on single DNA substrates or single RNAP molecules. Moreover, if a long reaction is to be observed (*e.g.* the entire transcription of a gene), it is likely that the fluorophores will bleach during the process, precluding conclusions about the last stages of the reaction. This is in sharp contrast to the long (>10 h) time-trajectories accessible to the magnetic-tweezers approach. We note, however, that recent advances in development of photostable fluorophores and in understanding and controlling fluorophore photophysics have increased the survival of single fluorophores, enabling measurements of FRET pairs for timescales longer than 10 min. This exciting development should open up experiments currently available only to label-free methodologies.

The need to measure distances between specific molecular parts within complexes also introduces the need for site-specific labelling. While site-specific labelling is trivial in the case of DNA fragments and relatively straightforward for σ^{70}, it is not simple for RNAP core, complicating transcription elongation experiments and calling for better strategies for site-specific labelling of large proteins such as RNAP. One also needs to demonstrate that fluorophores do not perturb the function under study. Thus, ensemble or (ideally) single-molecule functional assays have to be performed for all labelled biomolecules to demonstrate that their modification has little or no effect.

The flip-side of FRET sensitivity within the 3–10 nm range is that when distances remain out of the range, no distance changes can be detected, even in the presence of a conformational change. Therefore, one needs to strike a balance between labelling biomolecules close enough to the site of action to monitor a conformational change but far enough from critical interactions so that no function is significantly altered.

Finally, operating at > 1 nM concentrations of labelled biomolecule in solution (and > 50 nM concentrations of labelled biomolecule for immobilized molecules) is currently challenging since, above these concentrations, there are more than one molecule in the detection zone at a given moment (see also next section).

5.8 Future Prospects

Exciting prospects lie ahead in terms of both extending the current methodology, and applying existing ALEX methods to outstanding questions in transcription.

Initial ms-ALEX studies had limited temporal resolution; each frame corresponded to a 400-ms exposure of the immobilized molecules with the fluorophores and imaging system used. Thanks to improvements in fluorophores, imaging buffers (oxygen scavengers, reducing agents) and low-light-imaging electron-multiplying CCD cameras, the frame exposure times have been reduced to < 10 ms,[97,98] and should soon allow a more detailed view of abortive initiation and promoter escape.

One important application involves the use of three-color alternating-laser excitation (3c-ALEX) of single molecules to study transcription.[62] Three-color ALEX is an extension of ALEX that can measure up to three intramolecular distances and complex interaction stoichiometries in solution. 3c-ALEX was realized by adding a laser and a detector to the two-colour ALEX scheme (Figure 5.3), and by sorting molecules in multi-dimensional probe-stoichiometry and FRET-efficiency histograms. As with its 2c-ALEX counterpart, 3c-ALEX uses probe-stoichiometry histograms to perform analytical

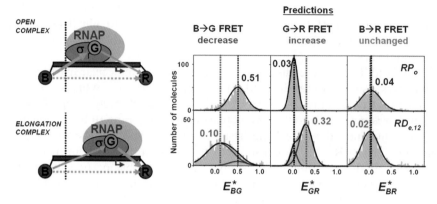

Figure 5.12 Multi-perspective monitoring of RNAP translocation on promoter DNA. Left-hand panel: concept and labelling scheme. Triply labelled transcription complexes are prepared; the G probe is on RNAP and the B and R probes upstream and downstream of the RNAP binding site, respectively. B→G: trailing-edge FRET pair; G→R: leading-edge FRET pair. The direction of transcription (and of downstream translocation) is denoted by the black arrow on DNA. Upon RNAP translocation downstream, trailing-edge FRET decreases and, correspondingly, leading-edge FRET increases, reporting on the direction of motion. Right-hand panels: Comparison of one-dimensional proximity-ratio histograms for RP_o and $RD_{e,12}$ show that for ~85% of the complexes, trailing-edge FRET decreases (as measured by E^*_{BG}) and leading-edge FRET correspondingly increases (as measured by E^*_{BG}), which is consistent with downstream RNAP translocation. The remaining 15% of the complexes are inactive.

sorting, identification, and selection of diffusing species; selected molecules are subsequently represented in FRET-efficiency histograms, generating up to three intramolecular distances. All these capabilities have been demonstrated using triply labelled DNA fragments.[62] Moreover, 3c-ALEX was used to study transcription complexes, and monitor the downstream translocation of RNA polymerase on DNA from two perspectives within the complex (Figure 5.12).

Briefly, triply labelled transcription complexes were prepared with probes denoted blue (B), green (G) and red (R). The G probe is located on RNAP and the B and R probes are located upstream and downstream of the RNAP binding site, respectively. In this labelling scheme, B can act as a FRET donor to G (*i.e.* B→G: a trailing-edge FRET pair) and G can act as a FRET donor to R (*i.e.* G→R: a leading-edge FRET pair); B and R are too far apart to participate in FRET. Upon RNAP translocation downstream, trailing-edge FRET is expected to decrease, and, correspondingly, leading-edge FRET is expected to increase, reporting on the direction of motion. Indeed, comparison of one-dimensional E^* histograms for the open complex and for $RD_{e,12}$ show that, for ~85% of the complexes, trailing-edge FRET decreases (as measured by E^*_{BG}) and leading-edge FRET correspondingly increases (as measured by E^*_{BG}), which is consistent with downstream RNAP translocation. This experiment validated the use of 3c-ALEX for transcription studies and opened up new possibilities of exciting experiments that allow real-time monitoring of conformational changes, especially after 3c-ALEX is adapted for use on immobilized molecules.[47]

A salient feature of ALEX is the improved distance measurement within single molecules.[26] Since the sensitivity of single-molecule fluorophore is high (requiring many orders of magnitude less material than crystallography or NMR), the accurate distance-measurement capability of ALEX[26] with fluorophores having adequate rotational freedom (see Section 5.2, discussion on orientation factor κ^2), combined with rapid preparation of small amounts of labelled biomolecules, may lead to solution-based, low-resolution structures of biomolecules and their complexes, as done using ensemble-FRET and distance-restrained docking.[35] Such an approach is suitable for large, multicomponent, heterogeneous complexes of cellular machinery (especially if inaccessible to conventional structural biology), and it will benefit from existing crystallographic or NMR-based structures.

As discussed earlier, one limitation of ALEX is the need to operate at subnanomolar concentrations of fluorescent biomolecule. The applicability of the method will be improved if the working concentration range can be extended to study fluorescent species in concentrations higher than 1 nM. This can be achieved by combining ALEX with methods that confine excitation volume[104–107] or by using nonfluorescent analytes, which can be present at any concentration and can exert an effect by altering the fluorescence observables of the labelled species.

In terms of applications to transcription, ALEX analysis of doubly labelled derivatives of σ^{70} will shed light on the conformational transitions occurring in σ^{70} at various steps of the transcription cycle. Labelling RNAP core, DNA and RNA will also allow the study of conformational changes in RNAP occurring during a single nucleotide addition cycle of transcription elongation, complementing the large body of work produced using optical tweezers[59,108] (see Part II of this book). ALEX methods can be used to study holoenzymes containing other σ factors such as σ^{54}.[109,110] ALEX and single-molecule FRET should also be useful for studies of eukaryotic RNAP, such as elongation complexes of polII (ref. 111). Finally, the holy grail of the field is to be able to monitor transcription at the single-molecule level in real-time and in living cells. Although this is an extremely challenging endeavour, recent success in monitoring the diffusion properties of single *lac* repressor molecules in single bacteria[112] suggests that elegant *in vivo* transcription experiments await and that ensuing discoveries may be closer than previously thought.

5.9 Summary

ALEX spectroscopy is a powerful tool that was recently added to the single-molecule detection toolbox; it can analyse complex samples without strict requirements for high purity or optimal distance range, facilitating reagent preparation. The first ALEX application to transcription provided convincing answers to questions related to σ^{70} release in elongation and formed an important link between ensemble and single-molecule FRET experiments. Moreover, FRET-based observation of abortive initiation occurring through a DNA-scrunching mechanism helped resolve a long-standing conundrum in the field and open up a whole new area of experimentation. Experiments on diffusing molecules are currently being supplemented by experiments with immobilized molecules, which provide real-time single-molecule views of the kinetics of initial transcription and promoter escape. Finally, additional extensions of the method are expected to continue advancing our understanding of basal and regulated transcription, especially as it occurs in its natural context, the chaotic but beautiful world of the living cell.

Acknowledgements

We are grateful to all the current and former members of the Weiss' group (especially Emmanuel Margeat, Nam Ki Lee, Xavier Michalet, Philip Tinnefeld, Soeren Doose, Sam Ho and You Wang) who contributed to studies of transcription mechanisms using ALEX methodologies. We also thank Richard Ebright, Jayanta Mukhopadhyay, Vladimir Mekler and Ekaterine Kortkhonjia for fruitful collaboration on transcription-related projects, and Ms Joanne Tang for editorial assistance. This work was funded by NIH Grant GM069709-01 to S.W. and A.N.K., DOE grants 02ER63339 and 04ER63938 to S.W., and EPSRC grant EP/D058775 and Marie Curie EU grant MIRG-CT-2005-031079 to A.N.K.

References

1. R. Landick, Shifting RNA polymerase into overdrive, *Science*, 1999, **284**, 598–9.
2. M. T. Record Jr., W. S. Reznikoff, M. L. Craig, K. L. McQuade and P. J. Schlax, *Escherichia coli* RNA Polymerase (Esigma70), Promoters, and the kinetics of the steps of transcription initiation, In *Escherichia coli and Salmonella*, ed. F. Neidhart, ASM Press, Washington, D.C., 1996, **Vol. 1**, pp. 792–820.
3. B. A. Young, T. M. Gruber and C. A. Gross, Views of transcription initiation, *Cell*, 2002, **109**, 417–20.
4. K. S. Murakami, S. Masuda, E. A. Campbell, O. Muzzin and S. A. Darst, Structural basis of transcription initiation: an RNA polymerase holoenzyme-DNA complex, *Science*, 2002, **296**, 1285–90.
5. K. S. Murakami, S. Masuda and S. A. Darst, Structural basis of transcription initiation: RNA polymerase holoenzyme at 4 Å resolution, *Science*, 2002, **296**, 1280–4.
6. D. G. Vassylyev, S. Sekine, O. Laptenko, J. Lee, M. N. Vassylyeva, S. Borukhov and S. Yokoyama, Crystal structure of a bacterial RNA polymerase holoenzyme at 2.6 Å resolution, *Nature*, 2002, **417**, 712–9.

7. D. G. Vassylyev, M. N. Vassylyeva, A. Perederina, T. H. Tahirov and I. Artsimovitch, Structural basis for transcription elongation by bacterial RNA polymerase, *Nature*, 2007, **448**, 157–62.
8. D. G. Vassylyev, M. N. Vassylyeva, J. Zhang, M. Palangat, I. Artsimovitch and R. Landick, Structural basis for substrate loading in bacterial RNA polymerase, *Nature*, 2007, **448**, 163–8.
9. H. Buc and W. R. McClure, Kinetics of open complex formation between *Escherichia coli* RNA polymerase and the lac UV5 promoter. Evidence for a sequential mechanism involving three steps, *Biochemistry*, 1985, **24**, 2712–23.
10. M. Buckle, I. K. Pemberton, M. A. Jacquet and H. Buc, The kinetics of sigma subunit directed promoter recognition by *E. coli* RNA polymerase, *J. Mol. Biol.*, 1999, **285**, 955–64.
11. P. L. deHaseth and J. D. Helmann, Open complex formation by *Escherichia coli* RNA polymerase: the mechanism of polymerase-induced strand separation of double helical DNA, *Mol. Microbiol.*, 1995, **16**, 817–24.
12. P. L. deHaseth, M. L. Zupancic and M. T. Record Jr., RNA polymerase-promoter interactions: the comings and goings of RNA polymerase, *J. Bacteriol.*, 1998, **180**, 3019–25.
13. J. D. Helmann and P. L. deHaseth, Protein–nucleic acid interactions during open complex formation investigated by systematic alteration of the protein and DNA binding partners, *Biochemistry*, 1999, **38**, 5959–67.
14. R. M. Saecker, O. V. Tsodikov, K. L. McQuade, P. E. Schlax Jr., M. W. Capp and M. T. Record Jr., Kinetic studies and structural models of the association of *E. coli* sigma(70) RNA polymerase with the lambdaP(R) promoter: large scale conformational changes in forming the kinetically significant intermediates, *J. Mol. Biol.*, 2002, **319**, 649–71.
15. B. Sclavi, E. Zaychikov, A. Rogozina, F. Walther, M. Buckle and H. Heumann, Real-time characterization of intermediates in the pathway to open complex formation by *Escherichia coli* RNA polymerase at the T7A1 promoter, *Proc. Natl. Acad. Sci., USA*, 2005, **102**, 4706–11.
16. A. N. Kapanidis, E. Margeat, S. O. Ho, E. Kortkhonjia, S. Weiss and R. H. Ebright, Initial transcription by RNA polymerase proceeds through a DNA-scrunching mechanism, *Science*, 2006, **314**, 1144–1147.
17. A. Revyakin, C. Liu, R. H. Ebright and T. R. Strick, Abortive initiation and productive initiation by RNA polymerase involve DNA scrunching, *Science*, 2006, **314**, 1139–43.
18. I. Artsimovitch and R. Landick, Pausing by bacterial RNA polymerase is mediated by mechanistically distincts classes of signals, *Proc. Natl. Acad. Sci. USA*, 2000, **97**, 7090–5.
19. D. A. Erie, The many conformational states of RNA polymerase elongation complexes and their roles in the regulation of transcription, *Biochim. Biophys. Acta*, 2002, **1577**, 224–39.
20. P. H. von Hippel and Z. Pasman, Reaction pathways in transcript elongation, *Biophys. Chem.*, 2002, **101–102**, 401–23.
21. T. Förster, Intramolecular energy migration and fluorescence, *Ann. Phys.*, 1948, **2**, 55–75.

22. R. M. Clegg, Fluorescence resonance energy transfer and nucleic acids, *Methods Enzymol.*, 1992, **211**, 353–88.
23. P. R. Selvin, Fluorescence resonance energy transfer, *Methods Enzymol.*, 1995, **246**, 300–34.
24. P. R. Selvin, The renaissance of fluorescence resonance energy transfer, *Nat. Struct. Biol.*, 2000, **7**, 730–4.
25. A. N. Kapanidis, Y. W. Ebright, R. D. Ludescher, S. Chan and R. H. Ebright, Mean DNA bend angle and distribution of DNA bend angles in the CAP–DNA complex in solution, *J. Mol. Biol.*, 2001, **312**, 453–68.
26. N. K. Lee, A. N. Kapanidis, Y. Wang, X. Michalet, J. Mukhopadhyay, R. H. Ebright and S. Weiss, Accurate FRET measurements within single diffusing biomolecules using alternating-laser excitation, *Biophys. J.*, 2005, **88**, 2939–53.
27. P. G. Wu and L. Brand, Orientation factor in steady-state and time-resolved resonance energy-transfer measurements, *Biochemistry*, 1992, **31**, 7939–7947.
28. J. N. Forkey, M. E. Quinlan, M. A. Shaw, J. E. Corrie and Y. E. Goldman, Three-dimensional structural dynamics of myosin V by single-molecule fluorescence polarization, *Nature*, 2003, **422**, 399–404.
29. E. Toprak, J. Enderlein, S. Syed, S. A. McKinney, R. G. Petschek, T. Ha, Y. E. Goldman and P. R. Selvin, Defocused orientation and position imaging (DOPI) of myosin V, *Proc. Natl. Acad. Sci. USA*, 2006, **103**, 6495–9.
30. S. Callaci, E. Heyduk and T. Heyduk, Conformational changes of *Escherichia coli* RNA polymerase sigma70 factor induced by binding to the core enzyme, *J. Biol. Chem.*, 1998, **273**, 32995–3001.
31. S. Callaci, E. Heyduk and T. Heyduk, Core RNA polymerase from *E. coli* induces a major change in the domain arrangement of the sigma 70 subunit, *Mol. Cell*, 1999, **3**, 229–38.
32. S. Callaci and T. Heyduk, Conformation and DNA binding properties of a single-stranded DNA binding region of sigma 70 subunit from *Escherichia coli* RNA polymerase are modulated by an interaction with the core enzyme, *Biochemistry*, 1998, **37**, 3312–20.
33. J. Mukhopadhyay, A. N. Kapanidis, V. Mekler, E. Kortkhonjia, Y. W. Ebright and R. H. Ebright, Translocation of sigma(70) with RNA polymerase during transcription: fluorescence resonance energy transfer assay for movement relative to DNA, *Cell*, 2001, **106**, 453–63.
34. R. N. Leach, C. Gell, S. Wigneshwaraj, M. Buck, A. Smith and P. G. Stockley, Mapping ATP-dependent activation at a sigma54 promoter, *J. Biol. Chem.*, 2006, **281**, 33717–26.
35. V. Mekler, E. Kortkhonjia, J. Mukhopadhyay, J. Knight, A. Revyakin, A. N. Kapanidis, W. Niu, Y. W. Ebright, R. Levy and R. H. Ebright, Structural organization of bacterial RNA polymerase holoenzyme and the RNA polymerase-promoter open complex, *Cell*, 2002, **108**, 599–614.
36. T. Ha, T. Enderle, D. F. Ogletree, D. S. Chemla, P. R. Selvin and S. Weiss, Probing the interaction between two single molecules:

fluorescence resonance energy transfer between a single donor and a single acceptor, *Proc. Natl. Acad. Sci. USA*, 1996, **93**, 6264–8.
37. T. Ha, Single-molecule fluorescence resonance energy transfer, *Methods*, 2001, **25**, 78–86.
38. T. Ha, Structural dynamics and processing of nucleic acids revealed by single-molecule spectroscopy, *Biochemistry*, 2004, **43**, 4055–63.
39. S. C. Blanchard, R. L. Gonzalez, H. D. Kim, S. Chu and J. D. Puglisi, tRNA selection and kinetic proofreading in translation, *Nat. Struct. Mol. Biol.*, 2004, **11**, 1008–14.
40. A. A. Deniz, T. A. Laurence, G. S. Beligere, M. Dahan, A. B. Martin, D. S. Chemla, P. E. Dawson, P. G. Schultz and S. Weiss, Single-molecule protein folding: diffusion fluorescence resonance energy transfer studies of the denaturation of chymotrypsin inhibitor 2, *Proc. Natl. Acad. Sci. USA*, 2000, **97**, 5179–84.
41. T. Ha, I. Rasnik, W. Cheng, H. P. Babcock, G. H. Gauss, T. M. Lohman and S. Chu, Initiation and re-initiation of DNA unwinding by the *Escherichia coli* Rep helicase, *Nature*, 2002, **419**, 638–41.
42. P. J. Rothwell, S. Berger, O. Kensch, S. Felekyan, M. Antonik, B. M. Wohrl, T. Restle, R. S. Goody and C. A. Seidel, Multiparameter single-molecule fluorescence spectroscopy reveals heterogeneity of HIV-1 reverse transcriptase:primer/template complexes, *Proc. Natl. Acad. Sci. USA*, 2003, **100**, 1655–60.
43. B. Schuler, E. A. Lipman and W. A. Eaton, Probing the free-energy surface for protein folding with single-molecule fluorescence spectroscopy, *Nature*, 2002, **419**, 743–7.
44. X. Zhuang, H. Kim, M. J. Pereira, H. P. Babcock, N. G. Walter and S. Chu, Correlating structural dynamics and function in single ribozyme molecules, *Science*, 2002, **296**, 1473–6.
45. M. Bates, T. R. Blosser and X. Zhuang, Short-range spectroscopic ruler based on a single-molecule optical switch, *Phys. Rev. Lett.*, 2005, **94**, 108101.
46. M. Heilemann, E. Margeat, R. Kasper, M. Sauer and P. Tinnefeld, Carbocyanine dyes as efficient reversible single-molecule optical switch, *J. Am. Chem. Soc.*, 2005, **127**, 3801–6.
47. E. Margeat, A. N. Kapanidis, P. Tinnefeld, Y. Wang, J. Mukhopadhyay, R. H. Ebright and S. Weiss, Direct observation of abortive initiation and promoter escape within single immobilized transcription complexes, *Biophys. J.*, 2006, **90**, 1419–31.
48. A. A. Deniz, M. Dahan, J. R. Grunwell, T. Ha, A. E. Faulhaber, D. S. Chemla, S. Weiss and P. G. Schultz, Single-pair fluorescence resonance energy transfer on freely diffusing molecules: observation of Förster distance dependence and subpopulations, *Proc. Natl. Acad. Sci. USA*, 1999, **96**, 3670–5.
49. I. Rasnik, S. A. McKinney and T. Ha, Nonblinking and longlasting single-molecule fluorescence imaging, *Nat. Methods*, 2006, **3**, 891–893.
50. A. N. Kapanidis, M. Heilemann, E. Margeat, X. Kong, E. Nir and S. Weiss, Alternating-laser excitation of single molecules. In *Single*

molecules *Techniques: A Laboratory Manual* (T. Ha and P. R. Selvin, eds.), Cold Spring Harbor Laboratory Press, 2008, pp. 85–119.
51. A. N. Kapanidis, T. A. Laurence, N. K. Lee, E. Margeat, X. Kong and S. Weiss, Alternating-laser excitation of single molecules, *Acc. Chem. Res.*, 2005, **38**, 523–33.
52. A. N. Kapanidis, N. K. Lee, T. A. Laurence, S. Doose, E. Margeat and S. Weiss, Fluorescence-aided molecule sorting: analysis of structure and interactions by alternating-laser excitation of single molecules, *Proc. Natl. Acad. Sci. USA*, 2004, **101**, 8936–41.
53. T. A. Laurence, X. X. Kong, M. Jager and S. Weiss, Probing structural heterogeneities and fluctuations of nucleic acids and denatured proteins, *Proc. Natl. Acad. Sci. USA*, 2005, **102**, 17348–17353.
54. E. Nir, X. Michalet, K. M. Hamadani, T. A. Laurence, D. Neuhauser, Y. Kovchegov and S. Weiss, Shot-noise limited single-molecule FRET histograms: comparison between theory and experiments, *J. Phys. Chem. B. Condens. Matter Mater. Surf. Interfaces Biophys.*, 2006, **110**, 22103–24.
55. A. N. Kapanidis, E. Margeat, T. A. Laurence, S. Doose, S. O. Ho, J. Mukhopadhyay, E. Kortkhonjia, V. Mekler, R. H. Ebright and S. Weiss, Retention of transcription initiation factor sigma(70) in transcription elongation: single-molecule analysis, *Mol. Cell*, 2005, **20**, 347–56.
56. B. K. Muller, E. Zaychikov, C. Brauchle and D. C. Lamb, Pulsed interleaved excitation, *Biophys. J.*, 2005, **89**, 3508–3522.
57. A. Revyakin, J. F. Allemand, V. Croquette, R. H. Ebright and T. R. Strick, Single-molecule DNA nanomanipulation: detection of promoter-unwinding events by RNA polymerase, *Methods Enzymol.*, 2003, **370**, 577–98.
58. A. Revyakin, R. H. Ebright and T. R. Strick, Promoter unwinding and promoter clearance by RNA polymerase: detection by single-molecule DNA nanomanipulation, *Proc. Natl. Acad. Sci. USA*, 2004, **101**, 4776–80.
59. E. A. Abbondanzieri, W. J. Greenleaf, J. W. Shaevitz, R. Landick and S. M. Block, Direct observation of base-pair stepping by RNA polymerase, *Nature*, 2005, **438**, 460–5.
60. J. W. Shaevitz, E. A. Abbondanzieri, R. Landick and S. M. Block, Backtracking by single RNA polymerase molecules observed at near-base-pair resolution, *Nature*, 2003, **426**, 684–7.
61. J. Mukhopadhyay, V. Mekler, E. Kortkhonjia, A. N. Kapanidis, Y. W. Ebright and R. H. Ebright, Fluorescence resonance energy transfer (FRET) in analysis of transcription-complex structure and function, *Methods Enzymol.*, 2003, **371**, 144–59.
62. N. K. Lee, A. N. Kapanidis, H. R. Koh, Y. Korlann, S. O. Ho, Y. Kim, N. Gassman, S. K. Kim and S. Weiss, Three-color alternating-laser excitation of single molecules: monitoring multiple interactions and distances, *Biophys. J.*, 2007, **92**, 303–12.
63. U. M. Hansen and W. R. McClure, Role of the sigma subunit of *Escherichia coli* RNA polymerase in initiation. II. Release of sigma from ternary complexes, *J. Biol. Chem.*, 1980, **255**, 9564–70.

64. B. Krummel and M. J. Chamberlin, RNA chain initiation by *Escherichia coli* RNA polymerase. Structural transitions of the enzyme in early ternary complexes, *Biochemistry*, 1989, **28**, 7829–42.
65. W. Metzger, P. Schickor, T. Meier, W. Werel and H. Heumann, Nucleation of RNA chain formation by *Escherichia coli* DNA-dependent RNA polymerase, *J. Mol. Biol.*, 1993, **232**, 35–49.
66. D. C. Straney and D. M. Crothers, Intermediates in transcription initiation from the E. coli lac UV5 promoter, *Cell*, 1985, **43**, 449–59.
67. A. A. Travers and R. R. Burgess, Cyclic re-use of the RNA polymerase sigma factor, *Nature*, 1969, **222**, 537–40.
68. G. Bar-Nahum and E. Nudler, Isolation and characterization of sigma(70)-retaining transcription elongation complexes from *Escherichia coli*, *Cell*, 2001, **106**, 443–51.
69. B. Z. Ring, W. S. Yarnell and J. W. Roberts, Function of *E. coli* RNA polymerase sigma factor sigma 70 in promoter-proximal pausing, *Cell*, 1996, **86**, 485–93.
70. P. A. Osumi-Davis, A. Y. Woody and R. W. Woody, Transcription initiation by *Escherichia coli* RNA polymerase at the gene II promoter of M13 phage: stability of ternary complex, direct photocrosslinking to nascent RNA, and retention of sigma subunit, *Biochim. Biophys. Acta*, 1987, **910**, 130–41.
71. N. Shimamoto, T. Kamigochi and H. Utiyama, Release of the sigma subunit of *Escherichia coli* DNA-dependent RNA polymerase depends mainly on time elapsed after the start of initiation, not on length of product RNA, *J. Biol. Chem.*, 1986, **261**, 11859–65.
72. K. Brodolin, N. Zenkin, A. Mustaev, D. Mamaeva and H. Heumann, The sigma 70 subunit of RNA polymerase induces lacUV5 promoter-proximal pausing of transcription, *Nat. Struct. Mol. Biol.*, 2004, **11**, 551–7.
73. B. E. Nickels, J. Mukhopadhyay, S. J. Garrity, R. H. Ebright and A. Hochschild, The sigma 70 subunit of RNA polymerase mediates a promoter-proximal pause at the lac promoter, *Nat. Struct. Mol. Biol.*, 2004, **11**, 544–50.
74. C. L. Lawson, D. Swigon, K. S. Murakami, S. A. Darst, H. M. Berman and R. H. Ebright, Catabolite activator protein: DNA binding and transcription activation, *Curr. Opin. Struct. Biol.*, 2004, **14**, 10–20.
75. J. T. Wade and K. Struhl, Association of RNA polymerase with transcribed regions in Escherichia coli, *Proc. Natl. Acad. Sci. USA*, 2004, **101**, 17777–82.
76. R. A. Mooney, S. A. Darst and R. Landick, Sigma and RNA polymerase: an on-again, off-again relationship?, *Mol. Cell*, 2005, **20**, 335–45.
77. M. Raffaelle, E. I. Kanin, J. Vogt, R. R. Burgess and A. Z. Ansari, Holoenzyme switching and stochastic release of sigma factors from RNA polymerase in vivo, *Mol. Cell*, 2005, **20**, 357–66.
78. L. M. Hsu, Promoter clearance and escape in prokaryotes, *Biochim. Biophys. Acta*, 2002, **1577**, 191–207.

79. K. S. Murakami and S. A. Darst, Bacterial RNA polymerases: the wholo story, *Curr. Opin. Struct. Biol.*, 2003, **13**, 31–9.
80. A. J. Carpousis and J. D. Gralla, Cycling of ribonucleic acid polymerase to produce oligonucleotides during initiation *in vitro* at the lac UV5 promoter, *Biochemistry*, 1980, **19**, 3245–53.
81. A. J. Carpousis and J. D. Gralla, Interaction of RNA polymerase with lacUV5 promoter DNA during mRNA initiation and elongation. Footprinting, methylation, and rifampicin-sensitivity changes accompanying transcription initiation, *J. Mol. Biol.*, 1985, **183**, 165–77.
82. J. D. Gralla, A. J. Carpousis and J. E. Stefano, Productive and abortive initiation of transcription in vitro at the lac UV5 promoter, *Biochemistry*, 1980, **19**, 5864–9.
83. D. C. Straney and D. M. Crothers, A stressed intermediate in the formation of stably initiated RNA chains at the *Escherichia coli* lac UV5 promoter, *J. Mol. Biol.*, 1987, **193**, 267–78.
84. M. Pal, A. S. Ponticelli and D. S. Luse, The role of the transcription bubble and TFIIB in promoter clearance by RNA polymerase II, *Mol. Cell*, 2005, **19**, 101–10.
85. E. A. Campbell, N. Korzheva, A. Mustaev, K. Murakami, S. Nair, A. Goldfarb and S. A. Darst, Structural mechanism for rifampicin inhibition of bacterial rna polymerase, *Cell*, 2001, **104**, 901–12.
86. N. Naryshkin, A. Revyakin, Y. Kim, V. Mekler and R. H. Ebright, Structural organization of the RNA polymerase-promoter open complex, *Cell*, 2000, **101**, 601–11.
87. N. Korzheva, A. Mustaev, M. Kozlov, A. Malhotra, V. Nikiforov, A. Goldfarb and S. A. Darst, A structural model of transcription elongation, *Science*, 2000, **289**, 619–25.
88. G. M. Cheetham, D. Jeruzalmi and T. A. Steitz, Transcription regulation, initiation, and "DNA scrunching" by T7 RNA polymerase, *Cold Spring Harb. Symp. Quant. Biol.*, 1998, **63**, 263–7.
89. G. M. Cheetham and T. A. Steitz, Structure of a transcribing T7 RNA polymerase initiation complex, *Science*, 1999, **286**, 2305–9.
90. S. Mukherjee, L. G. Brieba and R. Sousa, Structural transitions mediating transcription initiation by T7 RNA polymerase, *Cell*, 2002, **110**, 81–91.
91. C. B. Harley and R. P. Reynolds, Analysis of *E. coli* promoter sequences, *Nucleic Acids Res.*, 1987, **15**, 2343–61.
92. D. K. Hawley and W. R. McClure, Compilation and analysis of *Escherichia coli* promoter DNA sequences, *Nucleic Acids Res.*, 1983, **11**, 2237–55.
93. D. J. Hoffman and S. K. Niyogi, RNA initiation with dinucleoside monophosphates during transcription of bacteriophage T4 DNA with RNA polymerase of *Escherichia coli*, *Proc. Natl. Acad. Sci. USA*, 1973, **70**, 574–8.
94. E. G. Minkley and D. Pribnow, Transcription of the early region of bacteriophage T7: selective initiation with dinucleotides, *J. Mol. Biol.*, 1973, **77**, 255–77.

95. P. R. Cook, The organization of replication and transcription, *Science*, 1999, **284**, 1790–5.
96. T. Ha, A. Y. Ting, J. Liang, A. A. Deniz, D. S. Chemla, P. G. Schultz and S. Weiss, Temporal fluctuations of fluorescence resonance energy transfer between two dyes conjugated to a single protein, *Chem. Phys.*, 1999, **247**, 107–118.
97. S. Myong, I. Rasnik, C. Joo, T. M. Lohman and T. Ha, Repetitive shuttling of a motor protein on DNA, *Nature*, 2005, **437**, 1321–5.
98. W. J. Koopmans, A. Brehm, C. Logie, T. Schmidt and J. van Noort, Single-pair FRET microscopy reveals mononucleosome dynamics, *J. Fluoresc.*, 2007, **17**, 785–95.
99. H. D. Kim, G. U. Nienhaus, T. Ha, J. W. Orr, J. R. Williamson and S. Chu, Mg^{2+}-dependent conformational change of RNA studied by fluorescence correlation and FRET on immobilized single molecules, *Proc. Natl. Acad. Sci. USA*, 2002, **99**, 4284–9.
100. M. Margittai, J. Widengren, E. Schweinberger, G. F. Schroder, S. Felekyan, E. Haustein, M. Konig, D. Fasshauer, H. Grubmuller, R. Jahn and C. A. Seidel, Single-molecule fluorescence resonance energy transfer reveals a dynamic equilibrium between closed and open conformations of syntaxin 1, *Proc. Natl. Acad. Sci. USA*, 2003, **100**, 15516–21.
101. M. Sing-Zocchi, S. Dixit, V. Ivanov and G. Zocchi, *Proc. Natl. Acad. Sci. USA*, 2003, **94**, 7605–7610.
102. S. Dixit, M. Singh-Zocchi, J. Hanne and G. Zocchi, Mechanics of binding of a single intergration-host-factor protein to DNA, *Phys. Rev. Lett.*, 2005, **94**, 118101.
103. C. Danilowicz, V. W. Coljee, C. Bouzigues, D. K. Lubensky, D. R. Nelson and M. Prentiss, DNA unzipped under a constant force exhibits multiple metastable intermediates, *Proc. Natl. Acad. Sci. USA*, 2003, **100**, 1694–9.
104. M. Foquet, J. Korlach, W. Zipfel, W. W. Webb and H. G. Craighead, DNA fragment sizing by single molecule detection in submicrometer-sized closed fluidic channels, *Anal. Chem.*, 2002, **74**, 1415–22.
105. M. Foquet, J. Korlach, W. R. Zipfel, W. W. Webb and H. G. Craighead, Focal volume confinement by submicrometer-sized fluidic channels, *Anal. Chem.*, 2004, **76**, 1618–26.
106. T. A. Laurence and S. Weiss, Analytical chemistry. How to detect weak pairs, *Science*, 2003, **299**, 667–8.
107. M. J. Levene, J. Korlach, S. W. Turner, M. Foquet, H. G. Craighead and W. W. Webb, Zero-mode waveguides for single-molecule analysis at high concentrations, *Science*, 2003, **299**, 682–6.
108. M. D. Wang, M. J. Schnitzer, H. Yin, R. Landick, J. Gelles and S. M. Block, Force and velocity measured for single molecules of RNA polymerase, *Science*, 1998, **282**, 902–7.

109. M. Buck, M. T. Gallegos, D. J. Studholme, Y. Guo and J. D. Gralla, The bacterial enhancer-dependent sigma(54) (sigma(N)) transcription factor, *J. Bacteriol.*, 2000, **182**, 4129–36.
110. M. Heilemann, K. Lymperopoulos, S. Wigneshweraraj, M. Buck and A. N. Kapanidis, Studying sigma54-dependent transcription with alternating-laser excitation (ALEX) spectroscopy. *Proceedings of SPIE-The International Society for Optical Engineering*, 2007, **6633**, 66332K.
111. J. Andrecka, R. Lewis, F. Bruckner, E. Lehmann, P. Cramer and J. Michaelis, Single-molecule tracking of mRNA exiting from RNA polymerase II, *Proc. Natl. Acad. Sci. USA*, 2008, **105**, 135–40.
112. J. Elf, G. W. Li and X. S. Xie, Probing transcription factor dynamics at the single-molecule level in a living cell, *Science*, 2007, **316**, 1191–4.

CHAPTER 6

Real-time Detection of DNA Unwinding by Escherichia coli RNAP: From Transcription Initiation to Termination

TERENCE R. STRICK[a] AND ANDREY REVYAKIN[b]

[a] Centre National de la Recherche Scientifique, Institut Jacques Monod, and University of Paris Diderot – Paris 7, Paris, France; [b] University of California Berkeley, 431 Stanley Hall, Berkeley, CA 94720 USA

6.1 Introduction

During transcription initiation RNAP imposes specific torsional deformations on DNA, beginning with promoter DNA unwinding.[1] At each successive stage of transcription – abortive initiation and promoter escape,[2,3] processive elongation[4] and transcription termination[5] – RNAP imposes further torsional rearrangements on the template DNA. These torsional deformations are profoundly coupled to the specific mode of RNAP translocation along DNA during each of the stages of transcription. In the case of multi-subunit RNAPs this is highlighted by the central role of the sigma factor in forming specific RNAP–promoter interactions and simultaneously acting as a kind of "clutch" during transcription initiation (see Chapters 1, 2 and 5 of this book).[6–9] The topological nature of the RNAP–DNA interaction has made it an opportune candidate for real-time single-molecule nanomanipulation analysis, and such experiments complement the single-molecule studies of the mechanochemistry of RNAP translocation, transcription and termination described later in this book. In the light of structural analysis presented over the last few years, these kinetic studies of DNA twisting by RNAP help describe the role of mechanical forces in regulating transcription. In this chapter we focus on the crucial torsional interactions between RNAP and DNA, as well as the intimate coupling between torsional deformation and linear translocation by RNAP on substrate DNA.

A particularly compelling example of this coupling occurs at the stage of promoter escape where *E. coli* RNAP breaks free from the unwound promoter to begin processive transcription. During this process of promoter escape, RNAP begins transcribing without actually having broken the ties that bind it to the promoter site.[2,3,10,11] Instead of translocating along DNA, RNAP unwinds and reels-in ("scrunches") the downstream DNA: this coupling of DNA synthesis, unwinding and reeling-in was confirmed through a conjunction of single-molecule measurements[12,13] (Chapter 5 and here) and structural data on T7 RNAP (Chapter 4). During this process NTP hydrolysis drives synthesis, reeling-in and unwinding of downstream DNA into the active complex. The mechanical energy thus stored in the strained ternary complex (DNA, RNA and RNAP) progressively grows. Again σ is central here, as it not only binds to the DNA promoter –35 and –10 sites but also appears to obstruct the nascent RNA's path out of the active site cleft.[6,7] In the high energy state generated *via* strain on the various components of the machinery, RNAP can either escape from the promoter and begin processive transcription or release the short RNA transcript and excess unwound DNA and perform rapid resynthesis of RNA during another attempt at escape. During this abortive cycling, the size of the transcription bubble could be expected to increase and then collapse as abortive RNA is synthesized and released.

Another example of this torsional/translational coupling can be found during transcription elongation, in which forward translocation by a single nucleotide must be accompanied by unwinding of an additional downstream base-pair and simultaneous rewinding of an upstream base-pair. Finally, during transcription termination, rewinding of the transcription bubble and forward translocation of RNAP[5,14,15] along the terminator may both play a part in driving dissociation of the ternary RNAP/DNA/RNA complex. A detailed understanding of how RNAP couples unwinding of DNA and translocation along DNA during each of the stages of transcription, and in particular during transcription initiation, helps illustrate the remarkable mechano-chemical properties of this molecular motor.

6.2 Twist Deformations at the Promoter

Transcription initiation by RNA polymerase (RNAP) involves two major and successive protein-induced DNA deformations at the promoter site termed promoter unwinding and promoter scrunching. Although both deformations share a common structural feature, namely the topological unwinding of approximately twelve base-pairs of DNA, they are differentially coupled to translocation of RNAP along DNA.

Promoter unwinding occurs stochastically after initial binding of RNAP to the promoter site, and requires neither nucleotide hydrolysis nor RNAP translocation along DNA. As one of the earliest and most elementary events of transcription initiation, this stable RNAP/promoter open complex (RPo) has long been amenable to structural analysis. An impressive array of methods has been brought to bear on the σ^{70}-RNAP holoenzyme, including bulk biochemical techniques such as nuclease attack to measure the ~ 65 nt footprint of RNAP on promoter DNA,[16,17] and chemical attack by hydroxyl radicals,[18] potassium permanganate[19] and other reactive compounds[20,21] to map the ~ 12 unwound base-pairs at the promoter site (Chapter 2). Kinetic analysis of the formation and stability of RPo has traditionally been performed using nitrocellulose filter retention assays, and has shown that the lifetime of RPo in physiological conditions of salt and temperature is essentially well-described by single-exponential kinetics.[22] More recently, high-resolution structural studies of RPo have been performed using engineered nucleic acid substrates to trap the related *Thermus aquaticus* RNAP in well-defined states on the promoter DNA.[8,23]

When provided with nucleotides, RPo isomerizes into an RNAP/promoter initially-transcribing complex, RPitc. In RPitc, RNAP begins transcribing without breaking free from the promoter. Here additional DNA unwinding ("scrunching") occurs concomitantly with reeling-in of DNA through RNAP. The excess unwinding of downstream DNA prior to promoter escape has been much more difficult to characterize structurally and kinetically than simple promoter unwinding. For instance the broad distribution of lengths of abortive RNA produced by RPitc[11] implies that the extent of DNA unwinding in RPitc is variable from one abortive transcript to the next. Such heterogeneity could complicate high-resolution crystallographic studies (see Chapter 4). From a kinetic perspective, it has not been straightforward to synchronize a population of molecules in RPo, track it as it proceeds into RPitc and then emerges as a transcription elongation complex.

To sum up, these two central conformational states on the pathway to processive transcription share a common structural feature – DNA unwinding – but are otherwise very different. Whereas RPo is structurally homogeneous with accessible kinetics, RPitc is structurally heterogeneous and difficult to assess kinetically. While each process may be amenable to study on its own, the temporal relation between the two processes is more difficult to access. Because of their structural and kinetic features, RP_o and RPitc have constituted interesting candidates for single-molecule experimentation and analysis.

6.3 Magnetic Trapping and Supercoiling of a Single DNA Molecule

6.3.1 General Features of the Magnetic Trap

Single-molecule DNA nanomanipulation using a magnetic trap[24] has proved to be an extremely powerful approach to studying transcription initiation.[12,25]

Two features of the method are of interest here. First of all, the magnetic trapping approach makes it possible to measure with near base-pair resolution the topological unwinding of DNA by RNAP.[26] In addition, the real-time nature of the single-molecule measurement enables one to monitor changes in the protein–DNA interaction through time. This provides simultaneous structure–function data on the interaction, enabling one to identify specific states along the reaction pathway at the same time as one identifies the time it takes an individual complex to enter and exit these states along the reaction pathway. Finally, by providing information on individual interaction events between a single RNAP molecule and the nanomanipulated DNA, one can directly access structural or kinetic heterogeneity. As with any approach, the possibility of making such observations depends on the observable having a lifetime greater than the temporal resolution of the method.

In the magnetic trap experiment, a double-stranded, linear, unnicked DNA molecule in aqueous solution is tethered at multiple points at one extremity to a treated glass surface and at multiple points at the other extremity to a small (micron-scale) magnetic bead (Figure 6.1A).[24,16–28] Multiple attachment points to bead and surface ensure that the DNA is topologically constrained. The magnetic field generated by a pair of permanent magnets located above the sample is used to exert force on the bead and rotate the bead and hence the

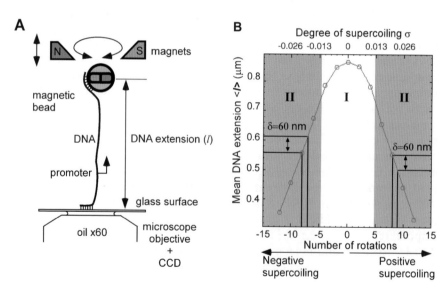

Figure 6.1 (A) Sketch of the experimental setup (see text for details). (B) Extension of DNA as a function of supercoiling. Extending force is constant, $F = 0.3$ pN, and $T = 34\,°C$ (as throughout) with 25 mM NaHepes (pH 7.9), 100 mM NaCl and 10 mM MgCl$_2$ as counter-ions. In white (regime I) twisting DNA causes the torque, Γ, to increase while extension varies nonlinearly. In grey (regime II), after a critical torque, Γ_b, has been reached twisting DNA causes it to buckle and form loops regularly at a rate of 1 loop per turn. Torque is roughly constant thereafter and loop circumference $\delta \sim 60$ nm.[25]

DNA molecule which tethers it to the surface. The magnets are mounted on computer-controlled translation and rotation stages, enabling quantitative and reversible control of the mechanical constraints – twisting and stretching – applied to the DNA. By moving the magnets closer to (respectively, farther away) from the sample, one can increase (respectively, decrease) the vertical extending force acting on the DNA. By rotating the magnets clockwise or counterclockwise turns (as seen from above) one causes the magnetic bead to rotate in perfect lock-step register as would a compass needle subjected to a magnetic field, enabling us to negatively or positively supercoil the DNA by the exact number of turns, n. An oil-immersion microscope objective located beneath the sample images the bead onto a CCD camera. Real-time computer-based tracking of the magnetic bead's image allows one to determine the position of the tethered bead in the three spatial directions.

Forces in the range of 10 fN (femto-newtons) to 100 pN (pico-newtons) can be exerted in this manner (1 pN = 10^{-12} N). By determining the position of the bead above the surface one obtains the end-to-end extension, l, of the nanomanipulated DNA molecule. This real-time extension measurement is the central readout of this type of single-molecule measurement, allowing one to determine the conformational state of the DNA molecule on the basis of initial mechanical calibration of its response to stretching and twisting. The spatial and temporal resolution of this measurement is limited by the Brownian motion of the magnetic bead, itself a function of the stiffness of the DNA tether (see below). At present the method's main limitation is a time resolution on the order of a 0.5 to 1 s, a result of the low stiffness of supercoiled DNA (see below). Finally, thermal regulation of the system is straightforward to implement, and guarantees environmental homogeneity of the sample to within 0.1 °C. This is particularly important given the strong temperature-dependence of promoter DNA unwinding by RNAP.[29]

6.3.2 Calibrating the DNA Sensor

A good understanding and calibration of the mechanical properties of the biopolymer "DNA sensor" used in single-molecule assays is central to these experiments. For studies of transcription initiation this involves not just calibration of linear deformation of DNA but also torsional deformation.

6.3.2.1 Extensibility of DNA

This essential calibration consists in determining the nanomanipulated DNA's stretching properties by measuring its force *vs.* extension response curve. In the magnetic trap system this is accomplished by measuring, for different distances between the magnets and the sample, the force applied to the bead and the time-averaged end-to-end extension, $<l>$, of the DNA. This curve was first measured in 1992 by C. Bustamante and colleagues,[30] and its theoretical description in the following years[31,32] has provided both a striking success of polymer physics and an invaluable tool for the experimental calibration of biopolymer "sensors" such as nucleic acids and proteins.[32–34] DNA is characterized locally by large bending rigidity, reflected in an exceptionally long persistence length, $\xi \sim 51$ nm (~ 150 bp).[32] Thus in solution (*i.e.*, in the absence of external force), DNA shorter than 150 bp is essentially rod-like. Longer DNA can "ball up" into a so-called random coil. The force–extension curve for DNA in the range of forces relevant to RNAP (*i.e.*, lower than ~ 30 pN) is characterized by two regimes: the so-called entropic and enthalpic regimes. In the entropic regime, extending the DNA causes the random coil it assumes in solution to unravel and align with the extending force, progressively reducing conformational entropy. This regime dominates at forces lower than ~ 0.5 pN, and it is important to point out that the DNA is not deformed locally by such low forces. However, once the DNA is completely extended by forces of about ~ 10 pN (*i.e.*, every bp projects 3.4 Å along the stretching direction), additional stretching tests the elastic mechanical features of the DNA double helix by reversibly and mildly straining the molecule. Thus up to a ~ 120 pN force the rise per base-pair extends by an additional $\sim 0.1\%$ for each additional pico-Newton of extending force, for a stretching elasticity of ~ 1000 pN. Although this latter regime is relevant to the study of RNAP elongation using optical trapping, it is the sub-pN force scale that is relevant to the study of topological interactions between RNAP and DNA.

6.3.2.2 Topological Properties of DNA

DNA topology is well described by two numbers, twist (Tw) and writhe (Wr), and their sum,[35] known as the linking number $Lk = Tw + Wr$ (Chapter 3). The twist measures the number of times the two strands of DNA cross each other in the double helix. Writhe counts the number of times the axis of the double helix crosses itself, for instance in plectonemic structures (looped structures similar to those commonly observed on a twisted phone cord). The linking number Lk thus measures the sum total number of times the two strands of the double helix cross each other. For torsionally relaxed DNA, $Lk_o = Tw_o = N/h$, where N is the number of base pairs and $h = 10.5$ bp per turn is the pitch of the double helix.[36] Wr is, on average, non-existent in the absence of supercoiling; $Wr_o = 0$. A change in linking number can be written as:

$$\Delta Lk = Lk - Lk_o = Tw - Tw_o + Wr = \Delta Tw + Wr$$

Thus, an excess or deficit of linking number must be stored as a change in average DNA helicity and formation of writhed structures such as plectonemic loops (or also solenoidal wraps; however, these are not favored on naked DNA). The degree of supercoiling, $\sigma = \Delta Lk/Lk_o$ allows one to compare supercoiling on DNA molecules of different lengths. For a topologically constrained DNA, Lk is constant (examples include a closed circular plasmid or magnetically trapped DNA) and any change in Tw must be compensated by an equal but opposite change in Wr.

The magnetic trap allows one to "dial in" any given Lk and observe how topology redistributes between twist and writhe. For a fixed magnet height one rotates the magnets in the horizontal plane, twisting the DNA incrementally while leaving the applied force unchanged. For each rotation point the time-averaged end-to-end extension, $<l>$, of the DNA is determined. Figure 6.1(B) gives an example of such a calibration curve obtained for a low extending force. The calibration curve is symmetric with respect to the relaxed position and presents two regimes. In regime I, over- or under-winding the DNA causes a small reduction in the DNA end-to-end extension, which is basically maximal for the torsionally relaxed state Lk_o of the DNA. In this regime, torque, Γ, acting on the DNA varies linearly with winding as the molecule accommodates torsional stress by small ($<1\%$) deformation of its helical pitch, resulting in $\Delta Tw = Tw - Tw_o$. Crossover to regime II occurs at n_b turns; here the torque has reached a critical threshold, Γ_b, at which point it is sufficient to cause DNA to buckle and form loops, i.e., writhe. As a result of loop formation, twist deformation of the DNA is converted into bend deformation and the torque acting on the DNA essentially ceases to grow. In regime II the DNA's end-to-end extension varies rapidly and linearly with supercoiling, with a change in DNA extension of about $\delta \sim 60$ nm for each turn (loop) added or removed from the DNA. If one increases the extending force the DNA buckles at a higher number of turns n_b corresponding to higher buckling torque, according to $\Gamma_b \sim F^{1/2}$,[37,38] but, again, once buckling takes place the torque thereafter is essentially constant.

Three caveats should be pointed out here. First, the symmetrical behavior of DNA to supercoiling is only true for low torque, which is experimentally only obtained at very low extending forces of $F < 0.3$–1 pN (depending on ionic and thermal conditions). The true control parameter is the torque acting on the DNA, not the force (as written above, the two are linked). Excess negative torque ($|\Gamma_-| > k_B T$) causes DNA to denature in A+T-rich regions,[27,39] and excess positive torque ($\Gamma_+ > 3k_B T$) causes DNA to hypertwist.[40] When the applied torque is greater than these values, Wr no longer forms upon unwinding and the extension vs. supercoiling curve is flat. When performing experiments with negatively supercoiled DNA as template for RNAP, care must therefore be taken to avoid excessive use of force, typically $F = 0.3$ pN. Positively supercoiled DNA template can be readily generated for extending forces below ~ 8 pN, but beyond this force overwound DNA also undergoes secondary structural transitions.

Second, supercoiled DNA in this assay is under fixed force and supercoiling, and the extension is the free variable. This is different from conditions under which bulk experiments using supercoiled plasmids are performed, for which the supercoiling is fixed and zero extension is imposed on the circular plasmid. In this latter case the thermodynamic force acting on the DNA is the free variable. To visualize this force, extend the branch of the curve for negative super coiling until the x-axis is reached. Here the DNA is just at zero extension, like a plasmid, and has the degree of supercoiling of the x-axis intercept. The DNA is held at a 0.3 pN force in this experiment. Thus, this is the thermodynamic force experienced by plasmid DNA supercoiled to the x-axis intercept value.

Third, magnetic trapping gives control over Lk but does not provide for direct control of the torque acting on the DNA. Again the torque varies with supercoiling until plectonemes form, at which point it saturates; to obtain a higher torque one must increase the force. The value of the torque can be inferred using theoretical models to account for the relation between plectoneme mechanical properties, DNA end-to-end extension, supercoiling and applied force.[27,38,41,42] More recently the use of optical trapping to supercoil DNA has been demonstrated, and it presents the advantage of providing a direct measurement of the torque acting on DNA.[43]

6.3.2.3 Using DNA Nanomanipulation to Detect Promoter Unwinding by RNAP: Advantages and Disadvantages

Consider now unwinding of ~ 12 bp of promoter DNA by RNAP. This corresponds to a localized decrease of Tw by about 1, which must be compensated topologically to keep Lk constant. When plectonemic supercoils are present, *i.e.*, in regime II (Figure 6.1B), this topological compensation takes place *via* the appearance of one unit of writhe with positive topology. For negatively supercoiled DNA, promoter unwinding by RNAP will titrate out a (negatively signed) plectoneme and the DNA extension will increase; for positively supercoiled DNA promoter unwinding will titrate in a (positively signed) plectoneme and the DNA extension will decrease (Figure 6.2). Note that intuitive predictions as to the effect of supercoiling on promoter unwinding can also be made here: one would expect that negative supercoiling will favor promoter unwinding, whereas positive supercoiling will disfavor promoter unwinding. The change in DNA extension is expected to be on the order of $\delta n/h$, where n is the number of base pairs unwound in RPo, h is the DNA pitch and δ is the size of a DNA supercoil with unit writhe. Of course, this assumes that unwinding of base pairs in RPo is complete, *i.e.*, that there is no local residual twisting of the two strands one about the other. Indeed formally one cannot distinguish in this method, as well as in other topological measurements, the full unwinding of 10 base-pairs and the half-unwinding of 20 base-pairs.

Fortunately, extensive biochemical knowledge accumulated over the last 40 years, and in particular the measurement of topological titration by RNAP of negatively supercoiled plasmids bearing a promoter site,[1,44] helps to address this ambiguity. These measurements indicated topological unwinding of slightly more than ~ 1 turn of DNA in RPo. At the same time the number of base pairs unwound at the promoter in RPo could be independently estimated using chemical reactivity assays such as permanganate- and oxygen radical-based cleavage of bases exposed to solvent and thus presumed to be unwound: these measurements pointed to unwinding of about 12 bp in RPo (see Introduction). Comparison of these two values suggests that topological measurements accurately reflect the extent of DNA unwinding in RPo, which fully covers approximately 12 to 13 base-pairs. Note that in an assay where the DNA

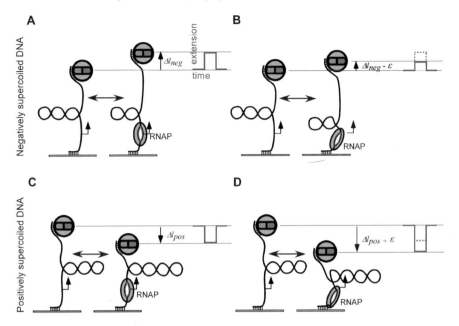

Figure 6.2 Topological predictions for RNAP–promoter interactions. For negatively supercoiled DNA, (A) promoter unwinding over ~10 bp removes ~1 DNA supercoil and extension increases by Δl_{neg} and (B) if there is concomitant promoter bending/compaction it causes a slight decrease in extension, subtracting a small amount ε from Δl_{neg}, leading to the observable $\Delta l_{obs,neg} = \Delta l_{neg} - \varepsilon$. For positively supercoiled DNA, (C) promoter unwinding over ~10 bp adds ~1 DNA supercoil and extension decreases by Δl_{pos} and (D) if there is concomitant promoter bending/compaction it causes a slight, additional decrease in extension, adding a small amount ε to Δl_{pos}, leading to the observable $\Delta l_{obs,pos} = \Delta l_{pos} + \varepsilon$. Δl_{pos} and Δl_{neg}, which have the same amplitude, can be recovered algebraically as described in the text.

is extended and supercoiled, the expected change in extension (~60 nm per titrated supercoil) is roughly 20-times the linear extent of DNA unwound in RPo (~3 nm per 10 bp); this is a result of the rigidity of DNA and the large DNA loops that result from it. This coupling of writhe deformation to twist deformation leads to "topological amplification" of signal by a factor of 20.

At the same time the fact that topological amplification and measurement requires the presence of plectonemic supercoils limits the spatio-temporal resolution of the experiment. Indeed, plectonemic supercoils are only present when the DNA is subjected to relatively weak extending forces. More specifically, in the context of physiologically-relevant negative supercoiling ($\sigma \sim -0.02$) the extending force cannot exceed ~0.3 pN or about 1/75 of the maximal force that a transcribing RNAP generates on its template DNA. Higher forces lead to DNA denaturation and thus a structurally inhomogeneous DNA.[24] In this low-force regime, the tethered bead's vertical Brownian fluctuations have a large amplitude, characterized by a Gaussian distribution with standard deviation σ_z of ~ 30 nm, and a relatively slow fluctuation timescale, $\tau \sim 0.125$ s (the inverse of the 8 Hz viscous cutoff frequency for a 2-kb long DNA molecule and a 1 μm-diameter bead, with $F = 0.3$ pN and $|\sigma| = 0.021^{26}$). Thus, as averaging the bead's z-position over 1 s only provides for ~ 8 independent measurements of the DNA end-to-end extension, the standard error on the mean extension (on the order of $\sigma_z/n^{-1/2}$) is about 10 nm. This compares favorably to the change in extension that results from unwinding of 1 base pair, i.e., 6 nm. Thus the number of base pairs unwound in RPo by any given RNAP (i.e., an individual single-molecule event) can be determined with a standard error of about ± 2 bp on a second timescale. Increasing time-averaging reduces this standard error, and for sufficiently long-lived events structural heterogeneity can even be observed if individual events display amplitude differences significantly greater than the standard error. Next, accumulating a dataset on the order of several dozens of events allows one to reduce statistical error. In such datasets it is sometimes possible to obtain the structural heterogeneity of the final state. Further improvement of the underlying spatio-temporal resolution requires using even shorter DNA molecules or smaller magnetic beads.[26]

6.4 Characterization of RPo at two Canonical Promoters

We first consider two canonical promoters: a consensus lac promoter, lac(cons), and one of the promoters driving transcription of ribosomal RNA, *rrnB* P1. Lac(cons) promoter has an ideal –35 sequence (TTGACA) and an ideal –10 sequence (TATAAT) separated by the ideal 17 bp spacing; ideal here is defined as forming the most stable σ^{70}-based open complex (Chapter 2). In the *rrnB* P1 promoter, non-consensus sequence features lead to lower stability of RPo but are partly offset by a direct interaction of the α-subunits' C-terminal domain with a DNA sequence known as the "UP" element (Chapters 1 and 2). Although reputed for forming a less stable and shorter-lived open complex, rrnBP1 is transcriptionally very active, responsible for transcription of a significant amount of RNA (the corresponding operons are responsible for generating $\sim 25\%$ of RNA in *E. coli*).

Figure 6.3 shows experimental time-traces of DNA extension *vs.* time $l(t)$; the DNA is a linear 4 kb GC-rich fragment containing either the lac(cons) or *rrnB* P1 promoter sites. As expected from the topological considerations presented in Figure 6.2, unwinding by RNAP of a negatively supercoiled promoter always leads to an *increase* in DNA end-to-end extension, while unwinding by RNAP of a positively supercoiled DNA always leads to a *decrease* in DNA end-to-end extension. These data confirm that both the rate of formation of RPo and the lifetime of RPo depend on the degree of supercoiling. For instance, at lac(cons) negative supercoiling extends the lifetime of RPo relative to that observed for positive supercoiling, where reversible cycles of promoter unwinding and

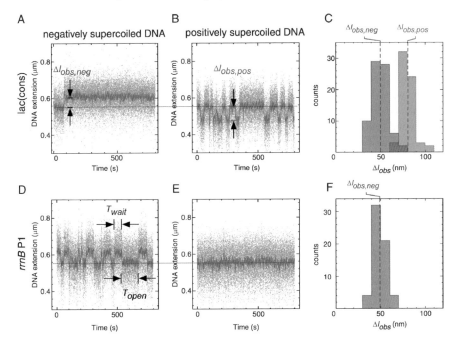

Figure 6.3 Real-time detection of promoter unwinding by a single RNAP on (A–C) lac(cons) and (D–F) rrnB P1 promoter under the same conditions as described Figure 6.1. DNA extension changes follow the pattern described in Figure 6.2(B) and (D), and can be characterized by both kinetic (T_{wait}, T_{open}) and structural ($\Delta l_{obs,neg}$, $\Delta l_{obs,\,pos}$) parameters. Amplitude histograms for $\Delta l_{obs,neg}$ and $\Delta l_{obs,\,pos}$ are shown to the right. DNA was supercoiled to $|\sigma| \sim 0.021$. Grey line indicates initial state. Data points (green) are collected at video rate (30 Hz) and time-averaged over ~ 1 s (red).[25]

rewinding are readily apparent. The data also confirm that promoter unwinding at *rrnB* P1 is significantly less stable than on lac(cons) under conditions of negative supercoiling, and is not even possible when positive supercoiling disfavors the reaction. Thus these data simultaneously encode both structural and functional features of the RNAP/promoter interaction. We will first consider the amplitude of the change in DNA end-to-end extension and relate this to the amplitude of promoter unwinding, and then go on to consider the kinetic analysis of the signal.

6.4.1 Structural Characterization of RPo

By analyzing the change in DNA extension that occurs for each promoter unwinding event, one can build a histogram of transition amplitude events. Figure 6.3 shows such histograms obtained on the lac(cons) and *rrnB* P1 promoters. From these Gaussian histograms the mean change in linking number is obtained as $\Delta Lk = <\Delta l_{obs}>/\delta$. Interestingly, at lac(cons) the mean amplitude observed using a positively supercoiled promoter is significantly larger than that observed with the negatively supercoiled promoter (80 nm rather than 50 nm).

This difference can be attributed to the formation of a protein-induced bend in promoter DNA as it is unwound by RNAP.[25,45] Indeed, bending at the promoter site will always *reduce* the end-to-end extension of the DNA by a small amount, ε (Figure 6.2B and D). Thus bending will subtract from the change in end-to-end extension due to promoter unwinding on negatively supercoiled DNA, but will add to the change in DNA end-to-end extension due to promoter unwinding on positively supercoiled DNA. If the (unsigned) change in DNA extension due to topological changes at the promoter site is Δl, then on positively supercoiled one will observe a change in DNA extension $\Delta l_{obs,pos} = \Delta l_{pos} + \varepsilon$ and on negatively supercoiled one will observe a change $\Delta l_{obs,neg} = \Delta l_{neg} - \varepsilon$. As Δl_{neg} and Δl_{pos} are expected to have the same amplitude, Δl, simple algebra allows one to extract both $\Delta l = (\Delta l_{obs,pos} + \Delta l_{obs,neg})/2$ and $\varepsilon = (\Delta l_{obs,pos} - \Delta l_{obs,neg})/2$. From the experimental data, one obtains $\Delta l = 65$ nm, corresponding to unwinding of about $\Delta Lk = 1.2$ turns, or about 13 ± 1 bp,[25] in excellent agreement with biochemical and crystallographic data. In addition, an estimate of the bend angle can be derived from the measure of $\varepsilon = 15$ nm by making the simplifying assumption that the bend is planar. Indeed, DNA has a persistence length of $\xi \sim 50$ nm, and at the extending force used in these experiments the molecule is $\sim 70\%$ extended, this implies that the angle between DNA entering and exiting the RNAP is on the order of $100°$, in qualitative agreement with results obtained, for instance, by Atomic Force Microscopy.[45]

Thus, structural analysis of the unwinding signal provides an estimate of the shift in DNA topology that results from promoter unwinding by RNAP. This analysis must be confronted to bulk biochemical results to estimate the number of base pairs affected by promoter unwinding. Similarly, differential analysis of promoter unwinding on positively and negatively supercoiled DNA provides an estimate for bending-mode deformation of DNA in RPo. However, to interpret this bend deformation several assumptions about its geometry must be made.

Finally, additional measurements[26] performed on shorter DNA (2 instead of 4 kb) provided for higher-resolution measurements. These show that much of the variance in the unwinding amplitude signal seen in the histograms of Figure 6.3 is due to experimental noise. They suggest that the great majority of RPo (roughly $\sim 70\%$) has no more than ± 1 bp of unwinding relative to the mean value, while the remaining $\sim 30\%$ are within less than ± 2 bp of the mean unwinding. As the corresponding difference in energy this represents is on the order of $k_B T$, the low level of heterogeneity is consistent with that to be expected from thermal fluctuations.

6.4.2 Kinetic Analysis of RPo

We now turn to analysis of the kinetics of the experimental signal. As shown in Figure 6.3, promoter unwinding events last a time $T_{unwound}$ and are separated by

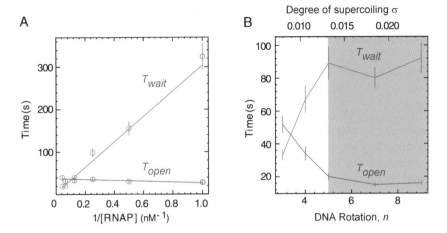

Figure 6.4 Dependence of the waiting time and lifetime of RPo for lac(cons) promoter on (A) RNAP concentration and (B) positive DNA supercoiling.[25]

an interval T_{wait} in which the promoter is in its closed configuration. Lifetime distribution histograms of each of these timescales display single-exponential statistics characteristic of processes dominated by a single kinetic barrier (at the temporal resolution of the experiment).[25] The main feature of this distribution is its mean lifetime $<T_{unwound}>$, which characterizes the height of the barrier according to $1/<T_{unwound}> \sim \exp(-\Delta G^{\neq}/k_B T)$.

As with the structural analysis of single-molecule data presented earlier, it is important to compare the kinetics of promoter unwinding as obtained in this assay with previous observations obtained thanks to bulk biochemistry. Determining the dependence of $<T_{unwound}>$ and $<T_{wait}>$ on RNAP concentration – i.e., performing a classical titration – is one way to do this. By measuring the half-life of each of these two kinetic parameters as a function of RNAP concentration ([RNAP]), one obtains the "tau plot" for the rate of formation of RPo presented in Figure 6.4(A). The "tau plot" (or Lineweaver–Burk plot) is a simple way to visualize the well-known Michaelis–Menten equation, usually written as:

$$k_{obs} = k_f[R]/([R] + K_m) \leftrightarrow k_{obs} = K_B k_f[R]/(1 + K_B[R])$$

for enzyme-driven formation of a biochemical product based on the reaction scheme:

$$R + P \underset{k_{-1}}{\overset{k_1}{\rightleftharpoons}} RPc \overset{k_f}{\longrightarrow} RPo$$

where [R] is RNAP concentration, k_f the rate of the reaction's forward step and $K_m = (k_{-1} + k_f)/k_1$ is the Michaelis constant. When $k_f << k_{-1}$ (i.e., when the forward step is rate-limiting) K_m is equal to the dissociation constant, $K_D = k_{-1}/k_1$ (itself the inverse of the binding constant K_B).

Rewriting the above equations using $T_{wait} = 1/k_{obs}$ and $T_f = 1/k_f$ gives:

$$T_{wait} = T_f + T_f/K_B[R]$$

At infinite RNAP concentration, where the promoter is saturated with polymerase, the (extrapolated) observed time $T_{wait} = T_f$ is the time required for an enzyme sitting on the promoter to form RPo. Knowing T_f, K_B can then be determined from the slope of the Lineweaver–Burk curve. From the experimental data collected using the lac(cons) promoter under conditions of positive supercoiling, we measured $T_f \sim 3\,s$ and $K_B \sim 10^7\,M^{-1}$. Although the conditions of positive supercoiling employed here preclude quantitative comparison with bulk results, the fact that the data are consistent with Michaelis–Menten kinetics is an important first kinetic benchmark.

We now consider the rewinding reaction characterized by the lifetime of RPo, $T_{unwound} = 1/k_r$:

$$RPc \xrightarrow{k_r} RPo$$

Analysis of the lifetime of RPo is more straightforward, as no titration is required to measure the rate of rewinding $k_r = 1/T_{unwound}$. Indeed as expected for a simple bimolecular species waiting to dissociate, $T_{unwound}$ is independent of RNAP concentration (Figure 6.4A). Moreover it is reasonable to assume that $T_{unwound}$ faithfully reflects k_r, and that promoter rewinding is accompanied by full RNAP dissociation: for instance were this not the case one would expect significant deviation from Michaelian kinetics for the forward process. Promoters with widely varying lifetimes in RPo [e.g., rrnB P1 vs. lac(cons)] can be analyzed in this way. Thus, under conditions of negative supercoiling the lifetime of RPo for rrnB P1 is on the order of 16 s, while for lac(cons) it is on the hour timescale.[25] This suggests that estimates of free-energy differences can be obtained for RPo on promoters of widely differing "strength" and under identical experimental conditions (in this case at least $\sim 5k_BT$, or $\sim 3\,kcal\,mol^{-1}$).

As we will see further on with specific examples, the concentration-independence of RPo lifetime contributes to making it a robust metric for further analysis of the effect of activators, inhibitors and effectors of transcription initiation. In contrast, T_{wait} is not as robust a metric, as it can vary with a range of experimental conditions. For instance the concentration of active enzyme molecules will decay over the course of several hours in the assay, leading to an artificial increase in T_{wait}. Similarly, the concentration of active enzyme molecules will also decay – albeit over longer times – under long-term storage conditions. Finally, the specific activity of a preparation of RNAP can also vary from one lot to another, leading to variability in the observed binding constant K_B.

6.4.3 Effect of Environmental Variables on Kinetics of RPo

It is now possible to use RPo lifetime data obtained in standard conditions to assess the effect of a range of environmental conditions (ionic strength and composition, temperature) and chemical or biological compounds. Experiments performed by varying ionic conditions and temperature agree well with previously published results, namely that reducing overall ionic strength[46,47] or increasing temperature[18,21,47–49] displaces the reaction equilibrium towards the promoter-open complex by increasing the rate of formation and/or decreasing the rate of dissociation of the open promoter complex RPo. However, although the rate of formation of RPo was seen in the single-molecule assay to follow these trends, the effect was not decomposed into changes in the binding or isomerization steps, as this requires complete RNAP titration data. Again the effect of temperature on lifetime of RPo is less ambiguous for the fact that it reflects bimolecular dissociation. Thus, a two-fold increase in lifetime of RPo formed on lac(cons), observed for a 4 °C increase from 30 to 34 °C, suggests a (negative) activation energy for promoter rewinding on the order of $50k_BT$, or 30 kcal mol^{-1} (the contribution of positive torque to this value is $\sim 10k_BT$). The choice of counter-ion is also important, as the use of glutamate counter-ions (rather than chloride) greatly enhances the affinity of RNAP for promoter sites[50] (A. Revyakin, R.H. Ebright and T.R. Strick, unpublished results).

The effect of supercoiling on formation of RPo is clear from the initial experimental data presented (Figure 6.3). As expected from topological and energetic considerations, as well as from extensive biochemical analysis,[1,51–54] promoter unwinding on positively supercoiled DNA occurs at a slower rate and reverses at a higher rate than for negatively supercoiled DNA. Further experiments carried out on positively supercoiled lac(cons) promoter provide more detail. Figure 6.4(B) shows how the rate of formation and rate of dissociation of RPo vary for increasing positive supercoiling at this promoter. A biphasic pattern is observed, with a supercoiling-dependent regime and a supercoiling-independent regime. Similar results were obtained using the *rrnB* P1 promoter under the same conditions, but for the negative supercoiling applied to the DNA template (*i.e.*, the lifetime of RPo increased as negative supercoiling increased) (A. Revyakin, R.H. Ebright and T.R. Strick, unpublished results).

The two regimes observed above coincide with regimes I and II, respectively, of the extension *vs.* supercoiling curve (Figure 6.1B) described previously. In regime I (where torque acting on the template grows with supercoiling) both rates of formation and dissociation of RPo change with supercoiling. In regime II (where torque acting on the template is constant as plectonemic supercoils have formed) both rates are unchanged. The fact that the rate of formation and lifetime of RPo depend on the torque experienced by the template indicates that the protein conformational changes that accompany unwinding or rewinding of the promoter are the rate-limiting steps of the promoter unwinding and promoter rewinding reactions. In addition the dependence of the dissociation rate on supercoiling in regime I is consistent with a simple Arrhenius model for torque mechanically biasing the promoter rewinding reaction.[25]

Overall, the equilibrium between RPo and free components is shifted 16-fold for an increase in supercoiling density of only 0.005 (*e.g.*, ~10% of the *in vivo* density of supercoiling). Extrapolating to the degree of supercoiling observed *in vivo* ($\sigma \sim -0.05$) suggests an 160-fold effect over the full range of physiological supercoiling (indeed unlike for these *in vitro* assays, *in vivo* torque is expected to vary continuously across the range of physiological supercoiling; see Section 6.3.2.2). When combined with the ~200-fold difference in stability of RPo at lac(cons) and *rrnB* P1, it is clear that simply the sequence- and supercoiling-dependence of RPo can account for a very wide range of behaviors. It is tempting to speculate that this degree of sensitivity to sequence and supercoiling may be buffered *in vivo* by transcription factors or events downstream of formation of RPo. Again, the lifetime of RPo provides for an experimentally more accessible metric than the rate of open promoter complex formation, which must be titrated against RNAP concentration to resolve into K_B and k_f.

A further comparison of classical results with single-molecule results is based on analysis of the impact of effector nucleotide ppGpp on the lifetime of RPo (see also Chapters 3 and 8). Both positively supercoiled lac(cons) promoter and negatively supercoiled *rrnB* P1 promoter were assayed, and showed the same threefold reduction in lifetime in the presence of ppGpp. This is consistent with bulk experiments that reported the same reduction in the lifetime of RPo for the same promoters.[55]

Because formation of RPo can be simply described as a relatively limited sequence of large-scale events, bulk biochemical measurements performed over the last 30 years have been very successful at quantitatively describing the kinetic and structural features of the process based on a relatively synchronous population of molecules. However, in such experiments the transcription reaction has typically been artificially limited to formation of RPo by withholding NTPs from the reaction. By using an initiating nucleotide and/or dinucleotide that complements the first few bases of the initially transcribed sequence and occupy the catalytic site, one can simulate the initial steps from open complex to initially-transcribing complex. Still, this is not sufficient for generating promoter escape. Thus, is it reasonable to assume that the kinetic features of RPo are the same in the absence as in the presence of nucleotides? This question becomes experimentally difficult to access in bulk assays, RPo being only a transient species under such conditions and rapidly evolving into the initially transcribing complex RPitc and then into the processively elongating complex RDe. Detection of such complex sequences of transient events becomes possible in single-molecule detection schemes.

6.5 Promoter Escape by DNA Scrunching

As pointed out in the Introduction, initiation of RNA synthesis and promoter escape are interesting to study from a single-molecule perspective because of their transient kinetics and the fact that such a measurement is immune to temporal desynchronization of sample molecules.

One question of particular importance has been to understand the transition from the open promoter complex to processively transcribing RNAP. This process, known as promoter escape, requires RNAP to release from the promoter to which it is stably bound in RPo. It has long been known that promoter escape is not a simple task for RNAP, as breaking the protein–DNA contacts that bind it *via* σ^{70} to open promoter involve energies on the scale of 10 kcal mol^{-1}. Moreover, rearrangements of RNAP–σ contacts also appear to take place during promoter escape, at an energetic cost of up to an additional 10 kcal mol^{-1}. Thus, escape from many promoters is preceded by the repetitive synthesis and release of a large number of short, 2–15 nt "abortive" RNA products[2,3,11,56,57] as RNAP repeatedly attempts to, but fails in, using translocation to break its contacts with the promoter –10 and –35 elements. In agreement with the energetic argument presented above, studies have shown that the greater the stability of RPo, the lower the likelihood of promoter escape.[58,59]

Mechanistically the process of promoter escape is complex, and several models have been proposed to explain the observed behavior of initially-transcribing RNAP (Chapter 5, Figure 5.8). Footprinting experiments do not show a shift in the upstream boundary of RNAP during abortive initiation,[3,10,60,61] suggesting that RNAP remains fixed relative to regions upstream of promoter DNA. Yet at the same time the active site of RNAP must obviously translocate relative to the gene's initially transcribed sequence,[2,62,63] suggesting that the RNAP active center moves relative to promoter DNA during the process. As described in Chapter 5, three major mechanisms have been proposed to explain these apparently contradictory observations (see Figure 5.8). The first invokes deformation of RNAP during abortive initiation as a means to translocate the active center along the initial template sequence. The second proposes deformation of DNA during abortive initiation, whereby reeling in and unwinding downstream DNA allows the RNAP to initially transcribe without leaving the promoter. The third suggests rapid and reversible excursions of RNAP from the promoter with no deformation of either. Hybrid processes could also be suggested. These mechanisms have in common the transient accumulation of strain energy in the complex up to a point where it may break the ties that bind it to the promoter. Whether the protein or the DNA template should support the main deformation was unknown, as well as the role of the deformation in the promoter escape reaction.

The combination of data generated by single-molecule fluorescence detection and data generated by single-molecule DNA nanomanipulation has provided a new way for assessing these proposals in detailed fashion. This example underscores the usefulness of such approaches in studying dynamic features of complex, transient molecular interactions.

6.5.1 Characterization of DNA Scrunching during Abortive Initiation

For these experiments the N25 early promoter of phage T5 was chosen for its ability to generate long abortive products.[11,64,65] In addition, its initially transcribed sequence of AUAAAUUUG provides a means for using subsets of NTPs to decompose transcription initiation into successive phases (formation of RPo, RPitc and RDe). At the wild-type N25 promoter, abortive products of up to 8 nt in length can be generated by providing RPo with only ATP and UTP. With the N25 A5C promoter mutant, abortive products of up to 4 nt in length can be generated by providing ATP an UTP, and abortive products of up to 8 nt in length can be obtained by providing ATP, UTP and CTP.

Furthermore, using the wild-type N25 sequence promoter escape can be induced by providing RPo with ATP, UTP and GTP, as promoter escape occurs at about position +15 (the maximal observed length of abortive products) and the first cytosine to be incorporated into the RNA only appears at position +30. For such experiments involving promoter escape, a terminator downstream of the transcription start site is required. Indeed, any RNAP molecule that escapes from the promoter forms an extremely stable ternary complex that cannot be easily removed from the DNA. Allowing RNAP to dissociate from the DNA upon accomplishing transcription termination is a straightforward way to "recycle" the DNA template, returning it to its initial RNAP-free state so that multiple successive rounds of transcription can be observed on the same DNA molecule. Thus, in the presence of all four NTPs, the full transcription cycle, including productive elongation and transcription termination, can be studied.

6.5.1.1 DNA Scrunching occurs during Abortive Initiation

The DNA scrunching mechanism for abortive initiation was tested by comparing the amplitude of DNA unwinding at the N25 promoter in the absence of nucleotides and in the presence of ATP and UTP (Figure 6.5). Under conditions in which abortive RNA products of up to 8 nt in length could be synthesized, the amplitude of DNA unwinding increased by about a half-turn of the double helix, or about 5 bp. This observation, consistent with the DNA scrunching model for abortive initiation, was controlled by performing experiments under conditions in which no abortive RNA was synthesized (i.e., in the presence of only ATP), and by performing experiments under conditions in which only dinucleotide abortive RNA was synthesized (i.e., in the presence of ATP, UTP and rifampicin, an antibiotic which prevents polymerization beyond the first phosphodiester bond[66,67]). No increase in the amplitude of DNA unwinding was observed in either of these control experiments. Thus no scrunching occurs upon binding of the first cognate NTP, nor upon synthesis of the first phosphodiester bond, but scrunching occurs upon synthesis of abortive RNA products greater than two nucleotides in length.

Note that these experiments were carried out under saturating NTP concentrations (500 μM each), conditions in which no intermediates such as RPo could be observed prior to formation of the expanded transcription bubble characteristic of abortive initiation. However, RNAP is expected to release multiple short RNAs prior to escape,[2] and so it is likely that after formation of RPitc there is repeated cycling between RPitc and RPo (on a timescale too brief

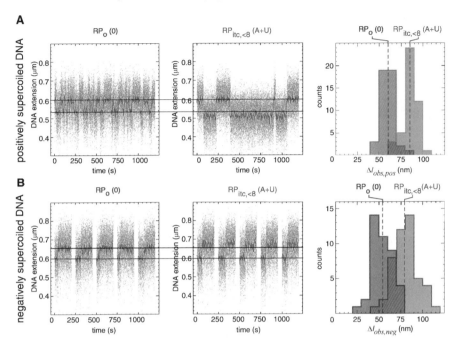

Figure 6.5 Real-time detection of DNA scrunching during abortive initiation on (A) positively supercoiled or (B) negatively supercoiled N25 promoter ($\sigma \sim 0.021$) in the absence (left) or presence (middle) of initiating nucleotides (ATP and UTP, allowing synthesis of up to 8-nt RNAs). Amplitude histograms (right) show an increase in transition amplitude, and thus DNA unwinding, during abortive synthesis.[12] Experimental conditions were as given in Figure 6.1, but for a slightly higher salt (NaCl) concentration (150 rather than 100 mM). (From ref. 12. Reprinted with permission from the AAAS.)

for this assay to detect) as the short abortive RNA is released and synthesized anew. This is supported by single-molecule fluorescence measurements that indicate that the rate-limiting step of abortive synthesis is product release, and that re-synthesis of RNA is rapid.[68]

6.5.1.2 Extent of DNA Scrunching is Proportional to the Length of Abortive Transcript

Next, by using the N25 A5C construct, it was possible to verify that the observed amplitude of promoter unwinding increased progressively as longer abortive RNA products could be synthesized. Under conditions in which abortive RNA products up to four bases long could be produced the amplitude of DNA unwinding was seen to increase by about 2 bp. Under conditions in which abortive RNA products up to 8 bases long could be produced the amplitude of DNA unwinding increased by about 5 bp, as observed with the wild-type N25 construct in the presence of ATP and UTP.

Taking these experiments together, a simple model for the relationship between length of abortive RNA and increase in amplitude of DNA unwinding was proposed. The model takes into account the fact that promoter unwinding in RPo extends from about −10 to +2, thus including the first two initially-transcribed nucleotides. Moreover at the initiation of phosphodiester catalysis, the RNAP active site must accommodate at least two NTPs, one in the so-called "P" site and one in the so-called "A" (or IS) site (Chapters 7 and 8), forcing the active site in the post-translocated configuration. In light of this it is not surprising that no additional DNA unwinding is required to bind a single NTP or to synthesize a dinucleotide. Beyond synthesis of the first two nucleotides, however, the simplest possible model would stipulate that further synthesis of RNA would require unwinding of downstream DNA sufficient to allow the next template base to line up with the active site – in other words, DNA unwinding would be proportional to the length of abortive RNA synthesized. These two elements lead to a model in which the amplitude of additional DNA unwinding (in bp) required to allow abortive synthesis goes as $N - 2$, where N is the length of the RNA product generated prior to escape.

6.5.2 Characterization of DNA Scrunching during Promoter Escape

6.5.2.1 *Assigning Molecular States by Single-molecule Pulse-chase Experiments*

To determine whether DNA scrunching occurs prior to promoter escape, experiments were performed using the N25 promoter and RNAP in the presence of all four NTPs. Under these conditions an RNAP can complete promoter escape as well as elongation and termination. Figure 6.6 shows time-traces collected under these conditions. The time-traces show a train of three "transcription pulses," each transcription pulse results from the action of a single RNAP as it goes from transcription initiation to elongation and to termination. The transcription pulses follow a remarkably similar pattern from one pulse to the next, and each pulse can be decomposed into four DNA unwinding/rewinding transitions: (1) formation of RPo, (2) formation of RPitc and concomitant DNA scrunching, (3) formation of RDe and concomitant DNA unscrunching and (4) transcription termination. The three consecutive states – RPo, RPitc and RDe – can be characterized both in terms of structure (amplitude of DNA unwinding) and function (*i.e.*, kinetics). Assignment of these states was verified by pulse-chase experiments on single transcription complexes (see Figure 6.7 for details).

6.5.2.2 *Structural Analysis of DNA Unwinding during and after Promoter Escape*

Structural analysis of the amplitude of DNA unwinding in RPo, RPitc and RDe provides information on topological changes in the transcription bubble

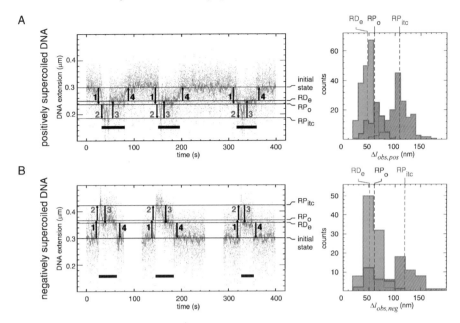

Figure 6.6 Real-time detection of the full transcription cycle in the presence of all four NTPs on (A) positively and (B) negatively supercoiled DNA. The promoter used was the N25 promoter. Three transcription pulses, corresponding to transcription by three successive RNAPs of the same DNA molecule, are underlined by black bars. Four unwinding/rewinding transitions (1–4) are labeled and discussed in the text and in Figure 6.7. Amplitude histograms (right) detail the change in DNA extension from the initial state in each of the three stages of transcription (RPo, RPitc and RDe).[12] Experimental conditions were as given in Figure 6.1, but for a slightly lower salt concentration (75 mM NaCl). (From ref. 12. Reprinted with permission from the AAAS.)

during the transcription cycle (Figure 6.6). In RPo, a transcription bubble of about 12 bp is formed, as expected. In RPitc, formation of the stressed, or "scrunched," intermediate causes the size of this unwound region to increase to about 21 bp on average. This is consistent with the $N - 2$ rule relating the amplitude of scrunching to the length of the abortive RNA. Indeed, the great majority of long abortive RNA products observed in radiolabeling experiments with this promoter are on the order of 10–11 nucleotides;[11] the $N - 2$ rule would imply that an additional 8–9 bp of DNA are unwound relative to RPo to form such lengths of RNA, and this is indeed what is observed experimentally. However, the observed distribution of transition amplitudes in RPitc is significantly broader than the distribution in RPo (Figure 6.6), with a very small fraction (<10%) of promoter-escape events occurring after "limited" scrunching (*e.g.*, a few base pairs) or "extensive" scrunching (*e.g.*, up to as many as 16–17 bp). This is also consistent with the distribution of abortive RNAs observed by gel electrophoresis, where abortive RNA products of up to 15 nt are still faintly detected.[11] Finally, the amplitude of DNA unwinding in

RDe is significantly smaller than in RPo, with only about 9–10 bp of unwound DNA. This is consistent with bulk measurements estimating the size of the RNA-DNA hybrid at about 8–9 base-pairs.[4]

At a typical promoter, escape occurs only after synthesis of an RNA product ~9 to 11 nt in length, and thus can be inferred to require scrunching of ~7 to 9 bp. Assuming an energetic cost of base-pair breakage of ~2 kcal mol^{-1} per bp, it can be inferred that, at a typical promoter, a total of ~14–18 kcal mol^{-1} of base-pair breakage energy is accumulated in the stressed intermediate. This free energy is high relative to the free energies for RNAP–promoter interaction (~7–9 kcal mol^{-1} for sequence-specific component of RNAP–promoter interaction) and for tight binding between enzyme and initiation factors (~13 kcal mol^{-1} for transcription initiation factor σ^{70}).

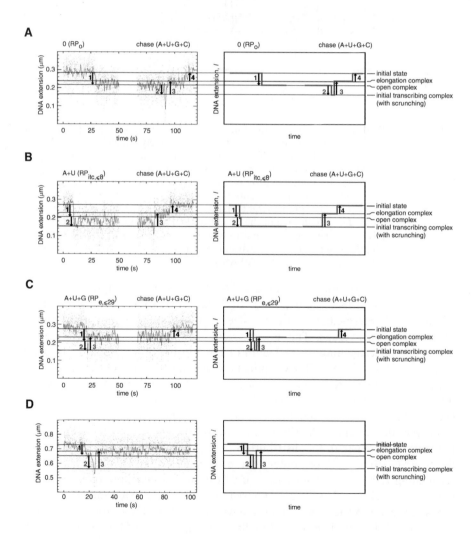

These observations strongly support the existence of an obligatory, mechanically-stressed intermediate, and the quantitation supports the view that the energy accumulated in that stressed intermediate is the energy that drives the transition from transcription initiation to elongation.

6.5.2.3 Kinetic Analysis of the Fate of the Transcription Bubble during Promoter Escape

We now consider the kinetic properties of the transcription bubble during promoter unwinding and promoter scrunching, beginning with formation of the open complex (Figure 6.6). At the N25 promoter, in the absence of NTPs RPo is long-lived even under conditions of positive supercoiling ($\sigma = 0.021$), with a lifetime of ~ 220 s at 34 °C in the moderate ionic conditions used here (essentially 25 mM Na Hepes, 75 mM NaCl and 10 mM $MgCl_2$). However, in the presence of NTPs (100 μM each), RPo only has a lifetime on the order of 3 s as NTP binding drives it to rapidly isomerize into RPitc. Moreover, comparison of the data shown in Figure 6.6 with those obtained in the presence of ATP and

---

Figure 6.7 Pulse-chase experiments confirming assignment of transitions (A) 1 as formation of RPo, (B) 2 as formation of RPitc, (C) 3 as promoter escape and formation of RDe and (D) 4 as transcription termination and closure of the transcription bubble. The left-hand column shows experimental time traces and the right-hand side gives idealized sketches of the transitions.[12] The pulse-chase experiments performed to confirm assignment of the above states were carried out by initially forming RNAP/DNA complexes in limiting conditions for NTP species, and then releasing them from the state they achieve by adding back the missing NTPs. (A) In initial conditions in which only RPo is allowed to form (*i.e.*, in the absence of any nucleotides), only transition (1) is observed. Upon adding back all four NTPs, the RNAP proceeds to cycle through the remaining three transitions. Transition (1) therefore corresponds to formation of RPo. (B) Next, under initial conditions in which RPitc is allowed to form (*i.e.*, in the presence of ATP and UTP), only transitions (1) and (2) are observed. Adding back the remaining NTPs allows the polymerase to complete the remaining two transitions. This confirms that transition (2) corresponds to formation of RPitc. (C) Next, under initial conditions in which RDe is allowed to form (*i.e.*, in the presence of ATP, UTP and GTP), only transitions (1)–(3) are observed. Upon add-back of the missing CTP, transition (4) is recovered. Transition (3) thus corresponds to promoter escape and formation of RDe. (D) Finally, transition (4) was assigned by performing experiments in the presence of all four NTPs but using a DNA template in which the tR2 terminator had been mutated and was no longer functional. Under these conditions, only transitions (1)–(3) were observed, confirming that transition (4) corresponds to final collapse of the transcription bubble during transcription termination. On a DNA substrate containing the tR2 terminator, multiple series of pulse-chase experiments could be performed in succession on the same DNA molecule as transcription termination during the chase cleared the DNA of any enzyme, returning it to its initial protein-free conformation. Experimental conditions were as given in Figure 6.6. (From ref. 12. Reprinted with permission from the AAAS.)

UTP at 500 μM each (where RPo is not detectable; see Figure 6.5) suggests that the lifetime of RPo continuously decreases with increasing NTP concentration until it is no longer detectable under physiological conditions. Furthermore at 100 μM NTP the rate of isomerization of RPo into RPitc is, within experimental error, the same for positively or negatively supercoiled templates (2–3 s). This can be interpreted as meaning that the transition state on the path of isomerization of RPo into RPitc has the same unwinding as in RPo, *i.e.*, is not torque- or supercoiling-sensitive. Thus, as pointed out earlier, under physiological conditions it is the rate-constant for isomerization into RPo that becomes the more important of the two forward rate constants involved in converting RPc into RPitc.[56,57,69]

We next consider the kinetic properties of DNA scrunching in RPitc. If DNA scrunching is an integral part of the pathway to promoter escape, we would expect that promoter escape is always preceded by a detectable amount of DNA scrunching. Analysis shows that 80% of transcription pulses display scrunching prior to escape. However, analysis of the lifetime distribution of scrunching events shows that not all scrunches are long-lived enough to be detected, so that the fraction of 80% is an underestimate. Indeed the lifetime distribution of scrunching events displays single-exponential statistics, with a mean scrunch duration of about 5 ± 1 s. In such a distribution, roughly 20% of events can be predicted to have a lifetime shorter than one second, which is precisely the lower detection limit of the instrument used. Thus all, or nearly all, promoter escape events are preceded by DNA scrunching, making this conformation an obligatory intermediate on the pathway to promoter escape.

Furthermore, as with RPo, the duration of the scrunched state prior to promoter escape is only weakly dependent on supercoiling – this despite the extensive accumulation of unwound DNA in the scrunched state. Thus the lifetime of RPitc on negatively supercoiled promoter ($\sigma = -0.021$) is about 3.5 s and that observed on positively supercoiled promoter ($\sigma = 0.021$) is about 5 s. The trend is to be expected, as the mechanical energy stored in negatively supercoiled DNA is still available to help drive the unwinding of the ultimate bp to be templated for NTP addition prior to promoter escape. Yet despite the difference in torque between positive and negative supercoiling the energy barrier to promoter escape is only $\sim 0.4\ k_B T$ lower for negative supercoiling, underlining the overall importance of translocation (relative to torsion) in undoing the stably unwound RPo complex.

We can consider these observations in light of the single-molecule fluorescence analysis described in Chapter 5. Indeed, as mentioned earlier, fluorescence analysis showed that during abortive initiation most complexes can be found in the "scrunched" state. In other words, the slow step in abortive cycling is neither initiation of RNA synthesis nor RNA synthesis itself but rather the ultimate stage of promoter escape where either RNA is released as an abortive product or RNAP is released from the promoter as an elongating complex.[68] Thus, once the last nucleotide has been added to the nascent RNA "*via*" the scrunching pathway, the system must pass over a final energy barrier to perform escape before kinetic competition with RNA release causes it to have to restart the entire process of nascent RNA synthesis.[70] From estimates of the torque and use of the Arrhenius law one can estimate that the transition state corresponding to this final energy barrier indeed has additional unwinding of less than a half base-pair relative to RPitc.

Structural studies of RNAP before and after promoter escape provide insight into how the σ subunit may be involved in abortive RNA release and what may constitute this final energy barrier[6,23] (see the Structural Atlas at the beginning of this book). First of all, the "linker" σ3.2 that connects σ domains 3 and 4 is seen as occupying the path of nascent RNA longer than two to three nucleotides. Thus, during initial synthesis this linker could provide a driving force for release of abortive RNA. Moreover, domain 4 of the σ^{70} subunit is observed obstructing the RNA exit channel (see Structural Atlas and Chapter 5) prior to escape, but no longer after escape. This could also constitute a way for RNAP to couple torsional deformation of DNA and linear translocation on DNA during promoter escape. By unwinding and reeling-in downstream DNA, RNAP grows an RNA chain in its central cleft that first displaces σ3.2 and then progressively presses against σ domain 4 until the latter releases, allowing the system to assume the structure of an elongating complex. This proposal merits testing in single-molecule experiments using mutants of σ mutated in the relevant domains.

6.5.2.4 Structural and Kinetic Analysis of the Fate of the Transcription Bubble during Transcription Elongation and Termination

After promoter escape, the transcription bubble persists until transcription termination takes place (Figure 6.7). The distribution of unwinding amplitudes in RDe is significantly more heterogeneous than for RPo (formed in the absence of nucleotides).[26] It is tempting to speculate about such heterogeneity, in particular to ask whether it could be static or dynamic, *i.e.*, could each RNAP molecule favor a hybrid with slightly different length, or could the hybrid length fluctuate slightly during transcription?

Kinetically, these measurements may also provide access to information concerning the rates of elongation and the rates and efficiencies of termination. By measuring the time interval between promoter escape and transcription termination one can determine $T_{el+term} = T_{el} + T_{term}$ the sum of times required to first elongate transcript (T_{el}) and then perform transcription termination (T_{ter}). As only T_{el} depends on the number of base pairs to transcribe, it can be obtained by measuring the variation of $T_{el+term}$ with transcript length. A rough estimate of $\sim 12\,\mathrm{nt\,s^{-1}}$ has been provided from measurements with 100 and 400 bp transcripts in the experimental conditions used here,[26] and is in good agreement with results from optical trapping experiments when NTP concentrations and temperatures are taken into account.[71] Determining T_{term} will require measurements of $T_{el+term}$ with many different transcript lengths if one extrapolates $T_{el+term}$ to transcripts of zero length.

Finally, the efficiency of termination can be estimated by comparing the number of transcription pulses where termination occurs to the number of transcription pulses that never return to baseline. For the tR2 terminator used in these experiments, and on positively supercoiled DNA, the termination efficiency was upwards of 90%. Although these experiments do not have the spatial and temporal resolutions of optical trapping experiments on elongation and termination,[72,73] they may still provide insight into topological effects during these phases of transcription.

6.6 Future Directions

There still remains much to be explored with the magnetic-tweezer approach to detecting RNAP–DNA interactions. For instance systematic analysis of the effect of promoter sequences and initially-transcribed sequences on the kinetics of promoter escape will be important to extend results obtained so far mostly on the N25 promoter (T.R. Strick, A. Revyakin and R.H. Ebright, unpublished results). More accurate determination of the torque acting on the DNA will also make it possible to obtain better measurements of free energy differences between states. Higher resolution measurements of DNA unwinding amplitude may help describe the role of structural heterogeneity in initiating and elongating RNAP. Improvements in resolution can still be achieved from the use of yet shorter DNA fragments or smaller magnetic beads, but this strategy may soon run its course as technical difficulties arise, for instance from unwanted, non-specific interactions between the bead and the surface.

Thus, one of the most pressing directions for future improvement of this approach involves its combination with single-molecule fluorescence methods such as those described in Chapter 5. Indeed, evanescent-wave microscopy has shown significant success in detecting transcription initiation, elongation and termination on immobilized complexes.[68,74] Fluorescence detection has a clear advantage over low-force mechanical manipulation of DNA, namely a much higher temporal resolution which could be pushed to the ~ 10 ms scale. Moreover, fluorophore lifetime can be brought into the range of timescales for a full transcription cycle, typically ~ 15–20 s on a 100 bp transcript in 100 mM NTP. Thus, for instance, using singly-labeled RNAP and nanomanipulated DNA one could directly determine the binding constant of RNAP for promoter as well as the time required to unwind promoter once bound to it. Using singly-labeled RNAP and singly-labeled, nanomanipulated DNA one could simultaneously measure relative motions of RNAP and template DNA and concomitant DNA unwinding/rewinding during transcription initiation or transcription termination. Using doubly-labeled RNAP and FRET it would be possible to directly measure the internal conformational changes at key residues involved in promoter escape, nucleotide addition and transcription termination. Using fluorescently-labeled nucleotides and nanomanipulation it may eventually be possible to measure the number of abortive RNA products released prior to promoter escape. The combination of single-molecule fluorescence and nanomanipulation approaches should therefore provide for a great many exciting opportunities in the study of RNAP as a molecular motor.

References

1. M. Amouyal and H. Buc, Topological unwinding of strong and weak promoters by RNA polymerase. A comparison between the lac wild-type and the uv5 sites of *Escherichia coli*, *J. Mol. Biol.*, 1987, **195**(4), 795–808.
2. A. J. Carpousis and J. D. Gralla, Cycling of ribonucleic acid polymerase to produce oligonucleotides during initiation in vitro at the lac uv5 promoter, *Biochemistry*, 1980, **19**(14), 3245–3253.
3. D. C. Straney and D. M. Crothers, A stressed intermediate in the formation of stably initiated RNA chains at the *Escherichia coli* lac uv5 promoter, *J. Mol. Biol.*, 1987, **193**(2), 267–278.
4. E. Nudler, A. Mustaev, E. Lukhtanov and A. Goldfarb, The RNA-DNA hybrid maintains the register of transcription by preventing backtracking of RNA polymerase, *Cell*, 1997, **89**(1), 33–41.
5. T. D. Yager and P. H. von Hippel, A thermodynamic analysis of RNA transcript elongation and termination in *Escherichia coli*, *Biochemistry*, 1991, **30**, 1097–1118.
6. K. S. Murakami, S. Masuda and S. A. Darst, Structural basis of transcription initiation: RNA polymerase holoenzyme at 4 A resolution, *Science*, 2002, **296**(5571), 1280–1284.
7. D. G. Vassylyev, S. Sekine, O. Laptenko, J. Lee, M. N. Vassylyeva, S. Borukhov and S. Yokoyama, Crystal structure of a bacterial RNA polymerase holoenzyme at 2.6 Å resolution, *Nature*, 2002, **417**(6890), 712–719.
8. K. S. Murakami and S. A. Darst, Bacterial RNA polymerases: the whole story, *Curr. Opin. Struct. Biol.*, 2003, **13**(1), 31–39.
9. B. E. Nickels, S. J. Garrity, V. Mekler, L. Minakhin, K. Severinov, R. H. Ebright and A. Hochschild, The interaction between sigma70 and the beta-flap of *Escherichia coli* RNA polymerase inhibits extension of nascent RNA during early elongation, *Proc. Natl. Acad. Sci. (USA)*, 2005, **102**(12), 4488–4493.
10. A. J. Carpousis and J. D. Gralla, Interaction of RNA polymerase with lacuv5 promoter DNA during mRNA initiation and elongation. Footprinting, methylation, and rifampicin-sensitivity changes accompanying transcription initiation, *J. Mol. Biol.*, 1985, **183**(2), 165–177.
11. L. M. Hsu, Promoter clearance and escape in prokaryotes, *Biochim. Biophys. Acta*, 2002, **1577**(2), 191–207.
12. A. Revyakin, C.-Y. Liu, R. H. Ebright and T. R. Strick, Abortive initiation and productive initiation by RNA polymerase involve DNA scrunching, *Science*, 2006, **314**, 1139–1143.
13. A. N. Kapanidis, E. Margeat, S. O. Ho, E. Kortkhonjia, S. Weiss and R. H. Ebright, Initial transcription by RNA polymerase proceeds through a DNA-scrunching mechanism, *Science*, 2006, **314**, 1144–1147.

14. W. S. Yarnell and J. W. Roberts, Mechanism of intrinsic transcription termination and antitermination, *Science*, 1999, **284**(5414), 611–615.
15. T. J. Santangelo and J. W. Roberts, Forward translocation is the natural pathway of RNA release at an intrinsic terminator, *Mol. Cell*, 2004, **14**(1), 117–126.
16. A. Schmitz and D. J. Galas, The interaction of RNA polymerase and lac repressor with the lac control region, *Nucleic Acids Res.*, 1979, **6**(1), 111–137.
17. M. L. Craig, O. V. Tsodikov, K. L. McQuade, P. E. Schlax, M. W. Capp, R. M. Saecker and M. T. Record, DNA footprints of the two kinetically significant intermediates in formation of an RNA polymerase-promoter open complex: evidence that interactions with start site and downstream DNA induce sequential conformational changes in polymerase and DNA, *J. Mol. Biol.*, 1998, **283**(4), 741–756.
18. E. Zaychikov, L. Denissova, T. Meier, M. Götte and H. Heumann, Influence of Mg^{2+} and temperature on formation of the transcription bubble, *J. Biol. Chem.*, 1997, **272**(4), 2259–2267.
19. S. Sasse-Dwight and J. D. Gralla, $KMnO_4$ as a probe for lac promoter DNA melting and mechanism in vivo, *J. Biol. Chem.*, 1989, **264**(14), 8074–8081.
20. U. Siebenlist and W. Gilbert, Contacts between *Escherichia coli* RNA polymerase and an early promoter of phage t7, *Proc. Natl. Acad. Sci. (USA)*, 1980, **77**(1), 122–126.
21. K. Kirkegaard, H. Buc, A. Spassky and J. C. Wang, Mapping of single-stranded regions in duplex DNA at the sequence level: single-strand-specific cytosine methylation in RNA polymerase-promoter complexes, *Proc. Natl. Acad. Sci. USA*, 1983, **80**(9), 2544–2548.
22. A. von Gabain and H. Bujard, Interaction of *Escherichia coli* RNA polymerase with promoters of several coliphage and plasmid DNAs, *Proc. Natl. Acad. Sci. USA*, 1979, **76**(1), 189–193.
23. K. S. Murakami, S. Masuda, E. A. Campbell, O. Muzzin and S. A. Darst, Structural basis of transcription initiation: an RNA polymerase holoenzyme-DNA complex, *Science*, 2002, **296**(5571), 1285–1290.
24. T. Strick, J. F. Allemand, D. Bensimon, A. Bensimon and V. Croquette, The elasticity of a single supercoiled DNA molecule, *Science*, 1996, **271**, 1835–1837.
25. A. Revyakin, R.H. Ebright and T.R. Strick. Promoter unwinding and promoter clearance by RNA polymerase: detection by single-molecule DNA nanomanipulation, *Proc. Natl. Acad. Sci. USA*, 2004, **101**, 4776–4780.
26. A. Revyakin, R. H. Ebright and T. R. Strick, Single-molecule DNA nanomanipulation: improved resolution through use of shorter DNA fragments, *Nat. Methods*, 2005, **2**(2), 127–138.
27. T. Strick, J.-F. Allemand, D. Bensimon and V. Croquette, The behavior of supercoiled DNA, *Biophys. J.*, 1998, **74**, 2016–2028.

28. P. R. Selvin and T. Ha (eds), *Single-Molecule Techniques: A Laboratory Manual.*, Cold Spring Harbor Laboratory Press, 2008.
29. A. Spassky, K. Kirkegaard and H. Buc, Changes in the DNA structure of the lac uv5 promoter during formation of an open complex with *Escherichia coli* RNA polymerase, *Biochemistry*, 1985, **24**(11), 2723–2731.
30. S. B. Smith, L. Finzi and C. Bustamante, Direct mechanical measurements of the elasticity of single DNA molecules by using magnetic beads, *Science*, 1992, **258**, 1122–1126.
31. C. Bustamante, J. F. Marko, E. D. Siggia and S. Smith, Entropic elasticity of λ-phage DNA, *Science*, 1994, **265**, 1599–1600.
32. C. Bouchiat, M. D. Wang, S. M. Block, J.-F. Allemand, T. R. Strick and V. Croquette, Estimating the persistence length of a worm-like chain molecule from force-extension measurements, *Biophys. J.*, 1999, **76**, 409–413.
33. M. Rief, J. M. Fernandez and H. E. Gaub, Elastically coupled two-level system as a model for biopolymer extensibility, *Phys. Rev. Lett.*, 1998, **81**, 4764–4767.
34. M. Rief, H. Clausen-Schaumann and H. E. Gaub, Sequence-dependent mechanics of single DNA molecules, *Nature Struct. Bio.*, 1999, **6**, 346–349.
35. J. H. White, Self linking and the gauss integral in higher dimensions, *Am. J. Math.*, 1969, **91**, 693–728.
36. J. C. Wang, Helical repeat of DNA in solution, *Proc. Natl. Acad. Sci. USA*, 1979, **76**, 200–203.
37. T. R. Strick, J.-F. Allemand, D. Bensimon and V. Croquette, Stress induced structural transitions in DNA and proteins, *Ann. Rev. Biophys. Biomol. Struct.*, 2000, **29**, 523–543.
38. G. Charvin, J.-F. Allemand, T. R. Strick, D. Bensimon and V. Croquette, Twisting DNA: single molecule studies, *Contemp. Phys.*, 2004, **45**, 385–403.
39. T. Strick, V. Croquette and D. Bensimon, Homologous pairing in streched supercoiled DNA, *Proc. Nat. Acad. Sci. (USA)*, 1998, **95**, 10579–10583.
40. J.-F. Allemand and D. Bensimon, R. Lavery and V. Croquette. Stretched and overwound DNA form a Pauling-like structure with exposed bases, *Proc. Natl. Acad. Sci. USA*, 1998, **95**, 14152–14157.
41. T. Strick, J.-F. Allemand, V. Croquette and D. Bensimon, Twisting and stretching single DNA molecules, *Prog. Biophys. Mol. Biol.*, 2000, **74**, 115.
42. J. Marko, *Phys. Rev. E.*, 2007, **76**, 021926.
43. S. Forth, C. Deufel, M. Y. Sheinin, B. Daniels, J. P. Sethna and M. D. Wang, Abrupt buckling transition observed during the plectoneme form of individual molecules, *Phy. Rev. Lett.*, 2008, **100**, 148301–148305.
44. J. C. Wang, J. H. Jacobsen and J. M. Saucier, Physiochemical studies on interactions between DNA and RNA polymerase. Unwinding of the DNA helix by *Escherichia coli* RNA polymerase, *Nucleic Acids Res.*, 1977, **4**(5), 1225–1241.
45. C. Rivetti, M. Guthold and C. Bustamante, Wrapping of DNA around the E. coli RNA polymerase open promoter complex, *EMBO J.*, 1999, **18**(16), 4464–4475.

46. J. H. Roe, R. R. Burgess and M. T. Record, Kinetics and mechanism of the interaction of *Escherichia coli* RNA polymerase with the lambda pr promoter, *J. Mol. Biol.*, 1984, **176**(4), 495–522.
47. J. H. Roe, R. R. Burgess and M. T. Record, Temperature dependence of the rate constants of the *Escherichia coli* RNA polymerase-lambda pr promoter interaction. assignment of the kinetic steps corresponding to protein conformational change and DNA opening, *J. Mol. Biol.*, 1985, **184**(3), 441–453.
48. K. Severinov and S. A. Darst, A mutant RNA polymerase that forms unusual open promoter complexes, *Proc. Natl. Acad. Sci. USA*, 1997, **94**(25), 13481–13486.
49. R. M. Saecker, O. V. Tsodikov, K. L. McQuade, P. E. Schlax, M. W. Capp and M. T. Record, Kinetic studies and structural models of the association of E. coli sigma(70) RNA polymerase with the lambdap(r) promoter: large scale conformational changes in forming the kinetically significant intermediates, *J. Mol. Biol.*, 2002, **319**(3), 649–671.
50. S. Leirmo, C. Harrison, D. S. Cayley, R. R. Burgess and M. T. Record, Replacement of potassium chloride by potassium glutamate dramatically enhances protein-DNA interactions *in vitro*, *Biochemistry*, 1987, **26**(8), 2095–2101.
51. J. A. Borowiec and J. D. Gralla, Supercoiling response of the lac ps promoter in vitro, *J. Mol. Biol.*, 1985, **184**(4), 587–598.
52. J. G. Brahms, O. Dargouge, S. Brahms, Y. Ohara and V. Vagner, Activation and inhibition of transcription by supercoiling, *J. Mol. Biol.*, 1985, **181**(4), 455–465.
53. S. Leirmo and R. L. Gourse, Factor-independent activation of *Escherichia coli* rRNA transcription. i. Kinetic analysis of the roles of the upstream activator region and supercoiling on transcription of the rrnb p1 promoter in vitro, *J. Mol. Biol.*, 1991, **220**(3), 555–568.
54. K. L. Ohlsen and J. D. Gralla, Interrelated effects of DNA supercoiling, ppgpp, and low salt on melting within the *Escherichia coli* ribosomal RNA rrnb p1 promoter, *Mol. Microbiol.*, 1992, **6**(16), 2243–2251.
55. M. M. Barker, T. Gaal, C. A. Josaitis and R. L. Gourse, Mechanism of regulation of transcription initiation by ppgpp. i. Effects of ppgpp on transcription initiation in vivo and in vitro, *J. Mol. Biol.*, 2001, **305**(4), 673–688.
56. W. R. McClure, C. L. Cech and D. E. Johnston, A steady state assay for the RNA polymerase initiation reaction, *J. Biol. Chem.*, 1978, **253**(24), 8941–8948.
57. J. D. Gralla, A. J. Carpousis and J. E. Stefano, Productive and abortive initiation of transcription in vitro at the lac uv5 promoter, *Biochemistry*, 1980, **19**(25), 5864–5869.
58. T. Ellinger, D. Behnke, H. Bujard and J. D. Gralla, Stalling of *Escherichia coli* RNA polymerase in the +6 to +12 region *in vivo* is associated with tight binding to consensus promoter elements, *J. Mol. Biol.*, 1994, **239**(4), 455–465.

59. N. V. Vo, L. M. Hsu, C. M. Kane and M. J. Chamberlin, In vitro studies of transcript initiation by Escherichia coli RNA polymerase. 3. Influences of individual DNA elements within the promoter recognition region on abortive initiation and promoter escape, Biochemistry, 2003, **42**(13), 3798–3811.
60. A. Spassky, Visualization of the movement of the Escherichia coli RNA polymerase along the lac uv5 promoter during the initiation of the transcription, J. Mol. Biol., 1986, **188**(1), 99–103.
61. B. Krummel and M. J. Chamberlin, RNA chain initiation by Escherichia coli RNA polymerase. Structural transitions of the enzyme in early ternary complexes, Biochemistry, 1989, **28**(19), 7829–7842.
62. M. A. Grachev and E. F. Zaychikov, Initiation by Escherichia coli RNA-polymerase: transformation of abortive to productive complex, FEBS Lett., 1980, **115**(1), 23–26.
63. L. M. Munson and W. S. Reznikoff, Abortive initiation and long ribonucleic acid synthesis, Biochemistry, 1981, **20**(8), 2081–2085.
64. R. Gentz and H. Bujard, Promoters recognized by Escherichia coli RNA polymerase selected by function: highly efficient promoters from bacteriophage t5, J. Bacteriol., 1985, **164**(1), 70–77.
65. W. Kammerer, U. Deuschle, R. Gentz and H. Bujard, Functional dissection of Escherichia coli promoters: information in the transcribed region is involved in late steps of the overall process, EMBO J., 1986, **5**(11), 2995–3000.
66. W. R. McClure and C. L. Cech, On the mechanism of rifampicin inhibition of RNA synthesis, J. Biol. Chem., 1978, **253**(24), 8949–8956.
67. E. A. Campbell, N. Korzheva, A. Mustaev, K. Murakami, S. Nair, A. Goldfarb and S. A. Darst, Structural mechanism for rifampicin inhibition of bacterial RNA polymerase, Cell, 2001, **104**(6), 901–912.
68. E. Margeat, A. N. Kapanidis, P. Tinnefeld, Y. Wang, J. Mukhopadhyay, R. H. Ebright and S. Weiss, Direct observation of abortive initiation and promoter escape within single immobilized transcription complexes, Biophys. J., 2006, **90**(4), 1419–1431.
69. W. R. McClure, Mechanism and control of transcription initiation in prokaryotes, Annu. Rev. Biochem., 1985, **54**, 171–204.
70. X. C. Xue, F. Liu and Z. C. Ou-Yang, A kinetic model of transcription initiation by RNA polymerase, J. Mol. Biol., 2008, **378**, 520–529.
71. E. A. Abbondanzieri, J. W. Shaevitz and S. M. Block, Picocalorimetry of transcription by RNA polymerase, Biophys. J., 2005, **89**, L61–L63.
72. E. A. Abbondanzieri, W. J. Greenleaf, J. W. Shaevitz, R. Landick and S. M. Block, Direct observation of base-pair stepping by RNA polymerase, Nature, 2005, **438**, 460–465.
73. M. H. Larson, W. J. Greenleaf, R. Landick and S. M. Block, Applied force reveals mechanistic and energetic details of transcription termination, Cell, 2008, **132**, 971–982.
74. L. J. Friedman, J. Chung and J. Gelles, Viewing dynamic assembly of molecular complexes by multi-wavelength single-molecule fluorescence, Biophys. J., 2006, **91**, 1023–1031.

Part II Transcription Elongation and Termination

INTERLUDE
The Engine and the Brake

HENRI BUC[a] AND TERENCE STRICK[b]

[a] CIS, Institut Pasteur, Paris, France; [b] Centre National de la Recherche Scientifique, Institut Jacques Monod and University of Paris Diderot-Paris 7, Paris, France

I.1 Introduction

Broadly defined, a molecular motor is a protein that uses the energy of hydrolysis of a small molecule such as a nucleoside triphosphate (NTP) to complete an enzymatic cycle during the course of which the protein performs directional motion. Molecular motors are therefore unusual machines that accomplish what man-made devices are unable to do: the direct and isothermal conversion of chemical energy into mechanical energy, without the need to rely on an intermediate energy carrier, heat or electricity.[1] This state of affairs raises a set of general questions that are relevant to any motor protein under consideration: How do these machines operate to realize this *tour de force*? How much fuel do they consume? What fraction of it is indeed converted into movement? How do they generate forces? How do they move and what path do they follow?

To answer these questions, one has to delve deeply into the details of their respective structures, the deformability of their various domains, and how these relative structural changes are properly coupled to achieve directional motion in a reasonable amount of time. Thus the answers are expected to be specific to a given enzymatic system. The primary goal of the following chapters of this book is to review these two problems: how does the RNA polymerase *engine* operate during elongation of the RNA message? What kind of *brake* comes into action to efficiently counteract this movement at specific termination signals?

Among molecular motors, nucleic acid polymerases have critical specificities, since they have to fulfill two particular and challenging tasks. First of all, RNA synthesis is amazingly processive, since the probability of dissociation of transcribing RNAPs from their templates is generally so low that it has not been accurately measured (P. von Hippel, personal communication). In addition, RNAP must also faithfully copy a DNA template of variable sequence. They successfully meet this second challenge, inserting an incorrect nucleotide not more than once every thousand positions. The copying device operating during elongation and termination remains essentially the same one characterized in the preceding chapters of this book as accurately recognizing and setting up promoter region during initiation of transcription.

RNA polymerases were discovered in the late 1950s and early 1960s (see Foreword). During an initial phase of research, much thinking was guided by the framework of enzymology, similar to the one prevailing at the time in the field of DNA polymerases, the characterization of the catalytic activities as well as the substrate and product specificities. As mentioned by Yager and von Hippel,[2] "although this approach provided valuable insights into the chemistry of RNA polymerization, it was blind to the topological aspects of transcription." Indeed during elongation the enzyme maintains unwound a constant region of the DNA template where nucleotide incorporation takes place, and in this active cleft a significant length of the product RNA remains hybridized with the template strand. Thus the chemistry taking place at the catalytic center, including the translocation of the catalytic site, and the maintenance throughout elongation of these topological features of the nucleic acid counterparts were recognized as intimately correlated. In 1986, after this paradigm shift, Yager and von Hippel could conclude, "the modern view of transcription therefore is a unification of the enzymological and the topological viewpoints."

At this time, no crystal structure of any RNA polymerase was yet available, and the movement of the enzyme relative to its template was only passively registered in bulk enzymatic assays. There was therefore no way to efficiently tackle the main questions listed above, though their importance was clearly recognized. A second shift in paradigm took place when it became possible to compare the free energy released during nucleotide incorporation with the product of the force developed by the molecular motor times its displacement, when conformational changes associated with this movement could be identified on accurate three-dimensional structures, and when the correlation between these various movements could be registered as the elongation process was taking place. We are now at a point in time when the catalytic and the topological viewpoints are more tightly linked through the comparison of their effects on several enzymes, and where mechanical forces, developed by the molecular motor and observed at the single-molecule level, appear as the third essential ingredient for understanding RNA polymerases at work.

The aim of this interlude is simply to give a general perspective of the interrelationship between these three points of view in today's research: catalysis during faithful transcription, the generation of force and motion, and the conservation of a specific, transient nucleic acid structure and topology.

Chapters 4–6 have already dealt with some of these aspects as RNA polymerases escape from their promoters. In particular, in Chapter 4, Tom Steitz has exposed what could be learned in this respect from the comparison of known structures of RNAP–DNA complexes of various origins. Most of the essential arguments on which this introduction is built are critically reviewed in the following four chapters; we simply intend here to provide considerations of general importance for novice readers. Finally, Carlos Bustamante and Jeffrey Moffitt will also carry this message in their concluding chapter, where they specifically point out important bottlenecks for such achievements.

I.2 The Engine

Even in the absence of any experiments involving assisting or hindering forces, inspection of the chemical energy provided by rupture at the α-β phosphodiester bond of the incoming NTP substrate into NMP (incorporated into the growing RNA chain) and pyrophosphate suggests that RNA polymerase has the potential of being a very powerful motor. Peter von Hippel and his colleagues accurately put the theoretical framework for this problem forward well before the experimental set-ups were available to perform the relevant experiments.[2,3]

Let's consider the simplest possible path describing the process of incorporation of a single nucleotide at position $i+1$:

$$E + S \Leftrightarrow ES \Leftrightarrow EP \Leftrightarrow E + P$$

where S is the incoming NTP and P the released pyrophosphate. In this formalism, the equilibrium constant of the reaction:

$$K_{eq} = [RNA_{i+1}][PPi]/[RNA_i][NTP]$$

can be rewritten under steady state conditions as the ratio:

$$K_{eq} = k_f K_m^{PPi} / k_r K_m^{NTP}$$

where k_f and k_r are phenomenological forward and backward rate constants, respectively, and K_m^{PPi} and K_m^{NTP} are the Michaelis constants, respectively, for inorganic pyrophosphate and NTP. The Gibbs free energy associated with the equilibrium constant K_{eq} is known for conditions close to the physiological ones, where the energy stored by the DNA–RNA enzyme complex at position i (before incorporation) and at position $i+1$ (after incorporation but when the translocation of the enzyme relative to the nucleic acid template has taken place to accommodate the next nucleotide) could be considered to be the same. Under these conditions, an equal amount of the i and $i+1$ species is obtained at equilibrium when the ratio of pyrophosphate to NTP concentration is equal to 100.

Erie et al. noticed, however, that after incorporation of a nucleoside monophosphate to the growing chain the situation encountered by the enzyme at positions i and $i+1$ were not strictly the same as the nucleic acid product has probably folded into a helical configuration containing one extra element. They estimated that incorporating a nucleoside monophosphate is as likely as removing a nucleotide via pyrophosphorolysis when the ratio between the pyrophosphate and NTP concentrations is around 30. In other words, the free energy gained by the system during nucleotide incorporation was estimated to be roughly 2 kcal mol^{-1} when substrate and product were present in equimolar concentrations.[3] This is equivalent to $3.4k_BT$, where $T=298$ K is the temperature in kelvin and $k_B = 1.38 \times 10^{-23}$ J K^{-1} is the Boltzmann constant, or also 14 pN nm in units of mechanical work.

In other motors like kinesin or myosin, the ATP fuel used is not incorporated into a growing product, but cleaved at the level of its β-γ phosphodiester bond. The free energy derived from ATP hydrolysis under cellular conditions is, in this case, much more important, of the order of 100 pN nm.[1] However, these other motors must translocate over large distances during one cycle of ATP hydrolysis, respectively 8 or 5 nm. By contrast the catalytic site of RNA polymerase only needs to travel a very short distance, 0.34 nm along the axis of a B-DNA helix, for each incorporation of an NMP at the 3' end of the message. The RNA polymerase engine therefore consumes a poorer fuel than kinesin or myosin but recharges its tank much more often during its vectorial movement. One can easily estimate the maximum force that the RNAP motor should develop if all of the fuel was efficiently used for translocation of the motor relative to its DNA template. This quantity, sometimes called the thermodynamic force, is equal to the chemical free energy divided by the step size. Equal to 12 or 20 pN for kinesin and myosin, respectively, it reaches here about 42 pN, a force roughly equivalent to the weight of 40 red blood cells.

I.2.1 Mechano-chemical Coupling at the Catalytic Site

As soon as it became clear that the movement of DNA relative to an immobilized enzyme could be followed during RNA elongation (Box 1), it became tempting to measure the force really developed by the engine, to compare it with the so-called thermodynamic force, and to obtain a rough estimate of the efficiency of the conversion of chemical energy into mechanical energy. A molecular motor can be experimentally characterized by a force–velocity curve, which specifies the velocity at which the motor advances when assisted or hindered by a specified force. The first experiments reported in Box 1 represent the zero-force extremum of this curve, i.e., the velocity of the unloaded motor. In the following few years, the development of optical tweezers as a means to generate an opposing or assisting load to transcribing RNAP allowed for further characterization of this F–V curve.

Box 1

Ingenious tethering strategies used to attach a motor to one surface (cover slip surface, micron sized polymer sphere) and its substrate to another (*e.g.*, micron-sized polymer sphere) make it possible to read out discrete motor activity from changes in bead-to-coverslip or bead-to-bead distance. Ångström-scale motor steps can now be directly visualized using such techniques. Along with these spatial features, temporal features of the individual reaction cycle (*i.e.*, kinetics) are also naturally recovered in such experiments. Furthermore, analysis and comparison of individual events from both a structural and a kinetic perspective enables one to estimate heterogeneity in the system, which is not possible when working with a bulk macroscopic sample.

The first optical experiments on RNA polymerase were described in 1991 by Schafer *et al.*[4] and given the name TPM (for Tethered Particle Motion) analysis. Researchers attached a small bead at the end of the DNA molecule and then had it transcribed by an RNA polymerase attached to a glass coverslip (Figure I.1A). As the RNA polymerase transcribed the DNA, it reeled the tethered bead in, progressively restricting its motion as the tether shortened. DIC microscopy of the bead provided for an estimate of the bead's range of motion and hence the rate of transcription by RNA polymerase. Although initial measurements were limited to timescales of several seconds, they led to the estimate that in the absence of an external force *E. coli* RNAP transcribed at a rate of approximately $10-20\,\text{nt}\,\text{s}^{-1}$, in agreement with bulk biochemical estimations.

Using a similar geometry Steven Block and co-workers next used an optical trap to manipulate the tethered bead (Figure I.1B).[5,6] The optical trap generates a harmonic potential well for the bead, and by a range of calibration methods (such as application of hydrodynamic flow or measurement of position density distribution) one can determine the linear relation between optical restoring force and position of the bead in the well. In the geometry used, transcribing RNAP pulls the bead progressively out of the optical trap. The bead thus experiences a force that opposes its motion towards the RNAP, progressively loading the motor until it stalls. Within this experimental geometry, relative drift of the optical trap and the sample stage meant that very low velocities could not be measured and base-pair scale displacement of the bead in the trap could not be directly determined.

To overcome this difficulty, Block and co-workers developed the so-called "dumbbell" geometry, which takes the RNAP off of the cover slip surface and attaches it to a second bead held in place by a second optical trap derived from the same laser beam as the first optical trap (Figure I.1C).[11] This helped to eliminate the mechanical drift of the stage relative to the optical trap and allowed for the direct observation of base pair stepping of RNAP along DNA, *i.e.*, a motion on the order of 3.4 Å.

It is then fascinating to follow from one publication to the next[4-11] the progresses made in the instrumentation to achieve this goal. They range from a proper immobilization of RNAP molecules on a suitable surface that did not significantly inhibit its enzymatic activity to the more and more accurate calibration of the optics and to the possibility of modulating the rate of elongation using the applied force, not at the weakest spots during a continuous elongation process, but at will, at any desired position selected at random on the DNA sequence. In parallel, it is important to stress that the data analysis was not trivial. In particular, for experiments performed with the *E. coli* enzyme, data analysis was made complex by the non-regular movement of RNAP along the template DNA sequence. Marked pauses were eliminated from the experimental time-trace by algorithms that differ from one laboratory to the next. The question of the homogeneity of the enzymatic preparations was another matter of serious concern,[8,9] at least during the first few years. Throughout these exciting developments, important qualitative results were already available from the earlier literature, although quantitatively convincing answers to the main questions stressed above rest on the most recent publications:

1. For *E. coli* RNAP working at saturating concentrations of substrate NTPs the average opposing force required to stall the progression of mRNA synthesis was initially reported to be on the order of 14 pN, a value raised to 25 pN when the trapping device was improved. Velocity appeared at the time to drop sharply within a very narrow range about this arrest force, and the distribution of values of this arrest force was rather broad.[6] Note that a stall force of 25 pN would represent 60% of the thermodynamic force estimated above, providing a rough estimate of the motor's efficiency. Thus the RNAP motor of *E. coli* is both powerful and efficient, much more than the yeast enzyme, where the average arrest force was more recently estimated to be of the order of 8 pN.[12]

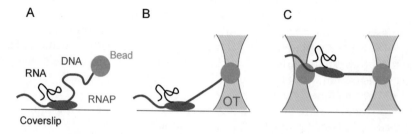

Figure I.1 Progression of single-molecule studies of transcription elongation and termination. (A) Tethered-Particle Motion assay. (B) Optical trapping using a single trap. (C) Dual optical trap, or "dumbbell," configuration. See Box 1 for a brief description of the setups.

2. When in trouble, RNAPs can always resort to an alternative strategy: to pause and to engage itself in a non-productive pathway such as backtracking, where dissociation still rarely occurs. Many questions here are still debated: What are the DNA sequences favoring this switch? How fast do the two pathways interconvert? What is the effect of force on the rate constants relevant to this second pathway (see refs 9, 13 and 14)? These issues are discussed in Chapter 9 as well as in the concluding chapter of this book. Without a clearer understanding of these points, the characterization of RNAPs as an engine will remain incomplete.
3. It is then possible to determine at which step of the pathway the applied force acts. For this the V vs. F curve was determined at different NTP concentrations and edited of pauses. The quality of the experimental device was now good enough to check two important points. The first was to assess that RNAP indeed advances by discrete steps, like any other motor. Analysis of transcription records obtained at low transcription velocity confirmed that the enzyme advances along the DNA by single base-pair nucleotide addition. Furthermore, at any nucleotide concentration the F–V relation was found to reasonably obey a simple Boltzmann relation:

$$v(F) = \frac{v_{\max}}{1 + \exp\left[-\frac{(F-F_{1/2})\delta}{k_B T}\right]}$$

where $F_{1/2}$ is the load for which transcription velocity is exactly one-half of $v_{\max}$. In this equation, the value of δ, the displacement of enzyme along DNA at each step, was found consistent with the rise of one base pair along the B-DNA double helix.[11]

From the dependence of the F–V curves on nucleotide concentration, and as explained in Chapter 9, it appeared that the only reaction steps markedly affected by the external mechanical load during NTP incorporation are the ones corresponding to substrate binding and not those linked to catalysis. This result is best accounted for through the formalism corresponding to the Brownian Ratchet model (introduced in Chapter 4, and discussed in more detail in Chapters 7 and 9) where, at each position on the template, the NTP-free catalytic site can oscillate between the pre- and the post-translocated register. This is true for the *E. coli* as well as for the T7 enzyme, although the likelihood of being in each register differs between the two enzymes. In the case of *E. coli* roughly 2/3 of the fluctuating population of catalytic sites is pre-translocated and can be rescued (made post-translocated) by an assisting force, while this proportion drops to 1/4 in the case of T7 RNAP.[10,11] Thus, after catalytic incorporation at position *i*, presumably as pyrophosphate is released, a major fraction (75% on average) of T7 RNAPs translocates "early" to establish the thermal equilibrium at the NTP-free catalytic site. For *E. coli* RNAP only ~1/3 of the molecules will do this. This prediction now requires a careful examination of the effect of force on the kinetic constants K_m^{PPi} and k_r, related to the reverse path, pyrophosphorolysis.

Having roughly assessed the energetic efficiency of these molecular motors and the steps at which translocation takes place during RNA synthesis does not mean that we have understood the way they work. Here, coupling is the key word, coupling between chemistry and movement. A reflection on ordinary motors, compared to molecular ones, may help explain our viewpoint.

There are wonderful books to explain how macroscopic motors work: take the example of a sewing machine. Sequential impulses on the pedal provide the energy, converted into a smooth rotary motion *via* the chain drive to the thread take-up lever. All of the following steps require the existence of this smooth and unidirectional drive. But an account of the way in which a sewing machine works requires much more, in particular an accurate description of the sequential movements of the various pieces of the machine: the descent of the needle through the fabric, the advance of the point of hook to meet the needle, the formation of a loop in the thread, its location in a recess of the bobbin, and so on. Understanding the mechanism requires one to grasp some essential points about the relative motion of the various parts of the machine: for example why, when the needle moves, the shuttle containing the bobbin is stationary. Or, conversely, why the needle stops moving when the shuttle returns to its initial position. Or why the loop formed within the thread is initially slack but pulled tight in a second step so that the two threads interlock just in the middle of the fabric.[15]

An explanation of "the way molecular motors work" progresses along the same lines of reasoning. Let us take the example of an enzyme that burns ATP to accomplish a vectorial motion on a template. For the process to be fast and selective enough it is first necessary that the energy provided by ATP cleavage be redistributed into numerous steps to facilitate their mutual interconversion. But this is just akin to explaining how an impulse on a pedal has to be converted into a smooth rotary motion and enumerating the pieces involved. A mechanistic explanation of the process needs much more: let us examine the simplest biochemical coupling process, the one occurring during catalysis. Rate enhancement in substrate conversion requires that some of the side chain residues at the catalytic sites are not yet mobilized during substrate binding but recruited at the transition state, with this extra set of interactions providing the free energy required to lower the corresponding activation barrier. In addition, for a process able to generate unidirectional movement, one needs also to understand similar but more elaborate rules: when is this domain at rest, when is it moving? Why is this movement required to generate the subsequent repositioning of another domain? And so on.

Coupling, therefore, has a very specific meaning in enzymology, as formulated long ago by Jencks:[16]

the driving force arises only from the Gibbs free energy change of the overall process minus the work performed; as long as the intermediates are present at a concentration large enough to permit all the steps to proceed at a reasonable rate the energy changes in the individual steps are not very important . . . those differences cannot explain the mechanism of coupled vectorial processes; they serve only to make the process proceed at a reasonable rate for enforcements of the [structural] rules that are responsible for the coupling itself.

Here a set of complex, interrelated questions must then be addressed. What are the conformational changes at the catalytic site that permit cycling of the engine? Are there more global conformational changes that are required for catalysis and cycling? At which specific steps of the cycle can the enzyme translocate (forward but also backward) relative to DNA, and why is motion prevented elsewhere? Are these motions in fact empirically sensitive to an applied external load? Although significant progress has been made in the last decade, none of these questions is definitively addressed yet, and it is our hope that keeping them in mind will make reading the following chapters specifically stimulating.

In fact, illustrations of how these questions can be addressed have been already given for T7 RNAP in Chapter 4. The same questions are discussed in the following chapters in the context of multi-subunit RNA polymerases. For example, crystallographic studies highlight the role of conformational cycling in specific protein domains, notably the trigger loop.

We should not hide, however, how incomplete our present understanding remains: the three-dimensional structures on which our comprehension rests do not necessarily represent the exact tertiary structure of intermediates present on the enzymatic pathway at the stage of substrate or product binding. Indeed, for many nucleic acid polymerases, further rearrangements occur past this point, but prior to catalysis, leading to a tighter grip of the enzyme on its template.[17,18] Furthermore, the passage from one structure to the next, though extremely suggestive, relies for the time being on simple modelizations. Strong divergences of opinion still exist between specialists about the amount of overall tertiary structural changes required to pass from the pre- to the post-translocated state. Structural comparisons advocate that translocation occurs as the T7 enzyme passes from an open to a closed conformation, as described in Figure 4.3. However, estimates of the equilibrium constants between the two alternative registers in the Michaelis complex show that their free energy difference is less than k_BT ($\sim 0.6\,\text{kcal}\,\text{mol}^{-1}$), a fluctuation that could possibly occur as a rapid transconformation within a conserved overall tertiary structure.[18]

This type of dilemma is typical of the molecular world, and completely foreign to the description of an ordinary machine. This is probably the main adjustment we have to make as we travel from one world to the other. k_BT, the amount of energy made available by the thermal bath, is irrelevant for any explanation of *the way things work* at our scale. But, in the case of translocating polymerases, k_BT represents the equivalent of a force of 12 pN acting on a distance of 0.34 nm, the rise of one base-pair of B-DNA.

In addition, 12 pN corresponds already to half the average force necessary to arrest *E. coli* RNAP. Molecular machines working in a Brownian world probably do not strictly follow one-dimensional pathways, so convenient to materialize causal relationships. Mother Nature has coped with this problem and has nevertheless reached a reasonable optimization in the angström and pico-Newton world, as testified by the excellent efficiency of RNAP motors.

I.2.2 Coupling between Translocation and Topology

As elongation proceeds at the catalytic site, the enzyme maintains a relatively fixed structure of nucleic acids both within itself and in its immediate vicinity. Indeed, a "transcription bubble" of apparently constant size accompanies RNAP during translocation. This results from specific structural features of the multi-subunit RNAP that allow it to unwind a downstream base pair and rewind an upstream base pair during translocation. It is coupled to elongation of RNA by one base at its 3' end and disruption of the DNA:RNA hybrid about 9 bp upstream (see Figure A.4). How are these events coupled during their respective movements? Do they evolve in perfect synchrony, or is it possible that thermal fluctuations can mildly uncouple translocation from unwinding/rewinding? The energetics of this problem and their implication in sequence-specific pausing are discussed in Chapter 9.

Three notable differences with respect to the elementary translocation of the catalytic site can be pointed out:

1. Mechano-chemistry operating at the catalytic site involves mainly short-range interactions. Here, however, the relevant length-scales cover entire nanometres (*e.g.*, dimensions of the DNA:RNA hybrid) and thus conformational changes across the breadth of the protein may be coupled to topological changes in the nucleic acid.
2. Advances at the catalytic site are necessarily stepwise, one base-pair rise per cycle. Here, one can imagine behaviors ranging between two extremes. In the case of "*strong*" coupling between all movements, the length of the RNA:DNA hybrid as well as the size of the unwound region remain strictly constant through the whole elongation process. In the alternative case of "*weak*" coupling those distances could fluctuate in such a way as to minimize the energetic cost of translocation. For instance, as RNAP advances from an A+T-rich region to a G+C-rich region this could result in a slight shortening of the unwound region.
3. At the catalytic site, translocation only implies translation by a fixed amount. Here, winding and unwinding events require the application of a transient torque by the enzyme on its template, as discussed in Chapter 6. The protein domains involved, their mode of action (*e.g.*, the way any entanglement of the exiting RNA with the rewinding part of the bubble is prevented) as well as the intensity of the exerted torque are still unclear and constitute important challenges for the future.

It is, however, possible that the situation could simplify itself, as it often happens in the case of other complex enzymatic problems. For instance, two major structures of the enzyme could operate in concert, allowing the enzyme to alternately bind the DNA template with a tight or relaxed grip. The study of the kinetics of binding of DNA polymerases on their templates and the subsequent realization of one step of elongation has led to the proposal that the template is recognized by a so-called "open" protein structure. Conversion of this entity that binds the substrate into a catalytically competent "closed" structure constitutes in this particular case the rate-limiting step of deoxynucleotide addition.[17,18] For RNAP, transient relaxation of crucial protein–DNA contacts at the points where topology has to be modified could perhaps involve such types of "breathing" modes of action.

Such considerations are generically relevant to protein-induced DNA deformations. For instance, the mode of recognition of the lac operator by the lac repressor involves a "search" conformation of the sliding protein, converted only into a tight structure when the proper sequence is encountered.[19] The sudden commitment of elongating RNAPs at pauses, their conversion from an active mode of synthesis into a state where backtracking and one-dimensional searching processes are made possible,[12] could perhaps rest on an analogous concerted transconformation.

I.3 The Brake

A priori, the situation encountered at terminators is somewhat symmetrical to the one previously described at promoters, where transcription starts. However, the detailed mechanisms through which RNAPs form a kinetically competent complex at the promoter and then escape from their initial target are known in much more detail than the converse processes at terminators. The crucial elements of the RNA sequence required to compose an efficient terminator are known: a hairpin structure followed by a seven to nine uridine-rich tract at the 3' end of the growing RNA. The role of these two elements as termination proceeds is still a matter of debate. In Chapter 10, E. Nudler details the path of the nucleic acids within the structure of the *E. coli* enzyme and reviews the evidence favoring a mechanism of progressive invasion by the growing hairpin of a crucial protein site, distal from the catalytic center.

In this controversial field, there is, however, a consensus on the minimal kinetic scheme leading to termination. It is given in Figure 10.1B. When translocating RNAP encounters a termination signal it can either pause or continue transcribing. Because the reverse processes are slow, this "decision" results from kinetic competition between the local rate of elongation, k_n, and the forward rate of entry into the pausing intermediate, k_2. The efficiency of the terminator is thus determined by the ratio $k_2/(k_2 + k_n)$. This paused intermediate can then either return to the main pathway or dissociate; kinetic competition between these two options generally favors dissociation over return.

Modern biophysical methodologies have validated this scheme, and contributed to allocate the role of the AU stretch and the hairpin at the two major steps of this process. Yin *et al.* observed the Brownian motion of a reporter bead tethered to an immobilized *E. coli* RNAP *via* a DNA segment containing a terminator. In agreement with the above kinetic scheme, only two classes of events were observed when RNAP reached the terminator: either the RNAP continued transcribing and "read through" the terminator sequence or it paused and released from the DNA tether.[20] The duration of the pause allowed them to estimate the dissociation rate, k_t, while the termination efficiency, similar to the one observed in parallel bulk assays, gave access to the ratio $k_2/(k_2+k_n)$.

More recent experiments performed in bulk, (see ref. 21 for example) and results obtained in S. Block's laboratory by pulling either on the DNA template or on the RNA transcript[22] together with the data analyzed in Chapter 10, have started to specify structural changes occurring along this scheme. We can then focus on two simple questions, central for the purpose of this book: As RNAP encounters a terminator, why and how is the power of the engine reduced, and what features force this intrinsically processive device not only to stop but also to dissociate?

Commitment involves the entry of RNAP into a pause. The presence of the U tract is sufficient to force RNAP to pause, but it is known that, at this step, the two nucleic acid elements act synergistically, increasing the duration of the pause and therefore providing a longer time window for termination. Also, the reverse rate constant appears to be extremely slow. We can therefore consider that, at this stage, polymerases that have entered this pathway are almost irreversibly committed to terminate transcription. At the structural level, synergy could result from a cross-talk between the two essential elements, the A:U hybrid slowing active elongation, while the hairpin would start its nucleation; formation of the first base-pair at the base of the hairpin would, in turn, further destabilize the AU hybrid.

In principle, long pauses could permit backtracking but, as fully described in Chapter 10, and in ref. 22, the movement of the sliding polymerase is very restricted at this locus, probably inhibited by the presence of a growing hairpin. The locally frozen polymerase could therefore neither return to the main pathway where elongation of the transcript proceeds nor avoid the growth of the RNA hairpin by backtracking. This dual mechanism is aptly summarized by E. Nudler in Chapter 10 as the "spider" strategy: "Trap first, kill later."

I.4 Conclusions

RNA polymerases appear as very particular motors indeed. As they transcribe, they produce an accurate copy of their DNA template by *incorporating* into their biopolymer product the very fuel that drives them. The engine is powerful and efficient, probably operating as a Brownian ratchet, where the binding of the substrate NTP appears to have the determining role. It is also possible that the type of fluctuation in conformation implied by this mechanism operates again during pyrophosphate release, a hypothesis that would permit reconciling schemes derived from the analysis of crystallographic data with single-molecule assays. These motors are not easily arrested by opposing forces because, when they are in trouble, they can always resort to a second mode of translation on their templates, i.e. backtracking. This is probably why the brake exerted by intrinsic terminators appears so sophisticated at the present level of our knowledge.

This present knowledge, fully discussed in the four following chapters, is still extremely scattered. We lack some crucial three-dimensional structures. We speak of pausing complexes but the mechanism by which various pausing sequences exert their effect is still somewhat mysterious. In addition, we do not know in detail the specific characteristics of the inactivated intermediate present at termination, and the crucial long-distance interactions operating at this level. More generally, if we accept Bruce Alberts' motto "ordered movements drive protein machines," we are in general very far from understanding here the generation of order.[23]

Throughout the book, emphasis has been placed on emerging biophysical techniques. We would not like to leave the impression that more classical methods are now outdated. It is still crucial to perform bulk measurements under the conditions used for single-molecule assays. One cannot simply assume or claim that the dataset assembled *via* examination of a series of isolated complexes is necessarily representative of the ensemble under study. Indeed it is important to know rigorously the asymptotic behavior expected from our assays when the size of our sample is increased. Thanks to Avogadro, this is a precious asset provided by bulk experiments.

We have not emphasized enough here the power of genetics. As any child knows, we really do not understand how a machine works, be it a toy or a molecular motor, until we have fully disassembled and in fact destroyed it. There is no better way to pinpoint the rules that, according to Jencks, preside over the functioning of a given motor, than by selecting specific mutants where the postulated coupling process has been perturbed or annihilated. As detailed in Chapter 8, analysis of the mode of action of specific antibiotics directed against RNAPs provides a precious complement to more classical studies.

We have argued that RNAPs constitute a class of their own among molecular engines, the enzymatic zoo operating in concert within the cell. In his famous paper published in 1998 in *Cell*, B. Alberts suggested that a rigorous comparative analysis of protein machines is now required in biology.[23] He wrote:

some of the methodologies that have been derived by the engineers who analyze the machines of our common experience, are likely to be relevant in our field. At the heart of such methods they recognize certain fundamental behaviors in nature and then create an idealized element to represent each of these behaviors...They classify elements as those that store kinetic energy, those that store potential energy, and those that dissipate energy. Any particular part of a machine might be modeled as consisting of one or more of these basic constituent elements.

It is also true, though, as emphasized by François Jacob, that evolution does not proceed like an engineer, working according to a pre-established blueprint, but like a tinkerer, operating through empirical modifications of the material at hand, the various gene products. Yet natural selection very often reaches very similar solutions because such artifacts all bear some common limitations imposed on them by similar physical and chemical constraints.[24] It will certainly be an extraordinary challenge to actively participate in the emergence, from the comparative study of various motors at work, of a classification taking into account the contrasted points of views of the tinkerer and of the engineer.

References

1. J. Howard, *Mechanics of Motor Proteins and the Cytoskeleton*, Sinauer Associates, 2001.
2. T. D. Yager and P. H. von Hippel, *Escherichia coli and Salmonella typhimurium*, pages 1241–1275, Amercian Society for Microbiology, Washington, 1987.
3. D. A. Erie, T. D. Yager and P. H. von Hippel, The single-nucleotide addition cycle in transcription: a biophysical and biochemical perspective, *Annu. Rev. Biophys. Biomol. Struct.*, 1992, **21**, 379–415.
4. D. A. Schafer, J. Gelles, M. P. Sheetz and R. Landick, Transcription by single molecules of RNA polymerase observed by light microscopy, *Nature*, 1991, **352**, 444–448.
5. H. Yin, M. D. Wang, K. Svoboda, R. Landick, S. Block and J. Gelles, Transcription against an applied force, *Science*, 1995, **270**, 1653–1657.
6. M. D. Wang, M. J. Schnitzer, H. Yin, R. Landick, J. Gelles and S. Block, Force and velocity measured for single molecules of RNA polymerase, *Science*, 1998, **282**, 902–907.
7. L. Bai, T. J. Santangelo and M. D. Wang, Single-molecule analysis of RNA polymerase transcription, *Annu. Rev. Biophys. Biomol. Struct.*, 2006, **35**, 343–360.
8. R. J. Davenport, G. J. Wuite, R. Landick and C. Bustamante, Single-molecule study of transcriptional pausing and arrest by *E. coli* RNA polymerase, *Science*, 2000, **287**(5462), 2497–2500.
9. K. Adelman, A. La Porta, T. J. Santangelo, J. T. Lis, J. W. Roberts and M. D. Wang, Single molecule analysis of RNA polymerase elongation reveals uniform kinetic behavior, *Proc. Natl. Acad. Sci. USA*, 2002, **99**(21), 13538–13543.
10. P. Thomen, P. J. Lopez and F. Heslot, Unravelling the mechanism of RNA-polymerase forward motion by using mechanical force, *Phys. Rev. Lett.*, 2005, **94**(12), 128102.
11. E. A. Abbondanzieri, W. J. Greenleaf, J. W. Shaevitz, R. Landick and S. M. Block, Direct observation of base-pair stepping by RNA polymerase, *Nature*, 2005, **438**, 460–465.

12. E. A. Galburt, S. W. Grill, A. Wiedmann, L. Lubkowska, J. Choy, E. Nogales, M. Kashlev and C. Bustamante, Backtracking determines the force sensitivity of RNAP ii in a factor-dependent manner, *Nature*, 2007, **446**, 820–823.
13. N. R. Forde, D. Izhaky, G. R. Woodcock, G. J. L. Wuite and C. Bustamante, Using mechanical force to probe the mechanism of pausing and arrest during continuous elongation by *Escherichia coli* RNA polymerase, *Proc. Natl. Acad. Sci. USA*, 2002, **99**(18), 11682–11687.
14. K. C. Neuman, E. A. Abbondanzieri, R. Landick, J. Gelles and S. M. Block, Ubiquitous transcriptional pausing is independent of RNA polymerase backtracking, *Cell*, 2003, **115**, 437–447.
15. *The way things work, An Illustrated Encyclopedia of Technology*, Simon and Schuster, Part 1, 1967, 241–242.
16. W. P. Jencks, The utilization of binding energy in coupled vectorial processes, *Adv. Enzymol. Relat. Areas Mol. Biol.*, 1980, **51**, 75–106.
17. S. S. Carroll and S. J. Benkovic, *Nucleic Acids and Molecular Biology*, pages 99–113, Springer-Verlag, 1991.
18. K. A. Johnson, Conformational coupling in DNA polymerase fidelity, *Annu. Rev. Biochem.*, 1993, **62**, 685–713.
19. P. H. von Hippel, Biochemistry. Completing the view of transcriptional regulation, *Science*, 2004, **305**(5682), 350–352.
20. H. Yin, I. Artsimovitch, R. Landick and J. Gelles, Nonequilibrium mechanism of transcription termination from observations of single RNA polymerase molecules, *Proc. Natl. Acad. Sci. USA*, 1999, **96**(23), 13124–13129.
21. S. J. Greive, S. E. Weitzel, J. P. Goodarzi, L. J. Main, Z. Pasman and P. H. von Hippel, Monitoring RNA transcription in real time by using surface plasmon resonance, *Proc. Natl. Acad. Sci. USA*, 2008, **105**(9), 3315–3320.
22. M. H. Larson, W. J. Greenleaf, R. Landick and S. M. Block, Applied force reveals mechanistic and energetic details of transcription termination, *Cell*, 2008, **132**, 971–982.
23. B. Alberts, The cell as a collection of protein machines: preparing the next generation of molecular biologists, *Cell*, 1998, **92**(3), 291–294.
24. F. Jacob, Evolution and tinkering, *Science*, 1977, **196**, 1161–1166.

CHAPTER 7

Substrate Loading, Nucleotide Addition, and Translocation by RNA Polymerase

JINWEI ZHANG[a] AND ROBERT LANDICK[b]

[a] Department of Biomolecular Chemistry, University of Wisconsin-Madison, 1550 Linden Drive, Madison, WI 53706, USA; [b] Departments of Biochemistry and of Bacteriology, University of Wisconsin-Madison, 1550 Linden Drive, Madison, WI 53706, USA

7.1 Basic Mechanisms of Transcript Elongation by RNA Polymerase

After the RNA polymerase (RNAP) holoenzyme commences RNA synthesis, the initiation complex isomerizes into the elongation complex (EC) in which the RNAP translocates the template DNA and synthesizes RNA transcripts of up to $\sim 10^4$ (bacterial) to $\sim 10^6$ (eukaryotic) nucleotides. At the ends of transcription units, the EC can be dissociated by either intrinsic or factor-dependent (e.g., rho) termination processes.[1–5] An elongating RNAP maintains a 12–14 base-pair transcription bubble and an 8–9 base-pair RNA-DNA hybrid, places the RNA 3' end in an active site near the center of the enzyme, and catalyzes nucleotide (nt) addition at 15–30 nt s^{-1} (for eukaryotic RNA Polymerase II, or Pol II) or 50–100 nt s^{-1} (for bacterial RNAPs).[6–8] Each round of nucleotide addition requires several steps collectively known as the nucleotide addition cycle.[9–12] Occasionally the nucleotide addition cycle can be interrupted by pause signals encoded in the DNA and RNA that play important regulatory roles such as ensuring synchronization of transcription and translation in bacteria.[1,13,14]

7.1.1 Active-site Features of an Elongation Complex

Recent crystallographic studies of bacterial and eukaryotic RNAPs and transcription complexes have produced a largely consistent picture of the RNAP active site (Figure 7.1).[15-28] Together with extensive biochemical and RNAP-nucleic-acid crosslinking data, these crystal structures suggest that multisubunit RNAPs from bacteria and eukaryotes evolved from a common ancestral core structure composed of two double psi-β barrels (DPBB) by accretion of both conserved and variable (i.e., lineage-specific) sequences.[29] The active site of RNAP resides in the center of the structural core, buried 30–40 Å deep from the enzyme surface (see Structural Atlas). Within the RNAP active site, an invariant aspartic acid triad (β' D460, β' D462, β' D464 in *E. coli* numbering, which will be used throughout this chapter) coordinates two catalytic Mg^{2+} ions (Figure 7.1A). Mg^{2+} I is usually bound in the RNAP active site due to its higher affinity ($K_d = \sim 100 \mu M$), whereas Mg^{2+} II is only loosely coordinated by an EC ($K_d > 10 mM$) and thus may only bind in complex with substrate nucleotide triphosphate (NTP) to participate in the two-metal-mediated catalysis of nucleotide addition.[30] Alteration of any of the three Asp residues dramatically reduces all known catalytic activities of the RNAP active site and strongly decreases Mg^{2+} binding to RNAP, which is consistent with the structural analysis of the active site.[31]

7.1.2 The Nucleotide Addition Cycle

First it is important to define the concept of register of the RNAP, which can be thought of as the position of the protein's catalytic site relative to the 3' end of the RNA:DNA hybrid (Figure 7.2A). We begin in the pre-translocated register: here a new nucleotide has been added to the nascent RNA and still fills the nucleotide insertion site (A site, also referred to as IS, or $i+1$). To add a new nucleotide to the RNA, RNAP must first move the RNA 3' nt from the insertion site to the product site (P site, also referred to as i site) to make room for the incoming nucleotide, a process termed translocation. Then in the post-translocated register the incoming nucleotide triphosphate (NTP) can enter the A site and align with the RNA 3' OH. Subsequently, catalysis of nucleotide addition occurs by S_N2-type nucleophilic attack of the RNA 3' OH on the NTP α-phosphorus atom and involves stabilization of a trigonal-bipyramidyl transition state by two Mg^{2+} ions.[11,32] This two-Mg^{2+}-mediated catalysis appears to be universal for all polynucleotide polymerases (Figure 7.1B).[33] The catalysis step covalently attaches one NMP to the RNA 3' end *via* a new phosphodiester bond, extending the RNA by one nucleotide and generating pyrophosphate (PPi) and a new RNA 3' end (PPi). Release of pyrophosphate from the active site completes the cycle and sets the stage for the next round of nucleotide addition (Figure 7.2A). A mechanistic understanding of each step in this cycle is essential to produce a complete molecular picture of the nucleotide addition process and to shed light on the complex regulatory mechanisms by which intrinsic signals in RNA and DNA and extrinsic factors modify and reprogram the transcribing complex.

7.1.3 Pyrophosphorolysis and Transcript Cleavage

The RNAP active site can catalyze two types of reverse reactions that shorten the RNA chain: pyrophosphorolysis and transcript cleavage.

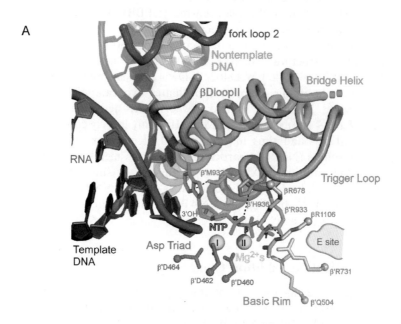

7.1.3.1 Pyrophosphorolysis

When RNAP is in the pretranslocated register the nucleotide addition reaction can be reversed in the presence of pyrophosphate (PPi), removing the 3′ nt in the form of NTP (Figure 7.2B, top panel). This chemical reversal of nucleotide addition is termed pyrophosphorolysis and can progressively shorten the RNA transcript at rates greater than ~1 min^{-1} in the presence of high concentrations of PPi and low concentrations of NTP.[34] Thus pyrophosphorolysis is one mechanism for removal of misincorporated nucleotides, helping ensure transcription fidelity.[12] However, pyrophosphorolysis cannot be responsible for most error correction in RNA *in vivo* for the following reasons. In the *E. coli* cell, PPi levels remain almost constant at ~0.5 mM due to the action of pyrophosphatase. However, the apparent K_d for pyrophosphorolysis is ~1–3 mM. In contrast, NTPs levels are at ~1–3 mM in cells whereas their apparent K_d for RNAP active site are ~50–200 μM.[12,34,35] Thus, the equilibrium constant for nucleotide addition *versus* pyrophosphorolysis is estimated to favor nucleotide addition over pyrophosphorolysis by more than 100-fold at cellular NTP and PPi concentrations.[12,36] Therefore, the pyrophosphorolysis reaction presumably does not occur to any appreciable extent in most cellular conditions.[36] Also as noted by Greive *et al.*, using pyrophosphorolysis to correct errors in RNA does not meet the "Hopfield Criterion," which posits that error corrections are only effective when using a different mechanism than the synthesis mechanism.[37] Finally, as the chemical reversal of nucleotide addition, pyrophosphorolysis

Figure 7.1 RNA polymerase active site and mechanism of catalysis. (A) A structural model of multi-subunit RNA polymerase active site and adjacent structures in the catalytically competent insertion state. Template DNA (black), nontemplate DNA (light orange), RNA (red), bridge helix (cyan), trigger loop (orange); βDloopII (salmon), fork loop 2 (blue), Asp triad (magenta), Mg^{2+} ions (yellow), incoming NTP substrate (green), basic rim (light blue) and E site (gray). Residues in the trigger loop that directly contact the incoming NTP, β′ Met 932, β′ Arg 933 and β′ His 936, are indicated and the interactions are shown as dashed lines. β′ Q504 of the basic rim in *Escherichia coli* corresponds to and is shown as β′ Arg 783 in *Thermus thermophilus*. The structural model is built based on Vassylyev *et al.*, 2007 PDB ID 2O5J,[15] and Wang *et al.*, 2006 PDB ID 2E2H.[17] All molecular graphics were prepared using Pymol (DeLano Scientific, Palo Alto, CA). (B) Mechanism for two-Mg^{2+}-mediated catalysis of nucleotide addition. Two subsites of the active site, P site and A site, are shown. NTP is shown in green. The red arrows illustrate the S_N2-type nucleophilic attack of the RNA 3′ OH on the NTP α-phosphorus atom and the reaction involves stabilization of a trigonal-bipyramidyl transition state by two Mg^{2+} ions.[30,33] (C) Comparison of the pre-insertion state (denoted PS) and the insertion state (A, also referred to sometimes as IS) during nucleotide addition. In the pre-insertion state, the trigger loop (blue) is partially unfolded away from the active site and the NTP is also shown in blue. In the insertion state, folding of the trigger loop (orange) shifts the NTP phosphates (green) towards the active site and facilitates catalysis. Note the differences in positioning of Mg^{2+} II in the pre-insertion state (blue sphere) and insertion state (green sphere).

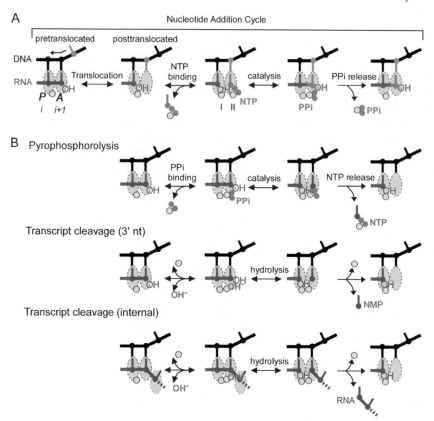

Figure 7.2 Nucleotide addition cycle and alternative reactions catalyzed by RNAP. (A) Steps of the nucleotide addition cycle include translocation, NTP-binding, catalysis and pyrophosphate (PPi) release. Template DNA (black), RNA (red) Mg^{2+} (yellow sphere) and NTP (green). (B) Alternative reactions catalyzed by RNAP include pyrophosphorolysis (top panel) and transcript cleavage at the 3′ nt (middle panel) or internal transcript cleavage (bottom panel). Pyrophosphorolysis occurs in the presence of high concentrations of PPi and low concentrations of NTP. RNAP enters the pretranslocated register and binds PPi at the A site. Then the non-bridging oxygen of PPi initiates nucleophilic attack on the scissile phosphorus atom, forming NTP which then exits from the active site. Transcript cleavage of the 3′ nt or internal transcript cleavage occurs by nucleophilic attack by the OH^- group on the scissile phosphate, hydrolysis of the phosphodiester bond and removal of NMP or polynucleotide RNA.

requires the same active-site geometry as nucleotide addition, and thus is inefficient for dealing with a misincorporated, misaligned RNA 3′ nt. A different mechanism, namely transcript cleavage, is likely utilized by the cell to remove misincorporated nucleotides.

7.1.3.2 Transcript Cleavage

The transcript cleavage reaction hydrolytically removes one or more nucleotides from the RNA 3′ end, rescues complexes arrested due to misincorporation, and contributes to transcription fidelity.[38,39] Transcript cleavage is mechanistically distinct from pyrophosphorolysis because in transcript cleavage the nucleophile that initiates the nucleophilic attack on the α-phosphorus is the OH^- group instead of pyrophosphate. Nevertheless, all three types of reactions – nucleotide addition, pyrophosphorolysis and transcript cleavage – require two catalytic Mg^{2+} ions to stabilize a trigonal-bipyramidal transition state during catalysis (Figure 7.1B).

Transcript cleavage is central to the rescue of arrested transcription complexes. During transcript elongation, incorporation of incorrect nucleotides (termed "misincorporation") into the RNA 3′ nt occurs at probabilities ranging from 10^{-3} to 10^{-5}, leading to formation of arrested transcription complexes.[12,40] In these arrested complexes, the incorrect RNA 3′ nt is unable to base pair with the template DNA base. Lack of base pairing at the 3′ end of the RNA:DNA hybrid can in turn cause the RNAP to slide back (upstream) relative to the DNA-RNA scaffold, placing the RNA 3′ proximal region containing the incorrect nucleotide(s) in the secondary channel, a process termed backtracking.[41] Backtracked transcription complexes are unable to incorporate nucleotides but can be rescued by hydrolytic removal of backtracked RNA *via* nucleophilic attack of OH^- on the scissile phosphate, placing a newly generated 3′ nt in the P site, ready to resume nucleotide addition (Figure 7.2B).[42] Such a reaction also requires Mg^{2+} II, which binds RNAP only weakly in the absence of NTP. Therefore, higher concentrations of Mg^{2+} or elevated pH (providing more OH^- as nucleophiles) accelerate transcript cleavage.[30,38,42]

In the presence of transcription cleavage factors such as GreA and GreB in bacteria and TFIIS in eukaryotes, the transcript cleavage reaction can be accelerated by more than 3000-fold.[24,39,43–45] These cleavage factors bind RNAP and insert a reactive domain into the RNAP secondary channel, placing several acidic residues in close proximity of the active site.[39,46,47] These factors can then induce transcript cleavage, presumably by stabilizing Mg^{2+} II binding.[30,39] In addition, the identity of the RNA 3′ nt affects its ability to stabilize Mg^{2+} II and to promote transcript cleavage. A nucleotide misincorporated at the 3′ end of the RNA appears to stimulate transcript cleavage at the penultimate position, thus providing an intrinsic proof-reading mechanism (Chapter 8).[48] In summary, transcript cleavage removes misincorporated nucleotides and converts arrested complexes back into active ECs.

7.1.4 Regulation of Transcript Elongation by Pauses

Occasionally the nucleotide addition cycle can be interrupted by pause signals encoded in the DNA and RNA that play important regulatory roles.[1,13,14] These pause signals embedded in the DNA and RNA sequence, coupled with the effect of extrinsic regulators, can transiently interrupt nucleotide addition in a fraction of elongating RNAPs and cause transcriptional pauses.[13,14,49,50] Transcriptional pauses ensure coupling of transcription and translation in bacteria, allow time for proper folding of RNA secondary and tertiary structures, facilitate properly-timed loading of elongation regulators, and precede intrinsic and regulator- (*e.g.*, Rho) dependent termination.[51] Although no general consensus sequence for pausing has emerged, transcriptional pauses are strongly sequence-dependent.[14,50,52,53]

Different types of pauses appear to result from a common initially paused state sometimes called the elemental pause, and are categorized based on different ways in which the elemental pause is extended into a stabilized pause. Thus one distinguishes hairpin-stabilized pauses (*e.g.*, the *his* pause from the histidine biosynthesis operon leader region), backtracking pauses (*e.g.*, the *ops* pause), and regulator-dependent pauses (*e.g.*, the promoter-proximal σ pause).[14,50] Structurally, the elemental pause results from a rearrangement of the protein and nucleic acid components in the active site, leading to inhibition of nucleotide addition. In the elemental pause the system is likely in the pre-translocated register.[13]

At the best understood hairpin-stabilized *his* pause site, interactions between RNAP and nucleic acid sequences (in the downstream DNA, in the active site, in the RNA:DNA hybrid, in the pause RNA hairpin, and in the 2–3 nt spacer between the hybrid and the hairpin) collectively contribute to the active site rearrangement in the elemental pause. These components plus the pause RNA hairpin subsequently stabilize the paused state, resulting in a long-lived pause.

Kinetically, pauses represent local free energy minima that precede relatively high free energy barriers to nucleotide addition. Kinetic models that seek to describe transcript elongation quantitatively and to predict pause sites based on sequence context are discussed in Chapter 9.

7.2 Structural Basis of NTP Loading and Nucleotide Addition

A single RNAP active site, deeply buried more than 30 Å from the RNAP surface, is responsible for all known activities of RNAP, including nucleotide addition, pyrophosphorolysis and transcript cleavage.[25,31] Key conformational changes in RNAP, involving protein secondary structure motifs, DNA, and RNA in the active site, mediate nucleotide loading and addition (Figure 7.1). Interference of these key movements by regulatory signals, extrinsic factors, and small molecules such as streptolydigin can dramatically impact or even completely shut off catalysis of nucleotide addition.[13,15,19,20] Therefore, it is crucial to understand the structural basis of nucleotide loading and addition before one can understand regulation of transcript elongation. However, detection of such small movements (on the order of 1–10 Å) buried 30–40 Å deep from the RNAP surface at 10^{-2}–10^{-3} s temporal resolution is no easy task. Nonetheless, recent high-resolution crystallographic studies and extensive biochemical work have yielded considerable insight and have begun to allow visualization of the nucleotide addition process in atomic detail.

An important question raised by these crystal structures is how substrate NTP initially gains access to the active site. The only evident route that connects the active site to the RNAP surface based on available RNAP crystal structures is a narrow funnel-shaped pore termed the secondary channel (see Figures A.2 and A.3).[27,28,54] Recent crystallographic, biochemical and simulation studies have identified an NTP "entry" site (E site) located at the inner pore of the secondary channel, lending support to the secondary-channel model of NTP entry.[22,54,55] However, based on biochemical studies of human Pol II, Burton and co-workers proposed an alternative NTP-entry model in which NTPs initially base-pair to downstream DNA template and load through the main channel.[23,56,57] These two models are further discussed in Chapter 8.

Within a ~15 Å radius of the active site, the following conserved RNAP structural features have been identified as important for RNAP catalytic activities and regulation: a long α-helical segment termed the bridge helix (β′780-815) that spans the main channel of RNAP and a flexible trigger loop (β′928-942 and β′1131-1145) just downstream of the active site. The bridge helix and the trigger loop have been observed to move toward and away from the active site or to fold and unfold near the active site, respectively, in structures containing or lacking nucleic acids. Additional structures identified as important for RNAP function and regulation include a basic rim (βR678, βR1106, β′Q504, β′R731, which contact substrate phosphates), a fork loop 2 (β533–541, termed βDloopI or streptolydigin-binding loop, proposed to bind NTP and control nucleotide addition) and a βDloopII (β558-575, which contacts the RNA:DNA hybrid) (Figure 7.1A).

These structures have led to several models for the crucial movements in the RNAP active site, which we will describe in the following sections.

7.2.1 Bridge-helix-centric Models of Nucleotide Addition and Translocation

Several structural models of nucleotide addition have postulated that a bridge helix oscillation between a uniform α-helical conformation (cyan, Figure 7.3A) and a locally unfolded conformation (magenta, Figure 7.3A) sterically drives translocation of DNA through RNAP and thereby controls the nucleotide addition cycle.[25,27,58,59] In these models, movements of the trigger loop located adjacent to the bridge helix are proposed to facilitate bridge helix oscillation and thus indirectly affect nucleotide addition *via* effects on bridge helix conformation.[58,59] These bridge-helix-centric models are based on the crystallographic observation of different conformations of the bridge helix in bacterial RNAP core/holoenzymes and in yeast Pol II. In the yeast Pol II structures, the bridge helix is uniformly alpha-helical[17,27,55,60,61] (Figure 7.3A, cyan), whereas in *Thermus aquaticus* core enzyme[62] and *Thermus thermophilus* holoenzyme[25] the bridge helix is locally unfolded or flipped out in the middle

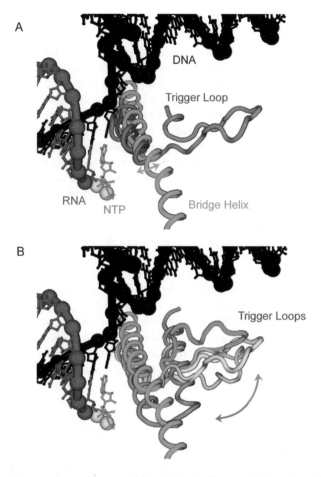

Figure 7.3 Proposed movements of the bridge helix and of the trigger loop during nucleotide addition. (A) Bridge-helix movements. The bridge helix is proposed to oscillate between an α-helical conformation ("straight," in cyan, modeled as in Gnatt et al.,[27] PDB ID 1I6H) and an unfolded conformation ("bent" or "flipped out," in magenta, modeled as in Vassylyev et al.,[25] PDB ID 1IW7) driving DNA translocation through RNAP. Alternatively, the bridge helix can assume a shifted conformation (in blue, modeled as in Wang et al.,[17] PDB ID 2E2H) and play roles in stabilizing trigger loop folding. Note that the unfolded bridge helix conformation (magenta) is sterically incompatible with the template DNA base in the $i+1$ site (black). The frequency of the bridge-helix oscillation (cyan arrow) therefore could determine the rate of nucleotide addition and control transcript elongation.[58] (B) Trigger-loop movements. The trigger loop is proposed to oscillate between a folded α-helical hairpin conformation [orange, modeled as in Vassylyev et al.[15] (PDB ID 2O5J) and Wang et al.[17] (PDB ID 2E2H)] and multiple unfolded loop conformations [green conformation is modeled as in Vassylyev et al.[25] (PDB ID 1IW7); yellow conformation is modeled as in Tuske et al.[19] (PDB ID 1ZYR); light orange conformation is modeled as in Kettenberger et al.[22] (PDB ID 1Y1V)]. Trigger loop folding brings its residues β' 933–936 into direct contact with the substrate NTP (Figure 7.1A), closing the active site, and aligning the RNA 3′ OH, NTP phosphates, and two Mg^{2+}s to facilitate catalysis.[15] Therefore, folding or movement of the trigger loop (orange–green arrow) controls nucleotide addition.

(Figure 7.3A, magenta). In this unfolded conformation the domain would sterically clash with the $i+1$ template DNA base based on these models and structures of the EC.[25]

These structural observations led to proposals that bridge helix unfolding pushes the $i+1$ template base to the i position, thus sterically driving DNA translocation through RNAP. Epshtein *et al.* reported that the RNA 3′ nt (substituted with the crosslinkable nucleotide analog 2′-deoxy-3′-isothiocyanate) crosslinks to β′ R789 of the bridge helix and β′ M932 of the trigger loop and postulated that an unfolded ("bent") bridge helix would place β′ R789 in the A site, evicting the pretranslocated RNA 3′ nt from the A site and preventing substrate NTP binding in the A site.[59] Therefore, it was suggested that the bridge helix oscillation involving the residues β′ R789, β′ T790, β′ A791, supported by a coordinated trigger loop movement involving β′ M932, could not only drive DNA/RNA translocation but could also define a "swinging gate" mechanism that controlled NTP entry.[59]

Further work by Bar-Nahum *et al.* isolated two dominant lethal substitutions in the trigger loop, G1136S and I1134V, which increase or decrease RNAP elongation rate, respectively. Exonuclease III footprinting with these mutants suggests that the fast G1136S RNAP is more likely to be in the post-translocated state than wild-type RNAP, whereas the slow I1134V RNAP is more likely to be in the pretranslocated state than wild type.[58] Therefore, these substitutions in the trigger loop were suggested to affect the folding/unfolding dynamics of the bridge helix and thus indirectly alter the translocation register and elongation rate of RNAP by affecting bridge-helix-controlled translocation. The RNAP inhibitor streptolydigin has been proposed to inhibit nucleotide addition by binding to and interfering with this conformational cycling of the bridge helix.[19]

The evidence supporting the bridge-helix-centric models faces certain objections. First, in recent yeast EC structures trapped in the post-translocated register, the observed displacement of the bridge helix towards the RNA:DNA hybrid (2.0 to 2.7 Å) takes place without unfolding.[17,55] Bacterial EC structures complexed with non-hydrolyzable NTP analogue AMPcPP or AMPcPP and streptolydigin did not reveal the unfolding (bending) of the bridge helix either.[15] To date, the unfolded bridge helix has only been observed in bacterial core or holoenzymes with no nucleic acid or substrate NTPs and has never been observed in yeast Pol II structures trapped in a number of states.[17,22,24,27,60] These structural findings cast doubt on the mechanistic relevance of an unfolded bridge helix.

Second, biochemical evidence supporting the bridge helix oscillation relies on the usage of artificial crosslinking groups such as 2′-deoxy-3′-isothiocyanate nucleotide analog, which lacks both 2′ OH and 3′ OH and is such a poor substrate for RNAP that Mn^{2+} rather than Mg^{2+} has to be used to boost incorporation efficiency.[58,59] Usage of Mn^{2+} is known to increase misincorporation and thus may cause distortions of the active-site structures and lead to artifactual crosslinking results.[63–65]

Third, it remains unclear how substitutions in the trigger loop isolated by Bar-Nahum *et al.*, which are located in the bridge-helix-distal side of the trigger loop, can dramatically alter the bridge-helix conformation. Indeed, to examine the possible level of involvement of bridge helix unfolding in nucleotide addition, Toulokhonov et al.[13] characterized a mutant RNAP carrying the A791G substitution in the center of the unfolded segment of the bridge helix. Predictions based on folding studies with isolated peptides suggested that such an alteration from helix stabilizer Ala to helix destabilizer Gly should destabilize the alpha-helical conformation by $\sim$20-fold.[66] However, the A791G mutation only slows nucleotide addition at a defined pause site by $\sim$30%[13] and supports *E. coli* growth.[19] Furthermore, a more drastic disruption of the bridge helix obtained by replacing the "oscillating" segment of five amino acids with a 3-glycine linker [$\Delta\beta'$K789-S793Ω(Gly)$_3$] gave a $\sim$90-fold reduction of the nucleotide addition rate at a pause site whereas deleting the trigger loop had a $\sim$460-fold effect.[13] Disrupting the bridge helix in an RNAP lacking a trigger-loop only reduced the rate of NTP incorporation four-fold. These results are at odds with the proposed pivotal role of bridge helix oscillations in nucleotide addition and pausing.[13,19,25,27,58]

7.2.2 Central Role of the Trigger Loop in Nucleotide Addition and Pausing

In contrast to the bridge-helix-centric structural models, recent biochemical and structural studies of bacterial and yeast ECs revealed that the trigger loop, located close to the bridge helix at the inner pore of the secondary channel, plays the key role in NTP loading, nucleotide addition and regulation of elongation (Figures 7.1A and 7.3B).[13,15,17] Movement of the trigger loop is proposed to be associated with the rate-limiting active-site rearrangement that aligns the substrate NTP and thus facilitates catalysis. It may also orchestrate the creation of paused transcription complexes when its movement is restricted.[13,15–17]

In earlier RNAP crystal structures, the conserved trigger loop is disordered, which is indicative of a high degree of mobility. Among the structures in which the trigger loop is modeled, different conformations have been observed depending on the particular enzyme, conditions of crystallization, and presence or absence of transcription factors and small molecules such as streptolydigin (Figure 7.3B).[15–17,19,20,22] Recent crystal structures of bacterial EC complexed with AMPcPP and of yeast Pol II EC with 3′deoxy RNA complexed with GTP both reveal a novel trigger loop conformation, a folded α-helical hairpin (Figures 7.1A and 7.3B) that is reminiscent of the O, O′ helices in DNA polymerases.[15,17] The folded trigger loop makes extensive interactions with the nearby bridge helix: together they form a three-helical bundle. Trigger loop folding brings its conserved residues β' Met932-His936 into direct contact with the incoming NTP (contacts occur both at the base and at the phosphates) and properly aligns the RNA 3′ nt OH, NTP triphosphates, and two catalytic Mg^{2+} ions for catalysis (Figures 7.1A and 7.3B).[13,15,17]

RNA–protein crosslinking experiments provided biochemical evidence that trigger loop folding occurs at least in some ECs. In non-backtracked, active transcription complexes halted at a nonpause site, the crosslinkable nucleotide analogue 4thioUMP incorporated at the RNA 3′ nt crosslinks to the A site.[67] This crosslink to the A site was shifted to the trigger loop when the halted EC gradually went into arrest, indicative of an active-site reconfiguration. More recent work showed that, at the non-backtracked *his* pause site, 4thioUMP incorporated at the RNA 3′ nt in the A site primarily crosslinks to the β′ Arg933-His936 segment of the trigger loop.[13] These results provide strong support for the existence of a folded trigger loop conformation and its close proximity to the RNA 3′ nt. The crosslinkable NTP analogue 4thioUMP used in these studies is superior to 2′-deoxy-3′-isothiocyanate used by other studies[58,59] in that it is a much better substrate for RNAP, does not require Mn^{2+} for efficient incorporation, and only moderately affects RNAP catalytic activity.[13]

Various alterations of the trigger loop establish its central role in nucleotide addition. Single amino-acid substitutions on the trigger loop profoundly alter elongation rate, transcriptional pausing, transcript cleavage, pyrophosphorolysis and termination.[13,15,58,59,68] Substitutions within the trigger loop that do not involve the NTP-interacting residues but that should specifically destabilize the folded conformation drastically reduce rates of nucleotide addition by as much as ∼10 000-fold with minimal effects on apparent NTP affinity.[13,15] Deletion of the trigger loop in either *E. coli* or in *T. aquaticus* also reduces nucleotide addition by a factor of ∼10 000 with minimal effects on apparent NTP affinity.[13,20] These biochemical results suggest that trigger loop folding is required for catalysis but not for NTP binding in the active site.

The trigger loop also plays important roles in transcriptional pausing. Various alterations of trigger loop residues greatly alter pausing. Importantly, deletion of the trigger loop from *E. coli* RNAP significantly compromised its ability to recognize a pause signal, and largely abolished the ∼100-fold difference in nucleotide addition rates at pause sites and non-pause sites.[13] These observations are consistent with the view that restriction of the trigger loop's movement contributes to formation of the paused state. Thus, the trigger loop not only serves as a facilitator of nucleotide addition, but also serves as a controller of the active site in response to regulatory signals such as pause signals embedded in the DNA and RNA.

7.2.3 A Trigger-loop Centric Mechanism for Substrate Loading and Catalysis

The picture that emerges from the aforementioned structural and biochemical studies supports the following trigger-loop model of nucleotide addition (Figures 7.3B and 7.4).[13,15,17] First, while the trigger loop is unfolded, NTP initially lands at an entry site (E site[55]) located at the junction of the main channel with the secondary channel. This initial binding can even occur in the absence of the trigger loop. After initial NTP binding, the trigger loop partially folds and NTP enters the A site. However, this first conformation of NTP in the A site is not aligned for catalysis and is termed a "pre-insertion" state

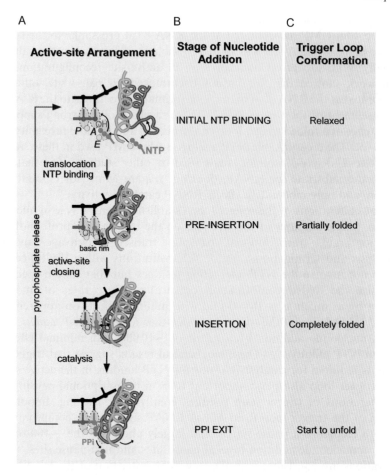

Figure 7.4 A trigger-loop-centric model of nucleotide loading and addition. The left-hand column depicts cartoon views of active-site geometry at each stage of the nucleotide addition cycle (middle column) and corresponding conformation of the trigger loop (right-hand column). Concomitant with or following translocation, incoming NTP initially binds at the E site when the trigger loop is in a relaxed loop conformation. NTP then moves into the A site, possibly inducing partial folding of the trigger loop and together constituting an open "pre-insertion" state, where NTP base-pairing with DNA template can be checked whereas catalysis is inhibited. The basic rim (blue) interacts extensively with the NTP phosphates at this stage (black lines) and likely prevents formation of a catalytically-competent insertion state. Incorrect NTPs can then be checked and rejected from the active site before a correct NTP induces complete folding of the trigger loop, closing the active site and aligning the RNA 3′ OH, NTP phosphates, and Mg^{2+} ions, facilitating catalysis of nucleotide addition at the insertion site. Following catalysis, PPi release from the active site removes the bridging interactions between the active site and the trigger loop. The trigger loop thus likely spontaneously unfolds in the absence of PPi and the active site is now ready to undergo another round of nucleotide addition.

(Figure 7.1c). Here the basic rim plays an important role: composed of multiple arginine residues that directly contact the NTP's phosphate groups (specifically, residues βR678, βR1106, β'Q504, β'R731) it is thought to interact with and prevent the phosphate groups from aligning with the RNA 3' OH and Mg^{2+} cations and undergoing premature catalysis. Thus the basic rim could play a role in proofreading the incoming NTP.

After binding of a correct NTP in the A site, the trigger loop completely folds into an α-helical hairpin, closing the active site and aligning the incoming NTP triphosphates with the RNA 3' OH and Mg^{2+} ions to facilitate catalysis (Figure 7.1A). This key conformational change of the trigger loop appears to be associated with the rate-limiting step in nucleotide addition. S_N2-type nucleophilic attack of the RNA 3' OH on the NTP α-phosphorus atom then ensues and the two properly-aligned Mg^{2+} ions stabilize a trigonal-bipyramidal transition state (Figure 7.1B). Lastly, PPi exits from the active site, the trigger loop reverts to its unfolded conformation, and the enzyme translocates DNA by one bp to await the next round of nucleotide addition. In this model, the bridge helix plays a secondary role by facilitating trigger loop folding *via* formation of a three-helical bundle.[13,15]

In conclusion, extensive structural and biochemical studies have revealed that trigger loop folding or movement, rather than the unfolding of the bridge helix, controls nucleotide loading and addition. A trigger-loop-centric model incorporating a two-step substrate loading mechanism appears to explain most experimental observations and is consistent with conservation of the fundamental mechanism of nucleotide addition throughout bacterial and eukaryotic RNAPs.[13,15]

The trigger loop has also been proposed to participate in ensuring transcription fidelity and preventing misincorporation.[17] It remains unclear what exact role the trigger loop plays in transcription fidelity, as RNAP with no trigger loop can still distinguish among different NTPs.[13] Quantitative analysis of transcription fidelity in mutant RNAPs with various trigger-loop alterations is necessary to examine the trigger loop's involvement in fidelity control. Specific mechanisms of fidelity control are discussed in more detail in Chapter 8.

7.3 Models of Translocation: Power-stroke *versus* Brownian Ratchet

RNAP is a powerful and processive mechanoenzyme that converts chemical energy stored in NTP into the mechanical work of threading DNA through RNAP. RNAP can generate forces as high as 30 pN[69] and can synthesize RNAs as long as 2000 kb.[70] Each nucleotide addition cycle involves a translocation step in which the DNA moves relative to RNAP by ~3.4 Å, the rise in B-DNA of a single base-pair. In the active site, the RNA 3' nt moves from the A site to the P site, one base pair is unwound in the downstream DNA duplex and another base pair is rewound in the upstream DNA. As already mentioned in Chapter 4, two types of models offer mechanistic explanations for the

A Power-stroke model

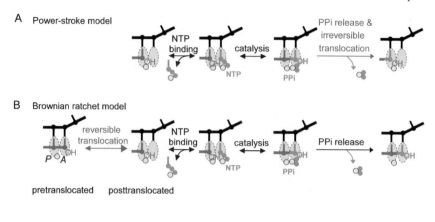

B Brownian ratchet model

C Modified Brownian ratchet model with second NTP binding site

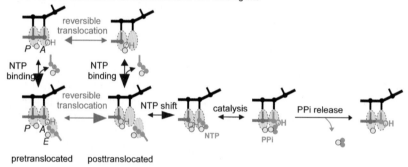

Figure 7.5 Power-stroke and Brownian ratchet models of RNAP translocation. (A) A power-stroke model where translocation is tightly coupled to irreversible PPi release. (B) A Brownian ratchet model where reversible interconversion between the pre- and post-translocated states occurs faster than nucleotide addition. NTP binding favors the post-translocated state and bias the EC towards forward translocation. (C) A modified Brownian ratchet model where translocation and NTP binding can occur in either order. This model requires the existence of a secondary NTP binding site to accommodate the possibility of nucleotide binding when the EC is in the pre-translocated state. Differences in the size of arrow heads illustrate the directionality of change in equilibrium.

translocation process: "power-stroke" models and "Brownian ratchet" models (Figure 7.5).[34,58,71–75]

7.3.1 Key Distinctions between Power-stroke and Brownian Ratchet Models

Power-stroke models posit that translocation occurs by directly coupling the chemical energy of triphosphate hydrolysis to irreversible mechanical translocation (Figure 7.5A). In contrast, Brownian ratchet models postulate that the pretranslocated and post-translocated states interconvert more rapidly than nucleotide is added and that NTP binding in the A site biases RNAP towards forward translocation (Figures 7.5B and 7.5C). The key distinction between these two types of models lies in whether the pretranslocated and post-translocated states interconvert fast enough with respect to nucleotide incorporation. The rate of interconversion is determined by the height of the energy barrier between these two states or translocation registers. This energy barrier has not been directly measured, although there is evidence that PPi biases the translocation register from post-translocated state to pretranslocated state and that NTP has the opposite effect.[34,58] Thus, definitive evidence that would unequivocally distinguish these models does not yet exist. This being said, biochemical, single-molecule and modeling efforts have been undertaken to address the question and have yielded considerable insights into the mechanism of translocation.

7.3.2 Power-stroke Models

A power-stroke model for RNAP translocation was proposed for T7 RNAP based on crystallographic snapshots of transcribing T7 RNAP ECs complexed with either incoming NTP or PPi.[71] The observation of a closed form of the O helix and fingers domain in the presence of PPi and of a more open form in the presence of NTP led to the proposal that irreversible PPi release destabilizes the closed conformation causing opening of the O helix, which then drives translocation by physically pushing the RNA:DNA hybrid from the A site to the P site (Chapter 4). The essence of this model is that irreversible PPi release is tightly coupled with translocation *via* the defined motion of the O-helix. This would ensure that the power-stroke operates unidirectionally, converting chemical energy into mechanical work of translocation (Figure 7.5A). Similar power-stroke models have been proposed for other molecular mechanical systems such as myosin-actin contraction and mitochondria F1-ATP synthase rotation. In such systems, the release of phosphates promotes certain well-defined conformational changes that are coupled to mechanical movement.[76,77]

7.3.3 Brownian Ratchet Models

Although the power-stroke model agrees well with the T7 RNAP EC structures, it has received little support from solution biochemistry or single-molecule studies. Instead, models analogous to Brownian ratchets have received general support from experimental and theoretical studies of both T7 and bacterial RNAPs.[34,58,72–75] Sousa and others lent biochemical support for Brownian ratchet models by showing that (1) halted ECs can slide on DNA, indicating that translocation is reversible; (2) roadblocks preventing RNAP movement reduce the apparent NTP affinity for RNAP, presumably by decreasing the population of post-translocated ECs (recall that it is in the post-translocated register that the unoccupied A site can bind NTPs); and (3) rapidly elongating and halted ECs are equally sensitive to PPi, suggesting that the occupancy of the pre- and post-translocated states are not significantly different in the two cases and, therefore, that interconversion can be uncoupled from nucleotide addition.[34,72,73]

Furthermore, by using single-molecule force-clamp transcription assays where the resolution reaches the level of a single base pair, Abbondanzieri et al. characterized the force–velocity behavior of E. coli RNAP. They reported results in favor of Brownian ratchet models over power-stroke models in which PPi release is tightly coupled with translocation.[78] Specifically, statistical analysis of multiple single-molecule transcription traces at various assisting or hindering forces suggested that higher hindering forces are required to slow elongation at increasing concentrations of NTP. This suggests that NTP binding is coupled to translocation, which is consistent with Brownian ratchet models. Power-stroke models, in which translocation is driven by PPi release but not by NTP binding, do not predict significant effects of NTP concentration on the force-dependence of translocation. In addition, the observed insensitivity of elongation rate to [PPi] also argues against a power-stroke model in which translocation is coupled to PPi release.[69] In fact, the measured force–velocity relationship is well-described by a modified Brownian ratchet model in which either translocation or NTP binding can occur first prior to NTP alignment in the A site and catalysis (Figure 7.5C). This modified model requires the existence of a secondary NTP binding site that can accommodate NTP binding in pretranslocated ECs, which fits well with the observation of initial NTP binding in the E site observed in yeast Pol II.[54,55] An allosteric site proposed by Erie and co-workers (Section 7.4.1) would also be consistent with this observation, although there is currently no structural evidence for an allosteric NTP-binding site (for further discussion of all these points see Chapter 9). Similar single-molecule, force-clamp studies conducted with T7 RNAP also supported the general Brownian ratchet model.[79] Altogether these results suggest that the fundamental Brownian ratchet mechanisms of RNAP translocation may be conserved between single-subunit T7 RNAP and multi-subunit RNAPs.

7.3.4 Technical Outlook in Detecting the Precise Translocation Register

A mechanistic understanding of the translocation process requires the ability to precisely and non-invasively measure the translocation register of RNAP. A major concern in interpreting experiments that try to probe the translocation register is that most techniques can perturb the translocation register one is trying to measure. For example, exonuclease III footprinting of ECs can be biased by exonuclease III interacting with RNAP and altering its original register. $KMnO_4$ and other chemical probing methods can suffer from insufficient spatial or temporal resolution since direct observation of single-base-pair translocation during rapid elongation would require $\sim 10^{-10}$ m spatial resolution and ~10^{-3} s temporal resolution or better and also may chemically damage RNAP and alter translocation register. Guo *et al.* have reported sub-Å resolution in measurement of the average position of halted T7 ECs using an RNAP-tethered chemical nuclease (Fe^{2+}-BABE) to generate detectable cleavage products of radiolabeled RNA/DNA strands.[34] This approach is rapid enough, precise (sub-Å spatial resolution) and, more importantly, does not perturb the translocation register. Future adaptation of this technique for multi-subunit RNAPs would be highly beneficial. The key technical challenge of utilizing such an assay is to guarantee that a single chemical nuclease is specifically tethered to the same amino acid (a reactive cysteine) for each and every RNAP molecule. To date, there have been no documented attempts to examine how many of the cysteines in multi-subunit RNAPs are sufficiently exposed for conjugation to Fe^{2+}-BABE.

Fluorescence-based assays offer another promising technique to measure rapid translocation in real time. Successful applications of fluorescent nucleotide analogues have enabled detection of opening or closing of a single base pair.[80] Extension of these studies could lead to direct observation of fluorescence changes during rapid translocation. Alternatively, FRET-based distance measurements in rapidly elongating RNAPs could allow for time-resolved measurements of translocation (Chapter 5).

7.4 Kinetic Models of Nucleotide Addition

Landmark studies of rapid nucleotide addition by the Erie and Burton groups have demonstrated that the rate of nucleotide addition by bacterial and human RNAPs exhibited a greater-than-hyperbolic dependency on substrate NTP concentration.[23,35,56,81] Furthermore, the rate of incorporation of a single-nucleotide displays a "biphasic" nature at intermediate NTP concentrations. This suggests the co-existence of two molecular species – a faster one and a slower one – that do not rapidly interconvert.[81] To explain these results, Erie and colleagues proposed a kinetic model for NTP addition termed the allosteric NTP binding model, whereas Burton and colleagues have proposed an alternative model termed NTP-driven translocation.

7.4.1 Allosteric NTP Binding Model

Erie and colleagues proposed that two NTPs can bind RNAP simultaneously at two distinct sites characterized by different affinities: one incoming NTP can bind the A site to be incorporated next, while a second NTP can bind an "allosteric" site on RNAP and switch the EC from an "unactivated" state where nucleotide addition is slow to an "activated" state, where nucleotide addition is fast (Figure 7.6A). This allosteric model can explain the empirical rates measured at various NTP concentrations and this mechanism is proposed to adjust the overall rate of nucleotide addition based on the availability of

NTPs.[35,81] It also has been observed that the identity of downstream template base at the $i+2$ position significantly affects nucleotide addition immediately upstream at the $i+1$ position.[81] It is suggested that a small conserved loop in the β subunit called fork loop 2 (β533–541, also termed βDloopI or streptolydigin-binding loop) is the allosteric NTP binding site. Fork loop 2 is located only ~5–6 Å away from the $i+2$ downstream template base, close enough to directly contact an NTP bound there. The fork loop 2 sequence resembles that of well-characterized "P-loops" found in various nucleotide binding proteins, suggesting it may be a good candidate for an NTP binding site. Additionally,

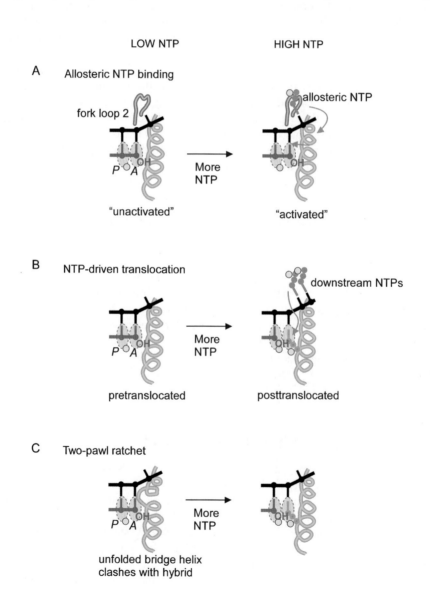

fork loop 2 assumes different conformations in *T. aquaticus* and *T. thermophilus* RNAPs.[25,62] Further work led to the proposal that an allosteric NTP bound at fork loop 2 can drive translocation *via* a "ratchet motion." Specifically, upon NTP binding at this allosteric site, the fork loop 2 would change its conformation and alter its direct contact with the bridge helix or the rifampicin-binding pocket residues (Figure 7.6A), both of which extensively contact the RNA:DNA hybrid and could directly affect the translocation process.[81]

This allosteric NTP binding model can explain the biphasic nature of the kinetics of nucleotide addition. It also suggests the existence of a regulatory pathway for relating the rate of transcript elongation to cellular NTP levels. Interestingly, recent yeast EC structures complexed with NTP have observed a 90° rotation of fork loop 2 between downstream DNA duplex strands, suggesting a role in unwinding downstream DNA,[17] supporting the proposal that repositioning of fork loop 2 is coupled to translocation. However, there is yet to be a direct demonstration that NTPs can bind to the proposed allosteric site, or that the conformation of fork loop 2 directly affects translocation. It seems at least possible that the fast and slow species of ECs simply reflect different semi-stable translocation registers of RNAP that interconvert more slowly than NTP addition but whose interconversion might be affected by NTP binding. Mutational analysis of the residues constituting the proposed

Figure 7.6 Kinetic models of nucleotide addition. Three kinetic models, (A)–(C), describe nucleotide addition at low (sub-saturating) and high (near-saturating or saturating) concentrations of NTPs, each providing mechanistic explanations for the empirical greater-than-hyperbolic NTP dependency of nucleotide addition. (A) Allosteric NTP binding model. In this model, two NTPs can bind simultaneously to two distinct sites on RNAP: one at the A site and the other at the allosteric site (fork loop 2, in blue). At low NTP, only the A site is populated by NTP, the EC is in an "unactivated" state, where nucleotide addition is slow; at high NTP, NTP binding at the allosteric site switches the EC to an "activated" state where nucleotide addition is fast, possibly by conformational changes in the fork loop 2 and consequent movements in the rifampicin-binding pocket (not shown) or the bridge helix (in cyan, green arrows).[35,81] (B) NTP-driven translocation model. In this model, at low NTPs, ECs are predominantly in the pretranslocated state. Therefore, nucleotide addition is slow. At high NTPs, increased NTP occupancy of the A site biases the equilibrium towards the post-translocated state and thus accelerates nucleotide addition by alleviating the inhibition due to pretranslocated states at low NTPs. This model distinguishes itself from the model in (A) by postulating that the NTP bound at downstream DNA is not "allosteric" or "non-essential," but rather is *en route* to the A site for incorporation.[23,56,57] (C) Two-pawl ratchet model. In this model, bridge helix oscillations occur at every nucleotide addition cycle. At low NTPs, an unfolded bridge helix can clash with the RNA 3' nt occupying the A site, leading to inhibition of nucleotide addition and even backtracking. At high NTPs, due to increased NTP occupancy of the A site and thus decreased RNA 3' nt occupancy of the same site, the probability for the unfolded bridge clashing with RNA 3' nt is reduced, leading to faster nucleotide addition.[58]

allosteric site and structural demonstration of NTP binding to the site will be crucial to test this model.

7.4.2 NTP-driven Translocation Model

A similar model termed "NTP-driven translocation" was proposed by Burton and co-workers. It is based on pre-steady-state kinetic studies of human Pol II. As opposed to the allosteric model, this model posits that the two kinetically distinct species identified in nucleotide addition simply correspond to the pre-translocated (slow species) and post-translocated (fast species) complexes. In this model, NTPs are proposed to line up base-paired to the downstream DNA template at positions $i+2$, $i+3$, and even $i+4$, favoring forward translocation of the enzyme by facilitating delivery of the NTP to the A site (Figure 7.6B).[23,56,57] This model is analogous to the aforementioned allosteric NTP binding model in that an NTP can base pair to the downstream $i+2$ position and accelerate the NTP incorporation at the A site.[57] However, it suggests that the NTP bound at downstream $i+2$ position is not "allosteric" but rather a substrate NTP *en route* to the A site (Figure 7.6B; see also Chapter 8).

7.4.3 Two-pawl Ratchet Model

Bar-Nahum *et al*. have proposed an alternative model to explain the greater-than-hyperbolic NTP dependency of nucleotide addition without invoking an additional NTP binding site.[58] In this model, nucleotide addition and unidirectional translocation are mediated by two pawls: NTP as a "stationary pawl," which binds to the A site and prevents reverse translocation of the RNA 3′ end, and the oscillating bridge helix as a "reciprocating pawl," which periodically unfolds in the middle, sterically pushing the 3′ end of the hybrid from the A site to the P site to drive translocation. Importantly, unfolding of the bridge helix can fail to translocate the hybrid, but instead peel the RNA 3′ nt away from the DNA template base, destabilizing the hybrid and allowing backtracking (Figure 7.6C).[58] The frequency of this occurring is affected by the NTP occupancy of the A site: at low NTP the RNA 3′ nt has a higher probability of occupying the A site. In this case nucleotide addition is thus limited by the oscillation frequency of the bridge helix (Figure 7.6C).

The same model could also explain the greater-than-hyperbolic NTP dependence of incorporation kinetics. NTP can bind the A site and diffuse back out faster than bridge helix oscillations, leaving the active site temporarily in an NTP-primed state (the post-translocated state), which can now incorporate a nucleotide more rapidly than an unprimed active site. As the NTP concentration increases, translocation is made easier and incorporation of NTP is in turn facilitated leading to apparent substrate cooperativity. In addition, the state in which the unfolded bridge-helix clashes with and peels off the pretranslocated hybrid could represent a "trapped" or "paused" state, which could be stabilized by an RNA hairpin or engage in further backtracking or even termination.[58] The peeling of a DNA base off of the 3' end of the hybrid is similar to the previously proposed fraying of RNA 3' nt away from the template DNA at a pause.[1,13]

The "two-pawl ratchet" model is subject to several concerns. First, the model centers on the idea that bridge helix oscillations sterically control translocation and therefore nucleotide addition. For such a model to work, translocation has to be associated with the rate-limiting step of nucleotide addition. However, structural and biochemical evidence suggests that trigger loop folding, rather than translocation, is likely the key conformational change that is rate-limiting for nucleotide addition.[13,15,17] Second, the model requires that oscillations of the bridge helix occur in the context of transcribing ECs and must occur on the same timescale as nucleotide addition; these timescales remain unexplored. Finally, as noted earlier, several structural and genetic data argue against a central role of the bridge helix in nucleotide addition (Section 7.2.1).

7.4.4 Biophysical Models for Transcript Elongation

Pioneering work from von Hippel and colleagues has offered a quantitative thermodynamic approach to understanding transcript elongation. These models consider the transcribing EC at any given template position as facing three competing energy barriers to three potential reaction pathways, including elongation, backtracking ("editing") and termination.[36,82,83]

By extending the work of von Hippel, as well as that of Sousa and colleagues,[36,73,83] Bai et al. proposed a simplified sequence-dependent kinetic model for transcript elongation and used it to predict backtrack pauses along different DNA templates.[74] A more recent kinetic model by Tadigotla et al. sought to improve upon the Bai et al. model by incorporating sequence-specific contributions of RNA secondary structures on suppression of backtrack pauses and accounting for thermal fluctuations of the transcription bubble.[75] The respective values of these two models are discussed in Chapter 9.

These thermodynamic and kinetic modeling efforts have allowed for a more quantitative understanding of the elongation process. However, they also reveal that our current knowledge remains largely insufficient to produce an accurate picture of the mechanical behavior of an elongating RNAP. These models can be further refined as we gain more mechanistic knowledge of the elongating process, for instance by taking into account the energy involved in forming and breaking RNA tertiary structures or the free energy associated with trigger loop/bridge helix movements. Conceivably, these models can be refined by varying the assisting or hindering load on RNAP in single-molecule transcription assays and examining how the thermodynamic model responds to external perturbations. A further step in the modeling efforts would be to consider the chromosomal topology, the density of RNAPs on DNA, cellular spatial constraints, and the numerous DNA modifications RNAP encounters in living cells.

7.5 Technological Advances in Studies of Transcript Elongation

The fundamental catalytic mechanisms and regulatory functions of transcript elongation have been extensively studied since the discovery of RNAPs, primarily using classical biochemistry and genetics as tools. The first high-resolution X-ray crystal structure of a multi-subunit RNAP in 1999[62] set the stage for the elucidation of a series of RNAP crystal structures ranging from core RNAP to transcription elongation complexes with bound NTP. These structures have produced a high-resolution picture of transcript elongation largely consistent with results obtained from classical biochemical and genetic approaches. Only in the past few years have several novel biophysical techniques such as single-molecule force clamps,[50,69,78] fluorescence-based spectroscopy and microscopy – such as fluorescence resonance energy transfer experiments (FRET)[84] – and chromatin immunoprecipitation followed by whole-genome microarrays (ChIP-chip)[85,86] been successfully applied to investigate the mechanisms and regulation of transcript elongation. These powerful and increasingly accessible tools, backed up by classical biochemical and genetic approaches, have greatly advanced the field and will allow investigators to begin unraveling the complex structural mechanisms of transcript elongation and its regulation. More importantly, these new technologies may allow us to seek answers to several fundamental questions inaccessible to classical experimental approaches. The characterization of force–velocity relationships revealed by single-molecule force-clamp assays[78] and the observation of DNA scrunching in initiating complex measured by FRET assays[84] and single-molecule DNA nanomanipulation[87] are excellent examples of such applications.

7.6 Concluding Remarks

In this chapter we have provided an overview of our current understanding of the mechanisms underlying transcript elongation and its control. Specifically, we have addressed research questions involving how NTP substrates enter the RNAP active site, of how key conformational changes near the active site involving the trigger loop and the bridge helix control nucleotide addition, of how catalysis of nucleotide addition, pyrophosphorolysis, and transcript cleavage occur, and of how RNAP threads DNA and RNA through the main channel *via* repeated translocation events.

Nonetheless, several fundamental questions concerning the molecular mechanisms of transcript elongation remain important challenges that await investigators. It remains a technical challenge to obtain direct evidence that distinguishes the secondary-channel and main-channel models of NTP entry, to directly observe and measure key movements of the trigger loop/bridge helix during nucleotide addition, and to accurately and non-invasively monitor translocation in real-time. We are hopeful that the burgeoning applications of novel biophysical and biochemical approaches in the transcription field can help provide answers to these important questions in the near future.

References

1. I. Toulokhonov and R. Landick, The flap domain is required for pause RNA hairpin inhibition of catalysis by RNA polymerase and can modulate intrinsic termination, *Mol. Cell*, 2003, **12**, 1125–36.
2. T. Santangelo, R. Mooney, R. Landick and J. W. Roberts, RNA polymerase mutations that impair reconfiguration to a termination resistant complex by Q antiterminator proteins, *Genes Dev.*, 2003, **17**, 1281–1292.
3. J. P. Richardson, Loading Rho to terminate transcription, *Cell*, 2003, **114**, 157–9.
4. P. H. von Hippel and Z. Pasman, Reaction pathways in transcript elongation, *Biophys. Chem.*, 2002, **101–102**, 401–23.
5. I. Gusarov and E. Nudler, Control of intrinsic transcription termination by N and NusA: the basic mechanisms, *Cell*, 2001, **107**, 437–49.
6. U. Vogel and K. F. Jensen, The RNA chain elongation rate in *Escherichia coli* depends on the growth rate, *J. Bacteriol.*, 1994, **176**, 2807–13.
7. B. R. Bochner and B. N. Ames, Complete analysis of cellular nucleotides by two-dimensional thin layer chromatography, *J. Biol. Chem.*, 1982, **257**, 9759–69.
8. A. M. Edwards, C. M. Kane, R. A. Young and R. D. Kornberg, Two dissociable subunits of yeast RNA polymerase II stimulate the initiation of transcription at a promoter in vitro, *J. Biol. Chem.*, 1991, **266**, 71–5.
9. G. Rhodes and M. J. Chamberlin, Ribonucleic acid chain elongation by *Escherichia coli* ribonucleic acid polymerase: Isolation of ternary complexes and the kinetics of elongation, *J. Biol. Chem.*, 1974, **249** 6675–6683.
10. S. A. Kumar and J. S. Krakow, Studies on the product binding site of the *Azotobacter vinelandii* ribonucleic acid polymerase, *J. Biol. Chem.*, 1975, **250**, 2878–2884.
11. V. W. Armstrong, D. Yee and F. Eckstein, Mechanistic studies on deoxyribonucleic acid dependent ribonucleic acid polymerase from *Escherichia coli* using phosphorothioate analogues. 2. The elongation reaction, *Biochemistry*, 1979, **18**, 4120–4123.
12. D. A. Erie, T. D. Yager and P. H. von Hippel, The single-nucleotide addition cycle in transcription: a biophysical and biochemical perspective, *Annu. Rev. Biophys. Biomol. Struct.*, 1992, **21**, 379–415.
13. I. Toulokhonov, J. Zhang, M. Palangat and R. Landick, A central role of the RNA polymerase trigger loop in active-site rearrangement during transcriptional pausing, *Mol. Cell*, 2007, **27**, 406–19.

14. I. Artsimovitch and R. Landick, Pausing by bacterial RNA polymerase is mediated by mechanistically distinct classes of signals, *Proc. Natl. Acad. Sci. USA*, 2000, **97**, 7090–5.
15. D. G. Vassylyev, M. N. Vassylyeva, J. Zhang, M. Palangat, I. Artsimovitch and R. Landick, Structural basis for substrate loading in bacterial RNA polymerase, *Nature*, 2007, **448**, 163–8.
16. D. G. Vassylyev, M. N. Vassylyeva, A. Perederina, T. H. Tahirov and I. Artsimovitch, Structural basis for transcription elongation by bacterial RNA polymerase, *Nature*, 2007, **448**, 157–62.
17. D. Wang, D. A. Bushnell, K. D. Westover, C. D. Kaplan and R. D. Kornberg, Structural basis of transcription: role of the trigger loop in substrate specificity and catalysis, *Cell*, 2006, **127**, 941–54.
18. H. Kettenberger, A. Eisenfuhr, F. Brueckner, M. Theis, M. Famulok and P. Cramer, Structure of an RNA polymerase II-RNA inhibitor complex elucidates transcription regulation by noncoding RNAs, *Nat. Struct. Mol. Biol.*, 2006, **13**, 44–8.
19. S. Tuske, S. G. Sarafianos, X. Wang, B. Hudson, E. Sineva, J. Mukhopadhyay, J. J. Birktoft, O. Leroy, S. Ismail, A. D. Clark Jr, C. Dharia, A. Napoli, O. Laptenko, J. Lee, S. Borukhov, R. H. Ebright and E. Arnold, Inhibition of bacterial RNA polymerase by streptolydigin: stabilization of a straight-bridge-helix active-center conformation, *Cell*, 2005, **122**, 541–52.
20. D. Temiakov, N. Zenkin, M. N. Vassylyeva, A. Perederina, T. H. Tahirov, E. Kashkina, M. Savkina, S. Zorov, V. Nikiforov, N. Igarashi, N. Matsugaki, S. Wakatsuki, K. Severinov and D. G. Vassylyev, Structural basis of transcription inhibition by antibiotic streptolydigin, *Mol. Cell*, 2005, **19**, 655–66.
21. K. D. Westover, D. A. Bushnell and R. D. Kornberg, Structural basis of transcription: separation of RNA from DNA by RNA polymerase II, *Science*, 2004, **303**, 1014–6.
22. H. Kettenberger, K. J. Armache and P. Cramer, Complete RNA polymerase II elongation complex structure and its interactions with NTP and TFIIS, *Mol. Cell*, 2004, **16**, 955–65.
23. Y. A. Nedialkov, X. Q. Gong, S. L. Hovde, Y. Yamaguchi, H. Handa, J. H. Geiger, H. Yan and Z. F. Burton, NTP-driven translocation by human RNA polymerase II, *J. Biol. Chem.*, 2003, **278**, 18303–12.
24. H. Kettenberger, K. J. Armache and P. Cramer, Architecture of the RNA polymerase II-TFIIS complex and implications for mRNA cleavage, *Cell*, 2003, **114**, 347–57.
25. D. G. Vassylyev, S. Sekine, O. Laptenko, J. Lee, M. N. Vassylyeva, S. Borukhov and S. Yokoyama, Crystal structure of a bacterial RNA polymerase holoenzyme at 2.6 A resolution, *Nature*, 2002, **417**, 712–9.
26. S. A. Darst, N. Opalka, P. Chacon, A. Polyakov, C. Richter, G. Zhang and W. Wriggers, Conformational flexibility of bacterial RNA polymerase, *Proc. Natl. Acad. Sci. USA.*, 2002, **99**, 4296–301.

27. A. Gnatt, P. Cramer, J. Fu, D. Bushnell and R. D. Kornberg, Structural basis of transcription: an RNA polymerase II elongation complex at 3.3 Å resolution, *Science*, 2001, **292**, 1876–82.
28. P. Cramer, D. Bushnell, J. Fu, A. Gnatt, B. Maier-Davis, N. Thompson, R. Burgess, A. Edwards, P. David and R. Kornberg, Architecture of RNA polymerase II and implications for the transcription mechanism, *Science*, 2000, **288**, 640–649.
29. L. M. Iyer, E. V. Koonin and L. Aravind, Evolution of bacterial RNA polymerase: implications for large-scale bacterial phylogeny, domain accretion, and horizontal gene transfer, *Gene*, 2004, **335**, 73–88.
30. V. Sosunov, E. Sosunova, A. Mustaev, I. Bass, V. Nikiforov and A. Goldfarb, Unified two-metal mechanism of RNA synthesis and degradation by RNA polymerase, *EMBO J.*, 2003, **22**, 2234–44.
31. V. Sosunov, S. Zorov, E. Sosunova, A. Nikolaev, I. Zakeyeva, I. Bass, A. Goldfarb, V. Nikiforov, K. Severinov and A. Mustaev, The involvement of the aspartate triad of the active center in all catalytic activities of multi-subunit RNA polymerase, *Nucleic Acids Res.*, 2005, **33**, 4202–11.
32. D. Yee, V. W. Armstrong and F. Eckstein, Mechanistic studies on deoxyribonucleic acid dependent ribonucleic acid polymerase from *Escherichia coli* using phosphorothioate analogues. 1. Initiation and pyrophosphate exchange reactions, *Biochemistry*, 1979, **18**, 4116–4120.
33. T. A. Steitz, A mechanism for all polymerases, *Nature*, 1998, **391**, 231–2.
34. Q. Guo and R. Sousa, Translocation by T7 RNA polymerase: a sensitively poised Brownian ratchet, *J. Mol. Biol*, 2006, **358**, 241–54.
35. J. E. Foster, S. F. Holmes and D. A. Erie, Allosteric binding of nucleoside triphosphates to RNA polymerase regulates transcription elongation, *Cell*, 2001, **106**, 243–52.
36. S. J. Greive and P. H. von Hippel, Thinking quantitatively about transcriptional regulation, *Nat. Rev. Mol. Cell Biol.*, 2005, **6**, 221–32.
37. J. J. Hopfield, Kinetic proofreading: a new mechanism for reducing errors in biosynthetic processes requiring high specificity, *Proc. Natl. Acad. Sci. USA*, 1974, **71**, 4135–9.
38. M. Orlova, J. Newlands, A. Das, A. Goldfarb and S. Borukhov, Intrinsic transcript cleavage activity of RNA polymerase, *Proc. Natl. Acad. Sci. USA*, 1995, **92**, 4596–4600.
39. E. Sosunova, V. Sosunov, M. Kozlov, V. Nikiforov, A. Goldfarb and A. Mustaev, Donation of catalytic residues to RNA polymerase active center by transcription factor Gre, *Proc. Natl. Acad. Sci. USA*, 2003, **100**, 15469–74.
40. D. A. Erie, O. Hajiseyedjavadi, M. C. Young and P. H. von Hippel, Multiple RNA polymerase conformations and GreA: control of fidelity of transcription, *Science*, 1993, **262**, 867–873.
41. N. Komissarova and M. Kashlev, Transcriptional arrest: *Escherichia coli* RNA polymerase translocates backward, leaving the 3′ end of the RNA intact and extruded, *Proc. Natl. Acad. Sci. USA*, 1997, **94**, 1755–1760.

42. C. K. Surratt, S. C. Milan and M. J. Chamberlin, Spontaneous cleavage of RNA in ternary complexes of *Escherichia coli* RNA polymerase and its significance for the mechanism of transcription, *Proc. Natl. Acad. Sci. USA.*, 1991, **88**, 7983–7987.
43. S. Borukhov, V. Sagitov and A. Goldfarb, Transcript cleavage factors from *E. coli*, *Cell*, 1993, **72**, 459–466.
44. S. Borukhov, A. Polyakov, V. Nikiforov and A. Goldfarb, GreA protein: a transcription elongation factor from *Escherichia coli*, *Proc. Natl. Acad. Sci. USA*, 1992, **89**, 8899–8902.
45. M. Izban and D. Luse, The RNA polymerase II ternary complex cleaves the nascent transcript in a $3' \rightarrow 5'$ direction in the presence of elongation factor SII, *Genes Dev.*, 1992, **6**, 1342–1356.
46. N. Opalka, M. Chlenov, P. Chacon, W. J. Rice, W. Wriggers and S. A. Darst, Structure and function of the transcription elongation factor GreB Bound to bacterial RNA polymerase, *Cell*, 2003, **114**, 335–45.
47. O. Laptenko, J. Lee, I. Lomakin and S. Borukhov, Transcript cleavage factors GreA and GreB act as transient catalytic components of RNA polymerase, *EMBO J.*, 2003, **22**, 6322–34.
48. N. Zenkin, Y. Yuzenkova and K. Severinov, Transcript-assisted transcriptional proofreading, *Science*, 2006, **313**, 518–20.
49. S. Kyzer, K. S. Ha, R. Landick and M. Palangat, Direct versus limited-step reconstitution reveals key features of an RNA hairpin-stabilized paused transcription complex, *J. Biol. Chem.*, 2007, **282**, 19020–8.
50. K. M. Herbert, A. La Porta, B. J. Wong, R. A. Mooney, K. C. Neuman, R. Landick and S. M. Block, Sequence-resolved detection of pausing by single RNA polymerase molecules, *Cell*, 2006, **125**, 1083–94.
51. R. Landick, The regulatory roles and mechanism of transcriptional pausing, *Biochem. Soc. Trans.*, 2006, **34**, 1062–6.
52. M. Palangat, C. T. Hittinger and R. Landick, Downstream DNA selectively affects a paused conformation of human RNA polymerase II, *J. Mol. Biol.*, 2004, **341**, 429–442.
53. C. Chan, D. Wang and R. Landick, Multiple interactions stabilize a single paused transcription intermediate in which hairpin to 3' end spacing distinguishes pause and termination pathways, *J. Mol. Biol.*, 1997, **268**, 54–68.
54. N. N. Batada, K. D. Westover, D. A. Bushnell, M. Levitt and R. D. Kornberg, Diffusion of nucleoside triphosphates and role of the entry site to the RNA polymerase II active center, *Proc. Natl. Acad. Sci. USA*, 2004, **101**, 17361–4.
55. K. D. Westover, D. A. Bushnell and R. D. Kornberg, Structural basis of transcription; nucleotide selection by rotation in the RNA polymerase II active center, *Cell*, 2004, **119**, 481–9.
56. X. Q. Gong, C. Zhang, M. Feig and Z. F. Burton, Dynamic error correction and regulation of downstream bubble opening by human RNA polymerase II, *Mol. Cell*, 2005, **18**, 461–70.

57. Z. F. Burton, M. Feig, X. Q. Gong, C. Zhang, Y. A. Nedialkov and Y. Xiong, NTP-driven translocation and regulation of downstream template opening by multi-subunit RNA polymerases, *Biochem. Cell Biol.*, 2005, **83**, 486–96.
58. G. Bar-Nahum, V. Epshtein, A. Ruckenstein, R. Rafikov, A. Mustaev and E. Nudler, A ratchet mechanism of transcription elongation and its control, *Cell*, 2005, **120**, 183–193.
59. V. Epshtein, A. Mustaev, V. Markovtsov, O. Bereshchenko, V. Nikiforov and A. Goldfarb, Swing-gate model of nucleotide entry into the RNA polymerase active center, *Mol. Cell*, 2002, **10**, 623–634.
60. P. Cramer, D. Bushnell and R. Kornberg, Structural basis of transcription: RNA polymerase II at 2.8 Å resolution, *Science*, 2001, **292**, 1863–76.
61. A. Gnatt, J. Fu and R. D. Kornberg, Formation and crystallization of yeast RNA polymerase II elongation complexes, *J. Biol. Chem.*, 1997, **272**, 30799–805.
62. G. Zhang, E. A. Campbell, L. Minakhin, C. Richter, K. Severinov and S. A. Darst, Crystal structure of *Thermus aquaticus core* RNA polymerase at 3.3 Å resolution, *Cell*, 1999, **98**, 811–24.
63. Y. Huang, A. Beaudry, J. McSwiggen and R. Sousa, Determinants of ribose specificity in RNA polymerization: effects of Mn^{2+} and deoxynucleoside monophosphate incorporation into transcripts, *Biochemistry*, 1997, **36**, 13718–28.
64. J. J. Arnold, D. W. Gohara and C. E. Cameron, Poliovirus RNA-dependent RNA polymerase (3Dpol): pre-steady-state kinetic analysis of ribonucleotide incorporation in the presence of Mn^{2+}, *Biochemistry*, 2004, **43**, 5138–48.
65. D. Pinto, M. T. Sarocchi-Landousy and W. Guschlbauer, 2'-Deoxy-2'-fluorouridine-5'-triphosphates: a possible substrate for *E. coli* RNA polymerase, *Nucleic. Acids Res.*, 1979, **6**, 1041–8.
66. A. Chakrabartty, T. Kortemme and R. L. Baldwin, Helix propensities of the amino acids measured in alanine-based peptides without helix-stabilizing side-chain interactions, *Protein Sci.*, 1994, **193**, 843–52.
67. V. Markovtsov, A. Mustaev and A. Goldfarb, Protein-RNA interactions in the active center of the transcription elongation complex, *Proc. Natl. Acad. Sci. USA*, 1996, **93**, 3221–3226.
68. R. Weilbaecher, C. Hebron, G. Feng and R. Landick, Termination-altering amino acid substitutions in the β' subunit of *Escherichia coli* RNA polymerase identify regions involved in RNA chain elongation, *Genes Dev.*, 1994, **8**, 2913–2917.
69. M. Wang, M. Schnitzer, H. Yin, R. Landick, J. Gelles and S. Block, Force and velocity measured for single molecules of RNA polymerase, *Science*, 1998, **282**, 902–907.
70. M. Burmeister, A. P. Monaco, E. F. Gillard, G. J. van Ommen, N. A. Affara, M. A. Ferguson-Smith, L. M. Kunkel and H. Lehrach, A 10-megabase physical map of human Xp21, including the Duchenne muscular dystrophy gene, *Genomics*, 1988, **2**, 189–202.

71. Y. W. Yin and T. A. Steitz, The structural mechanism of translocation and helicase activity in T7 RNA polymerase, *Cell*, 2004, **116**, 393–404.
72. R. Guajardo, P. Lopez, M. Dreyfus and R. Sousa, NTP concentration effects on initial transcription by T7 RNAP indicate that translocation occurs through passive sliding and reveal that divergent promoters have distinct NTP concentration requirements for productive initiation, *J. Mol. Biol.*, 1998, **281**, 777–92.
73. R. Guajardo and R. Sousa, A model for the mechanism of polymerase translocation, *J. Mol. Biol.*, 1997, **265**, 8–19.
74. L. Bai, A. Shundrovsky and M. D. Wang, Sequence-dependent kinetic model for transcription elongation by RNA polymerase, *J. Mol. Biol.*, 2004, **344**, 335–49.
75. V. R. Tadigotla, D. O. Maoiléidigh, A. M. Sengupta, V. Epshtein, R. H. Ebright, E. Nudler and A. E. Ruckenstein, Thermodynamic and kinetic modeling of transcriptional pausing, *Proc. Natl. Acad. Sci. USA*, 2006, **103**, 4439–44.
76. G. Oster and H. Wang, Rotary protein motors, *Trends Cell Biol.*, 2003, **13**, 114–21.
77. R. D. Vale and R. A. Milligan, The way things move: looking under the hood of molecular motor proteins, *Science*, 2000, **288**, 88–95.
78. E. A. Abbondanzieri, W. J. Greenleaf, J. W. Shaevitz, R. Landick and S. M. Block, Direct observation of base-pair stepping by RNA polymerase, *Nature*, 2005, **438**, 460–5.
79. P. Thomen, P. J. Lopez and F. Heslot, Unravelling the mechanism of RNA-polymerase forward motion by using mechanical force, *Phys. Rev. Lett.*, 2005, **94**, 128102.
80. E. Kashkina, M. Anikin, F. Brueckner, E. Lehmann, S. N. Kochetkov, W. T. McAllister, P. Cramer and D. Temiakov, Multisubunit RNA polymerases melt only a single DNA base pair downstream of the active site, *J. Biol. Chem.*, 2007, **282**, 21578–82.
81. S. F. Holmes and D. A. Erie, Downstream DNA sequence effects on transcription elongation. Allosteric binding of nucleoside triphosphates facilitates translocation via a ratchet motion, *J. Biol. Chem.*, 2003, **278**, 35597–608.
82. Z. Pasman and P. H. von Hippel, Active *Escherichia coli* transcription elongation complexes are functionally homogeneous, *J. Mol. Biol.*, 2002, **322**, 505–19.
83. T. D. Yager and P. H. von Hippel, A thermodynamic analysis of RNA transcript elongation and termination in *Escherichia coli*, *Biochemistry*, 1991, **30**, 1097–1118.
84. A. N. Kapanidis, E. Margeat, S. O. Ho, E. Kortkhonjia, S. Weiss and R. H. Ebright, Initial transcription by RNA polymerase proceeds through a DNA-scrunching mechanism, *Science*, 2006, **314**, 1144–7.
85. N. B. Reppas, J. T. Wade, G. M. Church and K. Struhl, The transition between transcriptional initiation and elongation in *E. coli* is highly variable and often rate limiting, *Mol. Cell*, 2006, **24**, 747–57.

86. C. D. Herring, M. Raffaelle, T. E. Allen, E. I. Kanin, R. Landick, A. Z. Ansari and B. O. Palsson, Immobilization of *Escherichia coli* RNA polymerase and location of binding sites by use of chromatin immunoprecipitation and microarrays, *J. Bacteriol.*, 2005, **187**, 6166–74.
87. A. Revyakin, C. Liu, R. H. Ebright and T. R. Strick, Abortive initiation and productive initiation by RNA polymerase involve DNA scrunching, *Science*, 2006, **314**, 1139–43.

CHAPTER 8
Regulation of RNA Polymerase through its Active Center

SERGEI NECHAEV,[a] NIKOLAY ZENKIN[b] AND KONSTANTIN SEVERINOV[c,d]

[a] Laboratory of Molecular Carcinogenesis, NIEHS/NIH, Research Triangle Park, NC, 27709, USA; [b] Institute for Cell and Molecular Biosciences, Newcastle University, Newcastle upon Tyne, NE2 4HH, UK; [c] Waksman Institute, Department of Molecular Biology and Biochemistry, Rutgers, The State University, Piscataway, NJ 08854, USA; [d] Institute of Molecular Genetics, Russian Academy of Sciences, Moscow 123182, Russia

8.1 Introduction

RNA polymerase (RNAP) is arguably the most highly regulated enzyme in the cell. The ability of RNAP to sense and integrate signals from various factors, including nucleic acid sequences, protein regulators, and small molecules largely determines the ability of the cell, and ultimately the entire organism, to execute its programmed life cycle and adjust to changes in the environment. Consistent with evolutionary conservation of the basic transcription mechanism, RNAP's active center, the part of the enzyme that performs catalysis, consists of elements whose sequence and structure are highly conserved in all living organisms. At the same time, the active center is highly sensitive to inputs from regulatory signals, which are exceedingly variable from organism to organism. In this chapter, we discuss properties of the RNAP active center that make this highly conserved region of RNAP a versatile regulatory target. We highlight the currently known modes of transcription regulation through the RNAP active center, with a special emphasis on recent advances in elucidating the fundamental mechanisms of RNAP active center function and regulation.

8.2 Regulatory Checkpoints of the RNAP Active Center

8.2.1 Versatility of the Active Center. How many Metals are Enough?

Enzymes that make nucleic acids (RNA and DNA polymerases) or break nucleic acids (*i.e.*, nucleases and transposases) share a common mechanism of catalysis, where two divalent cations, usually Mg^{2+}, drive the appropriate reaction by stabilizing a pentavalent phosphate intermediate.[1–4] In the case of RNA polymerases, the scissile phosphate is the β-phosphate of the incoming NTP during the RNA polymerization reaction, or a nascent RNA phosphate during exo- or endonucleolytic cleavage.[3] Several features of the multi-subunit RNAP active center make it well suited for sensing and responding to regulatory signals.

First, unlike DNA polymerases or single-subunit RNAPs, the active center of multi-subunit RNAPs can perform not only synthesis and pyrophosphorolysis but also hydrolysis (discussed in Chapter 7 and illustrated in Figure 7.2). In this last characteristic reaction, multi-subunit RNAPs can use water to cleave RNA. This efficient reaction generates a new catalytically competent 3′ end and can take place even when the original 3′ end of the transcript is moved far away from the active center. That a properly positioned, catalytically competent 3′ end of RNA can be restored after backtracking or arrest regardless of how far the RNAP has backtracked is an important property of the enzyme, essential for the ability of RNAP to recover from paused and arrested states, and enabling regulation of transcription at the level of transcriptional pausing.

Second, while the act of catalysis requires two Mg^{2+} ions, only one ion, called $Mg^{2+}I$, is tightly bound in the RNAP active center. The other Mg^{2+} ion, $Mg^{2+}II$, is weakly bound and is brought into the active center in an *ad hoc* fashion by substrates, which in effect themselves become part of the active center. During polymerization, $Mg^{2+}II$ is brought in by the incoming complementary NTP substrate. During 3′-5′ exonuclease hydrolysis or pyrophosphorolysis, $Mg^{2+}II$ is stabilized by, respectively, a non-cognate NTP or pyrophosphate.[3] During endonucleolytic cleavage, $Mg^{2+}II$ is stabilized by the penultimate 3′-end nucleotide of the transcript.[5] In this regard, factors that affect the interaction of $Mg^{2+}II$ with RNAP have a profound influence on transcription. For example, transcript cleavage factors such as GreA and GreB strongly stimulate the RNA cleavage activity of the RNAP catalytic center by stabilizing $Mg^{2+}II$.

Third, and finally, the intrinsic flexibility of the RNAP active center allows for regulation of its activities through rearrangement of Mg^{2+} ions. As discussed below, the complexity of this type of regulation is only now beginning to be appreciated. Very small changes in positions of Mg^{2+} ions within the RNAP active center lead to dramatic changes in RNAP activities.

8.2.2 Delivery of NTPs to the Active Center. How many Channels are Enough?

Pioneering work on the crystal structure of the RNAP core enzyme from *Thermus aquaticus*[6] provided the first detailed view of a multi-subunit RNAP (see Structural Atlas). Some of the most prominent features of the enzyme structure are two openings that expose the active center, which is enclosed inside the enzyme, to the outside. One is the main channel, which separates the enzyme into two main lobes; this channel contains DNA during transcription initiation and the DNA-RNA hybrid during RNA chain elongation.[6] The existence of the main channel had been previously proposed based on electron-microscopy studies.[7]

A second opening, the secondary channel, connects the surface of RNAP with the active center. The secondary channel is only wide enough ($\sim$10–15 Å) to accommodate single-stranded nucleic acid. A similarly positioned secondary channel was independently observed in the eukaryotic (yeast) Pol II and archaeal RNAP structures, where it was designated as the "pore."[8,9] Earlier biochemical work showed that the 3'-end of the nascent RNA can crosslink to different parts of RNAP depending on whether the RNAP is arrested or elongating.[10] It was then shown that in the so-called arrested complexes, the segment of RNA proximal to the 3'-end must be threaded through the enzyme.[11–13] Taking into account all the biochemical and structural evidence, it was finally proposed that in these arrested complexes the RNAP backtracks along the DNA, extruding the RNA through the secondary channel and exposing its 3'-end.[14] This proposal was supported by single-molecule fluorescence analysis of Pol II.[15] In addition, it was suggested that the secondary channel is the route of NTP entry into the active center during transcription (Figure 8.1A).[6,14] While there is currently no dispute that the RNAP secondary channel is *the* path for RNA extrusion during backtracking and arrest, the question of where the NTP substrate enters RNAP during transcription remains open. Below, we briefly consider two alternative models of NTP substrate entry into the active center.

8.2.2.1 NTP Entry through the RNAP Secondary Channel

A role for the RNAP secondary channel as the NTP entry route is supported by studies on the mechanism of RNAP inhibition by the antibiotic Microcin J25 (McJ). McJ, a threaded-lasso molecule produced by maturation of a ribosomally-synthesized peptide, inhibits transcription by binding to RNAP. Mutations in RNAP that cause resistance to McJ cluster around the circumference of the secondary channel, close to the surface of RNAP.[16,17] In biochemical studies, McJ was shown to inhibit elongation and factor-assisted transcript cleavage and also to prevent backtracking of RNAP.[16] Taken together, these data

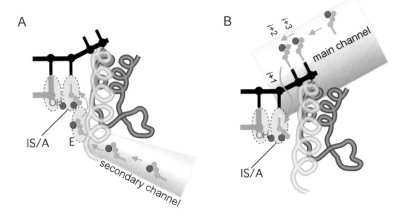

Figure 8.1 NTP entry routes into the RNAP active site. (A) Secondary-channel model of NTP entry. The incoming NTP enters through the secondary channel, initially landing at the E site (or at PS site, not shown), and finally arriving at IS for catalysis. (B) Main-channel model of NTP entry. The incoming NTPs bind and base-pair to the downstream template DNA at $i+2$, $i+3$, or even $i+4$ positions, travel over the bridge helix while remaining base-paired to DNA template, and finally land in the IS. DNA is shown in black and the 3′-proximal RNA nucleotide in orange; Bridge helix and trigger loop are shown in blue and dark orange, respectively. Incoming NTP is shown in green, with its phosphate moieties indicated with green circles; catalytic Mg^{2+} ions are in red. Green arrows indicate the proposed path of incoming NTP.

strongly argue that McJ binds in the secondary channel. This binding sterically prevents NTP substrates from accessing the catalytic center and similarly inhibits factors that act through the secondary channel. One could argue that if the secondary channel were the only route of NTP entry, then the binding of McJ and transcript elongation should be mutually exclusive. Single-molecule analysis of RNAP inhibition by McJ supports this prediction:[16] increasing the McJ concentration increased the probability of RNAP stalling but, importantly, did not affect the elongation velocity of RNAP between individual stalling events. Conversely, increasing the NTP concentration alleviated the inhibitory effect of McJ.[17]

Batada *et al.* performed computer simulations of NTP diffusion into the RNAP active center. The authors suggested that, while the narrow secondary channel dramatically reduces the effective diffusion rate of NTPs, the existence of a template-nonspecific NTP-binding site at the base of the channel (E site, see below) compensates for this reduction.[18] One can speculate that this steric decrease in NTP diffusion rate imposed by the secondary channel *increases* the effective residence time of NTPs in the E site, allowing more time for subsequent template-specific selection of NTP for incorporation. This might be especially important if nucleotide selection during the Nucleotide Addition Cycle (NAC) involves significant rearrangements in the RNAP active center that are relatively slow compared to the diffusion rate of free NTPs. Indeed, NTP selection might involve a rotation of the nucleotide within the active center,[19] or a closure of the active center.[20] *Together*, the secondary channel and the template-nonspecific NTP-binding "E" site might provide a reliable flow of NTP substrates under a wide range of conditions. At the high NTP concentrations that exist in the cell (~ 1 mM), the secondary channel limits the effective NTP diffusion rate to the active center (to approximately 200 NTP-binding events per second[18]), increasing the fidelity of transcription. On the other hand, at low NTP concentrations the existence of the E site is critical for NTP delivery.

The modeling work discussed above helps to explain why the secondary channel might be a good entry route for NTP substrates, but does not exclude alternative possibilities.

8.2.2.2 NTP Entry through the RNAP Main Channel

Available data do not exclude the possibility that rather than serving as an entry path for incoming NTPs, the secondary channel might instead be essential for another obligatory part of the NAC such as pyrophosphate (PPi) release. Indeed, the inner surface of the secondary channel is lined with negatively charged amino acid residues, which are conserved in all multisubunit RNAPs. One could argue that this negatively charged lining makes the secondary channel a poor route for the entry of bulky, negatively charged NTPs, but instead makes it better suited for removal of less bulky PPi. Moreover, some experimental evidence is not compatible with the idea that the secondary channel is the exclusive route of NTP delivery to the RNAP active center.

As mentioned in Chapter 7, work from the Burton laboratory on human Pol II showed that incorporation of NTP at register $+1$ appears to be influenced by NTPs specified by template DNA positions $+2$ and $+3$.[21] Since base-pairing of NTPs entering from the secondary channel to bases at DNA template positions $+2$ and $+3$ is impossible to reconcile with the existing crystal structure data, these results were interpreted as evidence that NTPs must be able to enter through a route other than the secondary channel (Figure 8.1B). The RNAP main channel is the obvious candidate for this alternative entry route. Indeed, structural modeling suggests that NTP entry through the main channel can in principle take place without significant sterical clashes with RNAP or nucleic acid.[21]

The main channel entry model implies that the DNA duplex remains single-stranded a few nucleotides downstream of the active center, and here the structural evidence is conflicting. In structures of eukaryotic RNAP elongation complexes, the DNA duplex indeed appears to be melted downstream of the active center. Note, however, that the observed melting might have been induced by mismatches in DNA at this position of the nucleic acid scaffold substrate used for co-crystallization. Conversely, available structures of bacterial RNAP elongation complexes formed on "fully matched" DNA templates indicate that the DNA becomes double-stranded immediately downstream of position +1, making the base pairing of NTP substrates with DNA bases at positions +2 and +3 impossible.[22] It remains to be determined whether differences in nucleic acid architecture observed in these studies stem from the technical constraints of crystallization, reflect structural plasticity of RNA polymerase or indeed show real differences between prokaryotic and eukaryotic transcription complexes. The first explanation appears to be corroborated by recent experiments by Kashkina *et al.*, who showed, using a fluorescent deoxynucleotide analogue in the template DNA strand, that DNA at position +2 is double-stranded both in bacterial and yeast RNAP transcription complexes.[23]

The model of NTP loading through the main channel remains attractive because it provides a simple mechanism ensuring correct nucleotide selection, which according to this model begins as many as three NACs before the incorporation event. As a result, RNAP has multiple opportunities to discriminate against incorrect NTPs.

One cannot exclude that, despite their high level of structural similarity, RNAPs of bacterial, yeast or human origin might differ in the ways in which substrates are predominantly delivered to their active centers. Alternatively, both pathways may be utilized in all enzymes, and a particular pathway might be predominant only under certain conditions, providing an additional mode of regulation. For example, one could speculate that the NTP delivery path might change in response to changes in NTP concentrations (with the main channel to be more prevalent at high NTP concentrations), during transition from initiation to elongation or escape from pauses, or in response to transcription factors.

8.2.3 Nucleotide Selection. How many Steps are Enough?

It is essential to minimize the number of errors introduced when making an RNA copy of a gene. Selectivity based solely on the difference in free energies between a mismatched and a proper Watson–Crick base pair only affords a 10–100-fold preference for binding of correct *versus* incorrect NTP. The observed fidelity of multi-subunit prokaryotic RNAP is higher, with estimated error rates on the order of 10^{-4}–10^{-5},[24] indicating that RNAP actively enforces high fidelity of transcription. In general, *fidelity* of transcription is determined at two principal steps: (i) through discrimination against incorrect NTPs prior to incorporation (*accuracy* of NTP incorporation) and (ii) through removal of incorrectly incorporated nucleotides (*proofreading*). Accuracy of NTP incorporation must be determined by interactions of nucleic acids with NTP substrates within the microenvironment of the RNAP active site. Proofreading is possible because of an intrinsic property of RNAP to slow down transcription subsequent to a misincorporation event, allowing extra time for removal of the incorporated NTP through one of several correction mechanisms. In the following section, we discuss mechanisms ensuring accuracy of NTP incorporation, while proofreading is discussed in Section 8.4.

8.2.3.1 Nucleotide Selection Intermediates: Pre-insertion Site or Check before you make a Bond

Several studies proposed the existence of a template-specified site on RNAP that binds the incoming NTP but is distinct from the Insertion Site (IS, also referred to as the A site) at which catalysis takes place (pathway following the top arrow in Figure 8.2A). This so-called Pre-insertion Site (PS) was first described in a study that analyzed a crystal structure of the elongation complex formed by single-subunit T7 RNAP.[25] In the same work, it was proposed that a functional analog of PS must also exist in multi-subunit RNAPs.[25] Indeed, PS was documented through the analysis of a crystal structure of eukaryotic Pol II elongation complex.[26] In PS the substrate NTP is base-paired with the template DNA $i+1$ base but is located too far from the 3′ end of the nascent transcript for catalysis of phosphodiester bond formation. NTP binding at this site could therefore provide an additional opportunity for discrimination.

However, one should bear in mind that, because of technical constraints, both in T7 RNAP and Pol II crystals the corresponding complexes were obtained in the presence of a non-hydrolyzable NTP analogue (AMP-cPP) in place of a natural NTP substrate. Therefore, the possibility remains that the substrate analog binds to the active center slightly differently than the natural NTP substrate. More recently, additional evidence for the existence of PS was obtained in a work that described a high-resolution crystal structure of *Thermus thermophilus* RNAP elongation complex in the presence of the substrate NTP analog AMP-cPP and RNAP inhibitor streptolydigin (see below).

8.2.3.2 Nucleotide Selection Intermediates: Insertion Site only or one Step with a Twist?

The existence of a template-nonspecific NTP entry (E) site that overlaps with but is distinct from IS was first proposed for the *E. coli* RNAP based on biochemical evidence (pathway following the bottom arrow in Figure 8.2A).[3] Evidence for the existence of the E site in Pol II is based largely on crystal structure analysis.

The Kornberg laboratory proposed a structure-based model, according to which the selection of the correct NTP takes place through (or during) a rotation of the incoming NTP from its initial position in the template-nonspecific E site to its final position in IS.[19] The authors obtained crystals of Pol II elongation complexes that contained a template-specified terminating nucleotide analog incorporated at the 3′ end of the transcript, as well as a free NTP, either correct or incorrect. The correct NTP in these crystals was found to be in the expected (IS) location, and would have been incorporated into the RNA chain if not for the use of the chain-terminating NTP analog. In contrast, the incorrect NTP, while bound at a site overlapping with IS, was oriented such that the base of the nucleotide did not make contact with either the template base or protein: it was held in place only through contacts with the sugar and phosphate. These findings led the authors to propose that rotation of an

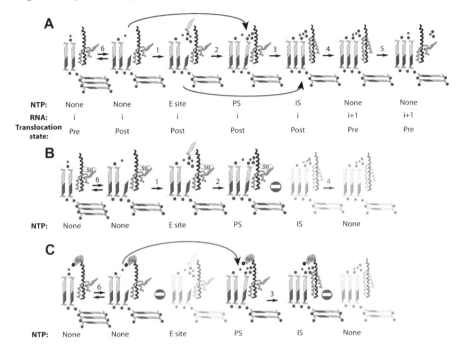

Figure 8.2 Nucleotide addition cycle and its inhibition by small molecule effectors. (A) Proposed NTP-binding events during NAC, shown on a scheme of the active center. DNA bases of the template strand are shown in blue and of the nontemplate strand in grey. RNA bases are indicated in orange. Nucleic acid sugar moieties are shown by small gray squares, with phosphates indicated by purple circles and Mg^{2+} ions by red circles. The base of incoming NTP is highlighted in green. BH and TL are indicated in purple and green, respectively. From left to right: the elongation complex exists in equilibrium between pre- and post-translocated states (discussed in Chapter 7). NTP binds to the elongation complex in E site (step 1), then rotates to bind in PS (step 2) and transitions to its place in IS (step 3), followed by the synthesis of a phosphodiester bond (step 4) and release of the pyrophosphate (step 5). The complex can then undergo translocation and bind the next NTP (step 6). Bent arrows indicate alternative models of NTP loading that exclude either E site (top arrow) or PS site (bottom arrow). (B) Proposed mechanism of transcription inhibition by streptolydigin. From left to right: binding of Stl (shown in yellow) does not affect translocation (step 6). Stl does not block NTP binding to E or PS sites (steps 1 and 2). Stl binding freezes the elongation complex in an inactive conformation and blocks the transition of incoming NTP from PS to IS (Step 3). (C) Proposed dual mechanism of transcription inhibition by tagetitoxin. Binding of a Tgt molecule (shown in salmon pink) to RNAP sterically interferes with NTP binding at the E site (step 1). The shunt provided by the upper arrow is not affected (see text). Furthermore, Tgt introduces an additional Mg^{2+} ion (tMg, highlighted with the black dot), which changes the relative positioning of catalytic cMgI and cMgII and blocks catalysis.

incoming NTP from the nonspecific E site to the IS position at the active center constitutes an important step of nucleotide selection. Furthermore, soaking the free Pol II crystals in solutions that contained natural NTP substrates resulted in binding of NTPs only in the E site orientation, indicating that the E site is the preferred place of NTP binding to RNAP and therefore is also a plausible site for preliminary substrate binding before template-specified selection takes place.

In summary, even though the molecular details of exact intermediates of nucleotide selection are not fully clear, available biochemical and crystallographic data argue that loading of an incoming NTP substrate into the RNAP active center takes place *via* multiple steps. Whether it is loading through the main channel, which is proposed to sense NTPs at registers +2 and +3, or through the secondary channel, which involves NTP selection at register +1 through prior binding at either a template-nonspecific E site or template-specific PS site, remains to be determined.

8.3 Regulators that Target the RNAP Active Center

8.3.1 Small-molecule Effectors of RNAP

Several small molecules that inhibit RNAP have been discovered and their mechanisms of action have been characterized to various degrees of detail. In this section, we discuss two such inhibitors, streptolydigin and tagetitoxin. We emphasize experimental and conceptual challenges facing researchers in this field, and discuss how conclusions derived from their analyses facilitated functional dissection of the RNAP active center.

8.3.1.1 Streptolydigin: Frozen at Last

Streptolydigin (Stl), a tetramic acid antibiotic produced by *Streptomyces lydigus*, is an inhibitor of eubacterial[27] but not eukaryotic RNAPs.[28] *In vitro*, Stl inhibits all catalytic functions of RNAP.[20,27] At low concentrations of Stl, which caused partial transcription inhibition, no change in the overall distribution of abortive RNA products was observed.[29] Thus, Stl interferes with substrate NTP incorporation at transcription initiation as well as elongation. The kinetic mechanism of RNAP inhibition by Stl was suggested to be complex, a combination of competitive and noncompetitive modes with respect to substrate NTPs, depending on the nature of the nucleotide substrate and of the template used.[29] Several mutations in *E. coli* RNAP that confer Stl resistance were obtained and mapped within the β and β' subunits of RNAP, mainly within their evolutionarily-conserved regions.[30–33]

Crystal structures of *Thermus thermophilus* RNAP in complex with Stl were obtained simultaneously and independently by two groups.[20,34] Both groups identified the same binding site for Stl on RNAP and showed that the antibiotic makes extensive contacts with critical RNAP structural elements including the β′ subunit's Bridge Helix (BH), the Trigger Loop (TL), and two β subunit loops (referred-to as STL1 and STL2) (see "fork loop 2" and "β D loop II" of Figure 7.1A, Figure 8.2B, and Structural Atlas Figure A.3).[20] The Stl-binding site is located ~20 Å away from the catalytic Mg^{2+} ions, suggesting that Stl is unlikely to directly interfere with substrate NTP binding or catalysis. Stl binding causes local yet significant structural changes in the vicinity of the catalytic center, such as a 2 Å shift of BH and a displacement of TL from its position in the apo-holoenzyme. Thus Stl might be an allosteric regulator.

Tuske et al. reported that the structures of the apo-holoenzyme and the holoenzyme–Stl complex differed in the conformations of BH.[34] The authors suggested that the binding of Stl freezes BH in the straight conformation, and that Stl inhibits transcription by blocking the cycling of BH between straight and bent conformation (for a discussion of BH cycling, see Chapter 7). To determine the effect of Stl on RNAP at different stages of NAC, the authors prepared stalled elongation complexes that existed in equilibrium between pre- and post-translocated states, as judged by Exo III footprinting. Stl changed the relative distribution of these complexes, increasing the proportion of complexes in the post-translocated state. However, in contrast to these findings, earlier work using crosslinkable substrate analogs showed that RNAP with a straight conformation of BH (the conformation presumably stabilized by Stl) favors the pre-translocated state.[35] This discrepancy had made it difficult to offer a plausible model of Stl mechanism of action, and so the authors had concluded that Stl inhibits BH cycling, which they state is "important for RNAP function."

Temiakov et al. also investigated the mechanism of Stl inhibition in several experiments.[20] First, they used rapid kinetic analysis and showed that Stl did not reduce the affinity of RNAP for the NTP substrate. This conclusion is consistent with the structural data obtained by both groups. They next investigated the effect of Stl on translocation. In one experiment, two types of elongation complexes that differed in their translocation states (as judged by their differential sensitivities to pyrophosphate) were assembled on synthetic scaffold templates.[20] The expectation was that if Stl were to inhibit translocation, it should have different effects on RNAP elongation complexes in different translocation states. However, the inhibitor showed no such bias and inhibited RNA extension by RNAP in the post-translocated complex and pyrophosphorolysis in the pre-translocation complex. Furthermore, Stl did not affect the equilibrium between pre- and post-translocated complexes assembled on scaffold templates, and had no effect on translocation induced by the addition of a non-hydrolysable NTP analog, as judged by Exo III footprinting. These observations contradict the above-mentioned findings of Tuske et al. It is possible that the differential effects of Stl on RNAP translocation obtained by Tuske et al. versus Temiakov et al. are due to the use of different RNAPs (E. coli and T. aquaticus, respectively) and/or differences in experimental design (promoter-initiated complexes stalled by incorporation of a chain terminating nucleotide versus artificially assembled elongation complexes). The differences might additionally reflect differential sensitivity of RNAP to Stl at different stages of transcription.

To further investigate the inhibition of RNAP by Stl, Temiakov et al. performed kinetic analysis of NTP incorporation in the presence of Stl. The authors used a slowly hydrolysable analog of NTP (alpha-phosphorothioate), which decreases the efficiency of nucleophilic attack of the RNA 3'-hydroxyl on the α phosphate of the NTP due to a decreased positive charge of the phosphorothioate group compared to the natural phosphate group. When the formation of a phosphodiester bond was thus made rate-limiting, the relative effect of Stl on incorporation was much less dramatic than that observed during incorporation of natural NTPs (20-fold *vs.* 1000-fold inhibition, respectively). The authors interpreted this difference as evidence that, during incorporation of the nucleotide analog, an early step that is distinct from formation of the phosphodiester bond becomes rate-limiting in the presence of Stl. From this, the authors reasoned that, in the presence of normal NTP, Stl should also target this early step that is distinct from catalysis. This conclusion is consistent with the observation that Stl is bound 20 Å away from the catalytic Mg^{2+}. This experiment has caveats, however: first, the observed absolute levels of NTP and α-S-NTP incorporation differed by three orders of magnitude, complicating direct comparisons of relative (fold-inhibition) effects of Stl on incorporation of the two substrates. Second, the phosphorothioate modification of the α phosphate may affect not only catalysis *per se* but might also alter the positioning of the modified NTP in the active center. For example, in the case of DNA polymerases, α-S-NTPs are readily incorporated into DNA, but incorrectly incorporated α-S-NTPs fail to be proofread.[36] Nevertheless, taken together, available structural and biochemical data indicate that catalysis *per se* is not directly targeted by Stl.

Having largely dismissed translocation and catalysis as steps that are directly inhibited by Stl, Temiakov et al. argued that Stl could interfere with NTP addition by "locking" the elongating RNAP complex with bound NTP at a step after NTP binding but before catalysis. To identify the most plausible intermediate, the authors superimposed the Stl-binding site with all previously observed and proposed incoming NTP-binding sites (E, PS and IS) on the crystal structure of *Thermus thermophilus* RNAP. The modeling showed that bound Stl should block the transition of the NTP substrate from PS to IS. The alternative model of NTP delivery invokes only the E site, the existence of which has been reasonably well demonstrated. Temiakov et al. offered a way to reconcile the E site *versus* PS models of NTP delivery to the active center by proposing that the E and PS complex intermediates are parts of the same pathway (Figure 8.2B). The authors also suggested that formation of the E complex is not an essential step of NAC, since in the post-translocated state the incoming NTP can enter PS directly. Thus, in the presence of Stl, the correct, template-specified NTP substrate can bind to RNAP in the pre- or post-translocated state either *via* the E site or directly to PS; the E complex can then isomerize to the PS complex. Because the resultant PS complex fails to isomerize into the IS complex, the incoming NTP remains bound to RNAP but cannot be properly positioned for catalysis.

More recently, structures of *Thermus thermophilus* RNAP elongation complexes with template-specified nonhydrolyzable AMP-cPP were obtained in the presence or the absence of Stl.[37] As expected, Stl and the incoming NTP bound to RNAP simultaneously. Moreover, Stl did not cause major alterations in the binding of the NTP substrate, which retained Watson-Crick base-pairing with the $i+1$ base of the template DNA strand. The authors proposed that the positions of the NTP substrate analog in complexes with and without Stl represent the PS and IS complexes, respectively. A caveat to this work is that the authors infer that the complex observed in the crystal structure in the presence of Stl reflects a true PS intermediate. In the absence of direct evidence for the PS complex, it is not possible to assert that the RNAP elongation complex containing NTP and Stl is identical to PS complex that is presumed to form in the absence of the drug. Nevertheless, that NTP in the complex with or without Stl contained approximately the same number of interactions with surrounding protein groups was taken as an argument supporting the biological relevance of both complexes.

Vassylyev et al.[37] observed that the main difference between RNAP complexes containing the NTP substrate in PS *versus* IS sites resides in the conformation of TL and not of BH, as had been inferred by earlier work.[34] The authors suggest that Stl interferes with the dramatic structural transition of TL that is essential for NAC. In support of this model, Temiakov et al. reported substitutions in TL that drastically reduced the rate of nucleotide addition but did not alter substrate binding (and thus mimicked the effect of Stl binding).[20] Similar structural rearrangements of TL were reported to take place in eukaryotic RNAP upon correct NTP binding.[38]

In addition, more recent work suggested that the effect of Stl on RNAP might be influenced by the local DNA context.[39] The authors showed that Stl failed to inhibit elongation complexes that escape from the hairpin-dependent *his* pause. Based on this surprising observation, the authors proposed that elongation complexes paused at this site have an altered conformation of the active center that prevents Stl binding altogether.

In conclusion, the studies of the mechanism of RNAP inhibition by antibiotic Stl would have been impossible without the advent and subsequent improvement in crystallization and analysis of various "frozen" complexes that are used to model different stages of NAC. This provides a compelling example of how feedback between genetics, biochemistry and structural analysis synergistically improves understanding of the structure, function and possible regulatory points of the RNAP active center.

8.3.1.2 *Tagetitoxin: Two Modes, three Magnesium Ions?*

Tagetitoxin (Tgt), an antibiotic produced by a bacterial plant pathogen *Pseudomonas syringae* pv. *tagetis*, inhibits chloroplast RNAP,[40] bacterial RNAP, as well as eukaryotic Pol III *in vivo* and *in vitro*.[40,41] Addition of Tgt to *in vitro* transcription reactions performed under conditions of multiple-round steady state transcription results in immediate cessation of transcription, indicating that Tgt is able to inhibit elongation. When initiation of transcription was assayed, Tgt strongly inhibited the synthesis of abortive dinucleotide pppApU from ATP and UTP, whereas the inhibitory effect was much weaker with AMP as the initiating substrate. These results suggested that Tgt does not occlude the RNAP active center but nevertheless binds close enough to interfere with the binding of initiating NTP phosphates. Transcription inhibition by Tgt could not be relieved by increasing the concentration of NTP substrates, pointing to a noncompetitive mechanism of inhibition. Using eukaryotic Pol III, it was shown that Tgt does not change the pattern of transcriptional pausing but instead increases the duration of intrinsic pauses.[42]

A crystal structure of *Thermus thermophilus* RNAP holoenzyme in complex with Tgt was obtained.[43] In this structure, a molecule of Tgt is bound at the base of the secondary channel, close to, but not overlapping with, the RNAP active center (Figure 8.2C). Substitutions of *E. coli* RNAP amino acids corresponding to Tth RNAP amino acids participating in Tgt binding were created and, consistent with expectations from the structure, were found to be resistant to Tgt *in vitro*. Since Tgt binds close to the RNAP active site, it was proposed that the mechanism of RNAP inhibition by Tgt might involve steric interference with NTP binding at one of the steps during NAC. Structure-based modeling of Tgt binding to transcription complexes with the NTP substrate in the E site, PS or IS was performed. The modeling suggested that binding of the incoming NTP to the E site is the most likely target of Tgt. Steric clashes between Tgt and NTP in PS were considered "slight," and no significant clashes between Tgt and NTP bound in IS were apparent.

Another important observation made from the structure analysis is that Tgt appears to bring an additional Mg^{2+} ion into the RNAP active center (Figure 8.2C). This Mg^{2+} ion, designated tMg^{2+}, is coordinated by the Tgt phosphate moiety and two residues in the RNAP active site.[43] This mode of coordination is similar to coordination of the catalytic Mg^{2+} II. However, structural modeling suggests that all three Mg^{2+} ions can bind at the active site simultaneously in the presence of Tgt. Since tMg^{2+} is located too far from the active center, coordination of the substrate by tMg^{2+} was deemed to be incompatible with catalysis. In addition, even if the NTP substrate in this triple-Mg^{2+} active center is coordinated by the catalytic Mg^{2+}II ion, an extra Mg^{2+} ion bound in the active center might perturb the relative positioning of Mg^{2+}I and Mg^{2+}II and thus alter the catalytic properties of RNAP.

Overall, it appears that Tgt binding leads to at least two different effects: first, Tgt directly interferes with NTP binding to the E site; second, by bringing in the additional tMg^{2+} ion, Tgt affects NTP coordination by the catalytic Mg^{2+} ions in the active center. Recruitment of an additional Mg^{2+} ion to the active center is a novel mechanism of RNAP regulation. A similar mechanism has been proposed for RNAse H.[44] Therefore, manipulation of catalytic Mg^{2+} ions might be a general strategy for active center regulation in enzymes that act on nucleic acids.

The proposed mechanism of RNAP inhibition by Tgt *via* interference with NTP binding is not easy to reconcile with the experimentally-observed noncompetitive mechanism of inhibition with regard to substrate NTP. One interpretation for this apparent paradox is that the E site might indeed not be an essential intermediate of NAC and might instead serve a regulatory role. In particular, from the observations of Steinberg and Burgess[42] it follows that the function of the E site in NAC might be more important during transcriptional pauses. One can speculate that during elongation by Pol III, "normal" catalysis involving Mg^{2+} I and Mg^{2+} II proceeds even in the presence of Tgt. However, when transcription complex reaches a pause site, the geometry of the active center changes, such that the NTP substrate is no longer able to coordinate catalytic Mg^{2+} II but instead coordinates inhibitory tMg^{2+}.

8.3.1.3 ppGpp: A Natural Effector of the Active Center, or does Orientation Matter?

ppGpp, a tetraphosphate regulatory nucleotide synthesized by the RelA protein in a ribosome-dependent manner,[45] is the principal mediator of stringent response in bacteria (see also Chapters 1–3). Dubbed "the magic spot" for historical reasons,[46,47] ppGpp acts on RNAP predominantly, but not exclusively, during transcription initiation. ppGpp directly represses transcription *in vivo* and *in vitro* from promoters responsible for synthesis of stable RNA, such as the ribosomal RNA promoters. Cells deficient in ppGpp production do not grow in media devoid of amino acids.[45,48] A direct stimulatory effect of ppGpp on promoters of amino acid biosynthesis genes was shown, although the mechanism of this stimulation is not completely clear.[49,50] Promoters that are negatively regulated by ppGpp are more likely to contain the so-called "discriminator" sequence and have a shortened spacer between their –10 and –35 motifs; they also tend to form unstable open complexes with RNAP.[51,52]

Identification of a ppGpp-binding site on RNAP by conventional methods has proven to be difficult, and so its binding site, yet again, was determined by crystallography.[53] In the structure of the *T. thermophilus* RNAP holoenzyme–ppGpp complex, ppGpp is bound deep within the RNAP secondary channel, overlapping with the Tgt binding site. Similar to the situation with Tgt, ppGpp binding does not appear to cause gross conformational changes in the RNAP molecule and appears to enable coordination of an extra Mg^{2+} ion in the active center. Unlike the situation with Tgt, no steric clashes between any of the substrate NTP binding sites and ppGpp are apparent.

In addition, crystal structure analysis suggested that ppGpp can bind to the same RNAP site in two different orientations.[53] In one orientation, ppGpp is positioned such that it can specifically base-pair with nontemplate DNA strand. This finding may be related to the frequent occurrence of GC-rich discriminator sequences in promoters regulated by ppGpp.[53] Indeed, the importance of nontemplate strand cytidines in the *downstream* part of the discriminator sequence for ppGpp response was experimentally confirmed on a derivative of the λPR promoter and on ribosomal promoters.[53] In addition, the authors presented evidence that, during elongation, ppGpp preferentially increases RNAP pausing at C residues, indicating that ppGpp inhibits transcription elongation, at least in part, through base-pairing with the DNA template.

The possible existence of two alternative, overlapping ppGpp binding sites in the RNAP suggests an intuitively attractive explanation for previously observed dual effects of ppGpp on transcription. The two orientations may correspond to the "stimulatory" and "repressive" modes of ppGpp action. However, the existence of the two orientations is based solely on crystal structure analysis, and so it remains unclear whether the same two orientations in fact exist *in vivo* and, if they do, whether these conformational states have regulatory roles.

A detailed biochemical analysis of the mechanism for ppGpp inhibition has recently been reported.[54] The authors found that their data not only directly contradicted the findings of Artsimovitch *et al.*[53] but they also offered alternative explanations of the earlier results. In particular, contrary to findings of Artsimovitch *et al.*,[53] ppGpp did not destabilize promoter complexes formed by *T. thermophilus* RNAP. Furthermore, the authors noted that ppGpp, while conserved in evolution, does not play the same roles in different organisms. Lastly, according to Vrentas *et al.*,[54] the binding site of ppGpp in the vicinity of the active site of *Thermus* RNAP might not be functionally relevant and in fact might be nonspecific.

8.3.2 Regulation of RNAP by Proteins that Bind in the Secondary Channel

The emergence of the RNAP secondary channel as *a* path for substrate NTP entry and *the* path for extrusion of 3′-end of RNA during backtracking leaves little doubt that this channel would be used for regulation by protein factors. Some of the proteins shown to bind to the secondary channel include the *E. coli* transcript cleavage factors GreA and GreB, a stringent response cofactor DksA, transcription inhibitor from *Thermus thermophilus* Gfh1, and eukaryotic transcript cleavage factor TFIIS (Figure 8.3). Below, we briefly discuss currently known mechanisms of regulation of the RNAP active center by proteins binding in the secondary channel.

8.3.2.1 Transcript Cleavage Factors GreA and GreB

GreA and GreB were isolated from *E. coli* cell lysates during purification of an activity that caused cleavage of RNA in transcription complexes.[55,56] The two factors are homologs and share extensive sequence and structural similarities (Figure 8.3).[56,57] The transcript cleavage activities of the two factors are not identical: GreA induces cleavage of 2–3 nt-long RNA whereas GreB releases longer cleavage products, up to 18 nucleotides long. Accordingly, while both proteins can prevent RNAP from entering a paused state, only GreB is capable of rescuing RNAP from the state of permanent arrest.[56,58,59] A low-resolution structure of RNAP–GreB complex was obtained.[60] GreB was shown to bind at the rim of the RNAP secondary channel and insert its coiled coil domain into the secondary channel, such that two conserved aspartate residues at the tip of

Regulation of RNA Polymerase through its Active Center

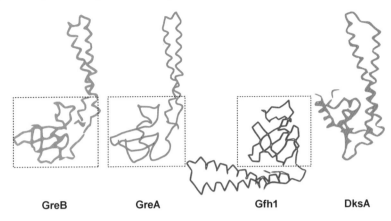

Figure 8.3 Structural comparison of bacterial transcript cleavage factors. A schematic representation of bacterial cleavage factors GreB, GreA, Gfh1 and DksA based on crystal structure analysis.[57,61,79,83] The structures are aligned by their N-terminal globular domains (boxes) to show the difference in relative orientations of the C-terminal coiled-coil domain of Gfh1 and Gre factors.

the coiled coil reach the RNAP active center (see Figure A.3). GreA is presumed to bind RNAP in a generally similar manner.[61] Structural modeling suggests that the aspartic acid residues at the tip of the coiled coil are positioned such that either GreA or GreB can stabilize Mg^{2+} II in the active center. This Gre-induced stabilization of Mg^{2+} II and its positioning with respect to Mg^{2+} I is believed to promote internal hydrolytic cleavage of the nascent transcript in the active center.[3,62,63] The RNAP catalytic center itself possesses an intrinsic transcript cleavage activity and the Gre factors merely stimulate this activity.[64] However, the intrinsic transcript cleavage activity requires harsh and unusual *in vitro* conditions (NTP starvation, high pH)[3] and is not sufficient to rescue elongation complexes from artificial arrest sites *in vivo*, in the absence of transcript cleavage factors.[65] However, we note, yet again, that while the active center may require additional factors to perform certain functions *in vivo*, it is its fundamental property to perform both RNA synthesis and degradation that makes regulation by these factors possible.

What is the cause of the differences between the transcript cleavage activities of GreA and GreB? One possibility is that GreA, is unable to rescue arrested elongation complexes because GreA, unlike GreB, is unable to bind to these complexes. An alternative possibility is that GreA and GreB differ in their ability to interact with nascent RNA in these complexes. A common structural feature of GreA and GreB, the so-called basic patch, is likely responsible for this interaction; the smaller patch of GreA might not be sufficient to bind to RNAP in the presence of long nascent RNA in a manner that would allow catalysis in the RNAP active center.[66] What is clear is that the presence of Asp residues at the tip of the coiled coil by itself is not sufficient to induce RNA cleavage, since DksA, which also contains these residues, does not support RNA cleavage (see below). These findings suggest that seemingly similar alterations of the RNAP active center by regulatory factors can lead to drastically different consequences, and indicate that the ultimate outcome of a regulatory signal is determined by the seemingly small changes in the structure of the active center.

8.3.2.2 TFIIS: at the Cutting Edge?

The eukaryotic Pol II transcription factor TFIIS is a functional analog of bacterial Gre factors.[59,67,68] Structural analysis suggests that TFIIS contains three domains. The C-terminal domains II and III bear a structural resemblance to bacterial Gre factors and are required for TFIIS RNA cleavage activity,[69] whereas its N-terminal domain I has no bacterial counterparts.[70] Like the two domains of Gre factors, TFIIS domains II and III are separated by a linker, and the tip of the TFIIS C-terminal domain contains a pair of acidic residues. Similar to the action of Gre factors on bacterial RNAP, TFIIS has been shown to induce RNA cleavage in Pol II transcription complexes[71,72] and to rescue Pol II from the paused state *in vitro* and *in vivo*.[16,73] In addition, TFIIS appears to increase the fidelity of Pol II transcription.[74,75] TFIIS-mediated RNA cleavage is similar to that induced by GreB: TFIIS is able to cleave off relatively long, 7–9 nucleotide, RNA products. Crystal structures of free Pol II with TFIIS[69] and Pol II elongation complex with TFIIS[26] were obtained. In both cases, TFIIS was found to bind at the rim of the secondary channel (pore) and insert its C-terminal domain into this pore such that the two acidic residues at the tip of the C-terminal domain reach the active center. Similar to the case of Gre factors, acidic residues of TFIIS are suggested to complement the Pol II active center by bringing in a Mg^{2+} ion (and a water molecule) are essential for hydrolysis of the RNA chain. However, the role of TFIIS in the cell and the mechanism of its action on Pol II might not be so simple: in addition to a GreB-like Mg^{2+}-coordinating activity, TFIIS appears to induce substantial changes in transcription complex structure, altering the alignment of nucleic acids in the active center.[69] The authors consider this as evidence that TFIIS may modify the general elongation properties of Pol II allosterically, independently of its ability to act as a cleavage factor.

8.3.2.3 Gfh1: Inhibition through the Secondary Channel

Gfh1 is one of two homologs of *E. coli* Gre proteins encoded in the *Thermus aquaticus* genome.[62,76] Unlike Gre factors, Gfh1 inhibits transcription *in vitro*. Crystal structures of Gfh1 have been independently obtained by several groups.[77–79] Analysis of these structures shows that the overall fold of Gfh1, although generally similar to that of GreA, bears two notable differences.

First, when the N-terminal globular domains of GreA and Gfh1 are superimposed, their C-terminal domains are positioned such that they appear to have been rotated by almost 180° relative to each other (Figure 8.3).[78] Structural considerations suggest that Gfh1 can not form a functional complex with

RNAP if Gfh assumes the conformation detected in the crystals. The authors therefore hypothesized that Gfh1 must exist in two alternative forms, and that the inactive form of Gfh1 must undergo a conformational change before it can bind to RNAP. Since the effects of Gfh1 on transcription *in vitro* are exquisitely sensitive to changes in pH, it has been suggested that Gfh1 might respond to intracellular pH changes in the cell.[78] Another important difference between Gfh1 and Gre is that Gfh1 has four acidic residues at the tip of the coiled-coil domain, compared to two such residues in Gre factors. This difference is likely responsible for the drastically different consequences of Gre factors and Gfh1 binding to RNAP. As with Gre, Gfh1 is thought to bring $Mg^{2+}II$ into the catalytic center. However, by providing an extra two acidic residues, Gfh1 effectively prevents $Mg^{2+}II$ coordination by the incoming NTP, leading to transcription inhibition. Consistent with this interpretation, a Gfh1 mutant lacking all four acidic residues had a greatly diminished ability to inhibit transcription (the residual inhibitory activity of this mutant was attributed to sterical occlusion of the secondary channel). An alternative model that is also consistent with the experimental data posits that Gfh1 inhibits transcription by introducing an additional Mg^{2+} ion into the active center, acting essentially like Tgt (above).[79] While additional experiments to distinguish between the two models are needed, it is clear that Gfh1 exemplifies yet another regulatory factor that exploits the amazing plasticity of the RNAP active center.

8.3.2.4 *DksA: A Potentiator of ppGpp*

The *in vitro* effects of ppGpp on transcription are much less pronounced than the magnitude of the *in vivo* ppGpp-dependent stringent response.[50,80] This discrepancy was resolved when DksA was identified as an essential cofactor of stringent response. DksA was originally isolated as a suppressor of *dnaK*⁻ mutants (hence the name DksA, *dnaK* suppressor A).[81] Subsequently, DksA was shown to be required for proper response of ribosomal promoters to ppGpp *in vivo*.[82] *In vitro*, DksA reduces the lifetime of open promoter complexes and amplifies the inhibitory effect of ppGpp on transcription.[82] *In vitro* activation of transcription from amino acid biosynthesis promoters by ppGpp is also dependent on DksA.

Despite the absence of a detectable sequence similarity, the structure of DksA closely resembles that of GreA (Figure 8.3).[83] Both proteins contain a globular domain and a coiled-coil domain with two acidic residues at its tip. Modeling shows that the acidic residues at the tip of the coiled-coil of DksA can reach the active center and coordinate catalytic Mg^{2+} ions when this domain is inserted into the secondary channel. Despite this structural similarity, DksA does not prevent RNAP pausing and is unable to induce RNA cleavage. However, like GreA, DksA prevents RNAP from backtracking and entering the arrested state. The different orientations of acidic residues in DksA and GreA with respect to the coiled-coil domains may account for the functional differences between these proteins.

While a crystal structure of RNAP in a complex with bound DksA is not yet available, structural modeling and biochemical analysis indicate that the acidic residues at the tip of the DksA coiled-coil may stabilize the binding of ppGpp. Indeed, substitution of these residues diminishes the ability of DksA to enhance the effect of ppGpp.[83]

In vivo tests aimed at determining whether GreA or GreB can substitute for DksA revealed that GreB (but not GreA) inhibited transcription from ribosomal promoters *in vivo*.[84] Unlike DksA, GreB did not upregulate *E. coli* promoters. While the mechanisms underlying the differences are not clear, these results have established an *in vivo* system that should facilitate functional dissection of factors that interact with the secondary channel of RNAP.

8.3.2.5 The ω Subunit of RNAP: A Remote Control for the Active Center?

The most obscure of the RNAP subunits, ω, is encoded by the *rpoZ* gene. While conserved in all organisms, ω is the only subunit that is dispensable for bacterial RNAP function *in vitro* and is non-essential for cell viability under normal growth conditions. It has been proposed that the main role of ω is to promote RNAP assembly,[85] indicating that ω might be more important at suboptimal growth conditions or during stress. Recent work demonstrated that RNAP that lacked the ω subunit was unable to respond to ppGpp *in vitro*, but the addition of DksA rescued the ability of ω-less RNAP to respond to ppGpp.[86] Subsequent work identified an *in vivo* defect of an *E. coli rpoZ* deletion, namely, a slow growth phenotype in the background of *dksA* deletion.[84]

At first sight, a direct interaction between ω and DksA appears impossible, since the two proteins bind on opposite sides of the RNAP molecule. However, structural analysis suggests that the binding sites for ω and ppGpp are connected to each other through two helical regions of β′ (β′ N458-L483 and N910-Q929 α-helices, the numbering refers to *E. coli* RNAP), opening up the possibility that the two proteins might communicate allosterically through these helices. It is interesting to consider whether additional factors that modulate transcription by interacting with ω exist in the cell or whether eukaryotic orthologs of ω, such as Rpb6 subunit of Pol II,[85] might serve as remote control points of the active center.

8.4 Transcript Proofreading

In the absence of proofreading, selection of nucleotides by a multi-subunit RNAP prior to incorporation was estimated to provide approximately 10^3-fold discrimination between the correct and incorrect NTP.[87] A value higher than that provided solely by Watson–Crick base pairing (10–100) presumably reflects active contribution of RNAP active center to this discrimination. High transcriptional fidelity is probably crucial for production of nontranslated

RNAs. On the other hand, because the frequency of translation errors is on the order of 10^{-4}, in the case of coding RNAs increases in transcriptional fidelity beyond this level would not bring substantial benefit to the overall rate of correct amino acid incorporation into proteins.[88,89]

The most crucial property of RNAP that permits proofreading is that the RNAP itself can sense a misincorporation event and respond by pausing.[5,74,90] RNA polymerase can also stall at sites where DNA is damaged, thus facilitating recruitment of the repair machinery.[91–93] The reduction of the rate of incorporation of correct NTP following a misincorporation event was estimated to be 50–500-fold for bacterial RNAPs[5,90] and 5–20-fold for eukaryotic RNAPs, depending on sequences of templates, the nature of misincorporated nucleotide, and the nature of NTP to be incorporated.[74,87,91] *In vitro*, archaeal RNAP failed to extend misincorporated nucleotides even after prolonged incubation times.[94] Recently, it was suggested that the reason why RNAP stalls upon misincorporation might be because the misincorporated 3′-end proximal NMP flips out of its position in IS and binds in (or in close vicinity to) the E site. This causes backtracking of the complex by 1 bp, such that the penultimate NMP of the transcript moves to the $i+1$ site of the active center and becomes amenable to cleavage (Figure 8.4).[5] Thus, the template-independent E site may play a role both in substrate selection during synthesis and in misincorporated substrate excision during proofreading.

In general, any process that allows removal of 3′-proximal nucleotide could contribute to transcriptional proofreading. Three such processes that are likely to be the most significant ones are discussed below.

8.4.1 Transcriptional Proofreading through Pyrophosphorolysis

The reaction of pyrophosphorolysis involves a nucleophilic attack of a phosphodiester bond by inorganic pyrophosphate.[95] Pyrophosphate can be fixed in the catalytic site more firmly than a water molecule due to a more extensive set of interactions with amino acids of the active center, which is probably the main reason why pyrophosphorolysis, unlike hydrolytic cleavage, can occur under physiological conditions without accessory factors. The pyrophosphate molecule likely enters the RNAP catalytic center in complex with a Mg^{2+} ion, just as the substrate NTP does during the synthesis reaction. As discussed in Chapter 7, the Hopfield Criterion dismisses the significance of proofreading through pyrophosphorolysis if one considers pyrophosphorolysis as cleavage of the misincorporated nucleotide, *i.e.*, a direct reversal of transcription reaction. However, pyrophosphorolysis of a penultimate phosphodiester bond in a backtracked, misincorporated complex is theoretically possible. Low concentrations of pyrophosphate (0.05 mM) were suggested to increase fidelity of bacterial RNAP 5- to 50-fold.[95] Since it is hard to determine pyrophosphate concentrations in the cell, the contribution of pyrophosphorolysis to transcription fidelity *in vivo* is not clear. Nevertheless, the estimated concentrations of PPi in the cell (0.1–2 mM) might be high enough to make a significant contribution.

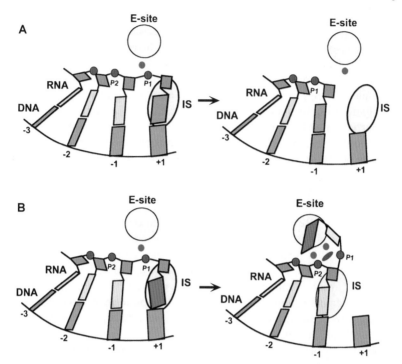

Figure 8.4 Transcript-mediated transcriptional proofreading. Nucleic acids in the RNAP active center are indicated as in Figure 8.2. Positions of the E site and IS are indicated by ovals; PS not shown. Positioning of RNAP IS with respect to the nascent RNA is indicated underneath the DNA. (A) Normal incorporation event. The template-specified NTP (indicated in green) is incorporated into the nascent RNA (left-hand panel). The resultant transcription complex can continue elongation, as indicated by its forward translocation (see Figure 8.2, step 6, and Chapter 7). (B) Transcript-mediated proofreading of a misincorporated NTP. Incorporation of an incorrect NTP (left-hand panel; misincorporated NTP base is shown in red) causes RNAP backtracking by 1 bp, such that the misincorporated NTP base binds in the E site (right panel) and can then be excised in the absence of additional factors with the exception of Mg^{2+}, which becomes cMgII (shown in red) and a molecule of water (blue oval).

8.4.2 Proofreading by Transcript Cleavage Factors

Proteins that assist RNAP in hydrolysis of phosphodiester bonds are present in all domains of life. They include eukaryotic TFIIS and Rpc11, archaeal TFIIS homolog TFS, and bacterial Gre factors. All of these factors have been shown to significantly increase proofreading *in vitro*.[74,90,94]

The significance of transcript cleavage for transcription fidelity *in vivo* might vary for different organisms. In bacteria, none of the cleavage factors are essential, and in *E. coli* a double *ΔgreA ΔgreB* mutant shows only moderate growth defects. In particular, the principal role of GreA in the cell might not be as a general RNA cleavage factor: a recent study suggested that, *in vivo*, GreA facilitates RNAP escape from promoters on a subset of promoters.[96] In eukaryotes, TFIIS does not appear to contribute significantly to Pol II fidelity *in vivo*,[97] although opposing views have also been presented.[75] In contrast, for Pol III whose active center lacks intrinsic proofreading activity, the transcript cleavage activity provided by the Rpc11 subunit is absolutely essential for the function of the enzyme.[98,99]

On the other hand, factors that do not appear to possess RNA cleavage activity have been shown to affect transcriptional fidelity *in vivo*. Examples include the Pol II Rpb9 subunit[97] and Hepatitis Delta Virus HDAg antigen.[100] Mechanisms through which these factors increase transcription fidelity are currently unknown.

8.4.3 Transcript-assisted Proofreading. A New Class of Ribozymes?

Recent work described a novel mechanism of transcriptional proofreading.[5] The authors postulated that upon incorporation of an incorrect NTP the geometry of the elongating transcription complex changes. The penultimate phosphodiester bond is repositioned such that it becomes amenable for cleavage. Remarkably, this cleavage does not require any additional external factors except for Mg^{2+} and is stimulated by the misincorporated nucleotide itself. This *transcript-assisted* transcription proofreading directly involves the nascent RNA transcript and thus was proposed to take place through a ribozyme-like reaction, in which the 3'-terminal nucleotide participates in cleaving the penultimate phosphodiester bond (Figure 8.4).

Evidence for RNA self-cleavage in misincorporated transcription complexes was obtained in an experiment that involved *Thermus aquaticus* RNAP elongation complexes assembled with synthetic DNA and RNA oligonucleotides. The complexes were identical except for the 3'-terminal RNA nucleotide. When preassembled complexes that contained an incorrect nucleotide at the 3'-terminus were supplemented with Mg^{2+}, the penultimate phosphodiester bond was efficiently cleaved, as judged by a two-nucleotide shortening of RNA. In contrast, RNA in complexes containing properly base-paired 3'-terminal nucleotide was cleaved much less efficiently, indicating that this self-cleavage is specific to misincorporation events. Importantly, this cleavage was distinct from the previously-described 3'-terminal phosphodiester bond cleavage in that the latter, but not the former, was stimulated by a free noncomplementary NTP.[3] Moreover, cleavage induced by incorrect free NTP was observed only in correctly incorporated complexes but not in misincorporated complexes.[5]

The results thus indicated that cleavage of the penultimate phosphodiester bond occurred through a novel mechanism that intimately involved the 3'-terminal misincorporated RNA nucleotide. In agreement with this proposal, the detailed kinetic parameters of this cleavage reaction were dependent on the exact nature of the 3′-terminal base. The authors dissected the contribution of individual groups within the bases by monitoring self-cleavage in elongation complexes assembled with 3′-terminal RNA nucleotides containing various modifications.[5]

It is also important to note that this ribozyme-like transcript self-cleavage requires that the 3-terminal nucleotide is relocated to the position that is normally occupied by the incoming NTP, the E site. Thus, the self-cleavage reaction can be regarded as a reversal of the incoming NTP entry in the active center during NAC.

While this ribozyme-like cleavage was efficient in an *in vitro* system, the generality of this mechanism and its potential role *in vivo* are yet to be established. As is also the case with other activities of the active center, self-cleavage of misincorporated nucleotides could be a regulatory target of as-yet unknown transcription factors.

8.5 Conclusions

Recent works have used a combination of incisive structural and biochemical approaches to produce high-resolution views of the multi-subunit RNAP at different stages of transcription. As a result, it is now possible to envision the molecular details of RNAP function and to begin to understand the very interactions that orchestrate individual atoms of substrates and individual atoms of amino acids in the active center to perform transcription. In particular, it is becoming increasingly clear that the active center of RNAP is not a rigid module dedicated exclusively to RNA synthesis, but is a highly flexible platform that can perform various reactions, such as transcript cleavage, in response to nucleotide misincorporation or regulatory factors. It is not an overstatement to say that these "extra" functions of the active center are what largely make transcription regulation possible. However, it is also clear that the current models of RNAP function are far from being complete and many aspects of RNAP function remain controversial today. In particular, many complexes from which the current assumptions about structures of transcriptional intermediates are being made have been successfully crystallized only with bacterial or only with eukaryotic RNAP, but not both. We also do not yet know whether the observed differences in the active center architecture between bacterial and eukaryotic RNAPs stem from differences in experimental conditions, or reflect genuine differences in the way these enzymes function. We fully expect, however, that the continuing refinement of structural and biochemical methods combined with a wider adoption and improvement of novel, including single-molecule, techniques will enable researches to ultimately understand the mechanisms of transcription and the principles of transcription regulation.

Acknowledgements

We are grateful to Robert Landick for generously sharing Figure 8.1 and to Sergei Borukhov for helpful discussion. This work was supported in part by the Intramural Research Program of the US National Institutes of Health, National Institute of Environmental Health Sciences. We also acknowledge support from BBSRC grants BB/F006462/1and BB/F013558/1 to N.Z., and from NIH as well as Russian Academy of Sciences Program in Molecular and Cellular Biology Grants to K.S.

References

1. T. A. Steitz and J. A. Steitz, *Proc. Natl. Acad. Sci. USA*, 1993, **90**, 6498.
2. T. A. Steitz, *Nature*, 1998, **391**, 231.
3. V. Sosunov, E. Sosunova, A. Mustaev, I. Bass, V. Nikiforov and A. Goldfarb, *EMBO J.*, 2003, **22**, 2234.
4. W. Yang, J. Y. Lee and M. Nowotny, *Mol. Cell*, 2006, **22**, 5.
5. N. Zenkin, Y. Yuzenkova and K. Severinov, *Science*, 2006, **313**, 518.
6. G. Zhang, E. A. Campbell, L. Minakhin, C. Richter, K. Severinov and S. A. Darst, *Cell*, 1999, **98**, 811.
7. S. A. Darst, E. W. Kubalek and R. D. Kornberg, *Nature*, 1989, **340**, 730.
8. P. Cramer, D. A. Bushnell and R. D. Kornberg, *Science*, 2001, **292**, 1863.
9. A. Hirata, B. J. Klein and K. S. Murakami, *Nature*, 2008, **451**, 851.
10. V. Markovtsov, A. Mustaev and A. Goldfarb, *Proc. Natl. Acad. Sci. USA*, 1996, **93**, 3221.
11. N. Komissarova and M. Kashlev, *Proc. Natl. Acad. Sci. USA*, 1997, **94**, 1755.
12. E. Nudler, A. Mustaev, E. Lukhtanov and A. Goldfarb, *Cell*, 1997, **89**, 33.
13. N. Komissarova and M. Kashlev, *J. Biol. Chem.*, 1997, **272**, 15329.
14. N. Korzheva, A. Mustaev, M. Kozlov, A. Malhotra, V. Nikiforov, A. Goldfarb and S. A. Darst, *Science*, 2000, **289**, 619.
15. J. Andrecka, R. Lewis, F. Bruckner, E. Lehmann, P. Cramer and J. Michaelis, *Proc. Natl. Acad. Sci. USA*, 2008, **105**, 135.
16. K. Adelman, J. Yuzenkova, A. La Porta, N. Zenkin, J. Lee, J. T. Lis, S. Borukhov, M. D. Wang and K. Severinov, *Mol. Cell*, 2004, **14**, 753.
17. J. Mukhopadhyay, E. Sineva, J. Knight, R. M. Levy and R. H. Ebright, *Mol. Cell*, 2004, **14**, 739.
18. N. N. Batada, K. D. Westover, D. A. Bushnell, M. Levitt and R. D. Kornberg, *Proc. Natl. Acad. Sci. USA*, 2004, **101**, 17361.
19. K. D. Westover, D. A. Bushnell and R. D. Kornberg, *Cell*, 2004, **119**, 481.
20. D. Temiakov, N. Zenkin, M. N. Vassylyeva, A. Perederina, T. H. Tahirov, E. Kashkina, M. Savkina, S. Zorov, V. Nikiforov, N. Igarashi, N. Matsugaki, S. Wakatsuki, K. Severinov and D. G. Vassylyev, *Mol. Cell*, 2005, **19**, 655.
21. X. Q. Gong, C. Zhang, M. Feig and Z. F. Burton, *Mol. Cell*, 2005, **18**, 461.

22. D. G. Vassylyev, M. N. Vassylyeva, A. Perederina, T. H. Tahirov and I. Artsimovitch, *Nature*, 2007, **448**, 157.
23. E. A. Kashkina, M. V. Anikin, W. T. McAllister, N. Kochetkov and D. E. Temyakov, *Dokl. Biochem. Biophys.*, 2007, **416**, 285.
24. A. Blank, J. A. Gallant, R. R. Burgess and L. A. Loeb, *Biochemistry*, 1986, **25**, 5920.
25. D. Temiakov, V. Patlan, M. Anikin, W. T. McAllister, S. Yokoyama and D. G. Vassylyev, *Cell*, 2004, **116**, 381.
26. H. Kettenberger, K. J. Armache and P. Cramer, *Mol. Cell*, 2004, **16**, 955.
27. C. Siddhikol, J. W. Erbstoeszer and B. Weisblum, *J. Bacteriol.*, 1969, **99**, 151.
28. K. Logan, J. Zhang, E. A. Davis and S. Ackerman, *DNA*, 1989, **8**, 595.
29. W. R. McClure, *J. Biol. Chem.*, 1980, **255**, 1610.
30. Y. Iwakura, A. Ishihama and T. Yura, *Mol. Gen. Genet.*, 1973, **121**, 181.
31. T. O. Morrow and S. A. Harmon, *J. Bacteriol.*, 1979, **137**, 374.
32. K. Severinov, M. Soushko, A. Goldfarb and V. Nikiforov, *J. Biol. Chem.*, 1993, **268**, 14820.
33. K. Severinov, D. Markov, E. Severinova, V. Nikiforov, R. Landick, S. A. Darst and A. Goldfarb, *J. Biol. Chem.*, 1995, **270**, 23926.
34. S. Tuske, S. G. Sarafianos, X. Wang, B. Hudson, E. Sineva, J. Mukhopadhyay, J. J. Birktoft, O. Leroy, S. Ismail, A. D. J. Clark, C. Dharia, A. Napoli, O. Laptenko, J. Lee, S. Borukhov, R. H. Ebright and E. Arnold, *Cell*, 2005, **122**, 541.
35. V. Epshtein, A. Mustaev, V. Markovtsov, O. Bereshchenko, V. Nikiforov and A. Goldfarb, *Mol. Cell*, 2002, **10**, 623.
36. T. A. Kunkel, F. Eckstein, A. S. Mildvan, R. M. Koplitz and L. A. Loeb, *Proc. Natl. Acad. Sci. USA*, 1981, **78**, 6734.
37. D. G. Vassylyev, M. N. Vassylyeva, J. Zhang, M. Palangat, I. Artsimovitch and R. Landick, *Nature*, 2007, **448**, 163.
38. D. Wang, D. A. Bushnell, K. D. Westover, C. D. Kaplan and R. D. Kornberg, *Cell*, 2006, **127**, 941.
39. I. Toulokhonov, J. Zhang, M. Palangat and R. Landick, *Mol. Cell*, 2007, **27**, 406.
40. D. E. Mathews and R. D. Durbin, *J. Biol. Chem.*, 1990, **265**, 493.
41. D. E. Mathews and R. D. Durbin, *Biochemistry*, 1994, **33**, 11987.
42. T. H. Steinberg and R. R. Burgess, *J. Biol. Chem.*, 1992, **267**, 20204.
43. D. G. Vassylyev, V. Svetlov, M. N. Vassylyeva, A. Perederina, N. Igarashi, N. Matsugaki, S. Wakatsuki and I. Artsimovitch, *Nat. Struct. Mol. Biol.*, 2005, **12**, 1086.
44. M. Nowotny, S. A. Gaidamakov, R. J. Crouch and W. Yang, *Cell*, 2005, **121**, 1005.
45. G. D. Cashel, M. V. Hernandez and D. Vinella, in *Escherichia coli and Salmonella*, ed. F.C. Neidhart, ASM Press, Washington, D. C., 1996, p. 1458.
46. K. Glazier and D. Schlessinger, *J. Bacteriol.*, 1974, **117**, 1195.
47. K. Potrykus, D. Vinella, H. Murphy, A. Szalewska-Palasz, R. D'Ari and M. Cashel, *J. Biol. Chem.*, 2006, **281**, 15238.

48. A. Maitra, I. Shulgina and V. J. Hernandez, *Mol. Cell*, 2005, **17**, 817.
49. B. J. Paul, M. B. Berkmen and R. L. Gourse, *Proc. Natl. Acad. Sci. USA*, 2005, **102**, 7823.
50. H. E. Choy, *J. Biol. Chem.*, 2000, **275**, 6783.
51. M. Cashel, *J. Biol. Chem.*, 1969, **244**, 3133.
52. B. J. Paul, W. Ross, T. Gaal and R. L. Gourse, *Annu. Rev. Genet.*, 2004, **38**, 749.
53. I. Artsimovitch, V. Patlan, S. Sekine, M. N. Vassylyeva, T. Hosaka, K. Ochi, S. Yokoyama and D. G. Vassylyev, *Cell*, 2004, **117**, 299.
54. C. E. Vrentas, T. Gaal, M. B. Berkmen, S. T. Rutherford, S. P. Haugen, W. Ross and R. L. Gourse, *J. Mol. Biol.*, 2008, **377**, 551.
55. S. Borukhov, A. Polyakov, V. Nikiforov and A. Goldfarb, *Proc. Natl. Acad. Sci. USA*, 1992, **89**, 8899.
56. S. Borukhov, V. Sagitov and A. Goldfarb, *Cell*, 1993, **72**, 459.
57. C. E. Stebbins, S. Borukhov, M. Orlova, A. Polyakov, A. Goldfarb and S. A. Darst, *Nature*, 1995, **373**, 636.
58. G. H. Feng, D. N. Lee, D. Wang, C. L. Chan and R. Landick, *J. Biol. Chem.*, 1994, **269**, 22282.
59. R. N. Fish and C. M. Kane, *Biochim. Biophys. Acta.*, 2002, **1577**, 287.
60. N. Opalka, M. Chlenov, P. Chacon, W. J. Rice, W. Wriggers and S. A. Darst, *Cell*, 2003, **114**, 335.
61. M. N. Vassylyeva, V. Svetlov, A. D. Dearborn, S. Klyuyev, I. Artsimovitch and D. G. Vassylyev, *EMBO Rep.*, 2007, **8**, 1038.
62. O. Laptenko, J. Lee, I. Lomakin and S. Borukhov, *EMBO J.*, 2003, **22**, 6322.
63. E. Sosunova, V. Sosunov, M. Kozlov, V. Nikiforov, A. Goldfarb and A. Mustaev, *Proc. Natl. Acad. Sci. USA*, 2003, **100**, 15469.
64. M. Orlova, J. Newlands, A. Das, A. Goldfarb and S. Borukhov, *Proc. Natl. Acad. Sci. USA*, 1995, **92**, 4596.
65. F. Toulme, C. Mosrin-Huaman, J. Sparkowski, A. Das, M. Leng and A. R. Rahmouni, *EMBO J.*, 2000, **19**, 6853.
66. D. Kulish, J. Lee, I. Lomakin, B. Nowicka, A. Das, S. Darst, K. Normet and S. Borukhov, *J. Biol. Chem.*, 2000, **275**, 12789.
67. M. G. Izban and D. S. Luse, *Genes Dev.*, 1992, **6**, 1342.
68. D. Reines, M. J. Chamberlin and C. M. Kane, *J. Biol. Chem.*, 1989, **264**, 10799.
69. H. Kettenberger, K. J. Armache and P. Cramer, *Cell*, 2003, **114**, 347.
70. V. Booth, C. M. Koth, A. M. Edwards and C. H. Arrowsmith, *J. Biol. Chem.*, 2000, **275**, 31266.
71. D. Wang and D. K. Hawley, *Proc. Natl. Acad. Sci. USA*, 1993, **90**, 843.
72. W. Gu and D. Reines, *J. Biol. Chem.*, 1995, **270**, 30441.
73. E. A. Galburt, S. W. Grill, A. Wiedmann, L. Lubkowska, J. Choy, E. Nogales, M. Kashlev and C. Bustamante, *Nature*, 2007, **446**, 820.
74. M. J. Thomas, A. A. Platas and D. K. Hawley, *Cell*, 1998, **93**, 627.
75. H. Koyama, T. Ito, T. Nakanishi and K. Sekimizu, *Genes Cells*, 2007 **12**, 547.

76. B. P. Hogan, T. Hartsch and D. A. Erie, *J. Biol. Chem.*, 2002, **277**, 967.
77. V. Lamour, B. P. Hogan, D. A. Erie and S. A. Darst, *J. Mol. Biol.*, 2006, **356**, 179.
78. O. Laptenko, S. S. Kim, J. Lee, M. Starodubtseva, F. Cava, J. Berenguer, X. P. Kong and S. Borukhov, *EMBO J.*, 2006, **25**, 2131.
79. J. Symersky, A. Perederina, M. N. Vassylyeva, V. Svetlov, I. Artsimovitch and D. G. Vassylyev, *J. Biol. Chem.*, 2006, **281**, 1309.
80. M. M. Barker, T. Gaal, C. A. Josaitis and R. L. Gourse, *J. Mol. Biol.*, 2001, **305**, 673.
81. P. J. Kang and E. A. Craig, *J. Bacteriol.*, 1990, **172**, 2055.
82. B. J. Paul, M. M. Barker, W. Ross, D. A. Schneider, C. Webb, J. W. Foster and R. L. Gourse, *Cell*, 2004, **118**, 311.
83. A. Perederina, V. Svetlov, M. N. Vassylyeva, T. H. Tahirov, S. Yokoyama, I. Artsimovitch and D. G. Vassylyev, *Cell*, 2004, **118**, 297.
84. S. T. Rutherford, J. J. Lemke, C. E. Vrentas, T. Gaal, W. Ross and R. L. Gourse, *J. Mol. Biol.*, 2007, **366**, 1243.
85. L. Minakhin, S. Bhagat, A. Brunning, E. A. Campbell, S. A. Darst, R. H. Ebright and K. Severinov, *Proc. Natl. Acad. Sci. USA*, 2001, **98**, 892.
86. C. E. Vrentas, T. Gaal, W. Ross, R. H. Ebright and R. L. Gourse, *Genes Dev*, 2005, **19**, 2378.
87. N. Alic, N. Ayoub, E. Landrieux, E. Favry, P. Baudouin-Cornu, M. Riva and C. Carles, *Proc. Natl. Acad. Sci. USA*, 2007, **104**, 10400.
88. J. Ninio, *Biochimie*, 1991, **73**, 1517.
89. R. J. Shaw, N. D. Bonawitz and D. Reines, *J. Biol. Chem.*, 2002, **277**, 24420.
90. D. A. Erie, O. Hajiseyedjavadi, M. C. Young and P. H. von Hippel, *Science*, 1993, **262**, 867.
91. G. E. Damsma, A. Alt, F. Brueckner, T. Carell and P. Cramer, *Nat. Struct. Mol. Biol.*, 2007, **14**, 1127.
92. F. Brueckner and P. Cramer, *FEBS Lett.*, 2007, **581**, 2757.
93. F. Brueckner, U. Hennecke, T. Carell and P. Cramer, *Science*, 2007 **315**, 859.
94. U. Lange and W. Hausner, *Mol. Microbiol*, 2004, **52**, 1133.
95. J. D. Kahn and J. E. Hearst, *J. Mol. Biol.*, 1989, **205**, 291.
96. E. Stepanova, J. Lee, M. Ozerova, E. Semenova, K. Datsenko, B. L. Wanner, K. Severinov and S. Borukhov, *J. Bacteriol.*, 2007, **189**, 8772.
97. N. K. Nesser, D. O. Peterson and D. K. Hawley, *Proc. Natl. Acad. Sci. USA*, 2006, **103**, 3268.
98. S. Chedin, M. Riva, P. Schultz, A. Sentenac and C. Carles, *Genes Dev.*, 1998, **12**, 3857.
99. E. Landrieux, N. Alic, C. Ducrot, J. Acker, M. Riva and C. Carles, *EMBO J.*, 2006, **25**, 118.
100. Y. Yamaguchi, T. Mura, S. Chanarat, S. Okamoto and H. Handa, *Genes Cells*, 2007, **12**, 863.

CHAPTER 9
Kinetic Modeling of Transcription Elongation

LU BAI,[a,b] ALLA SHUNDROVSKY[a,c] AND MICHELLE D. WANG[a]

[a] Department of Physics, Laboratory of Atomic and Solid State Physics, Cornell University, Ithaca, NY 14853, USA; [b] Current address: The Rockefeller University, New York, NY 10065, USA; [c] Current address: Department of Mechanical Engineering, Yale University, New Haven, CT 06511, USA

9.1 Introduction

Transcription elongation is the process by which RNA polymerase (RNAP) moves along template DNA and synthesizes a complementary RNA. During elongation, RNAP carries out a highly processive and directional net motion, even in the presence of a large external load.[1–3] From an energetics point of view, RNAP is a molecular motor capable of converting chemical energy derived from NTP hydrolysis into mechanical work.[4] Unlike other molecular motors such as kinesin or myosin that move along a uniform track, RNAP transcribes on a DNA substrate with varying sequence content. This variability of sequences significantly affects the kinetics of RNAP motion and results in non-uniform elongation rates.[5,6] Theoretical modeling of transcription has so far focused on two main aspects of elongation: (1) how RNAP couples chemical catalysis energy to its translocation and mechanical work and (2) how its motion is regulated by the DNA sequence.

The first question concerns the mechano-chemical coupling mechanism of RNAP and applies to any molecular motor, or more generally, to any motor. However, there is a major difference between molecular and macroscopic motors: molecular motors function on such a small scale that they are significantly affected by thermal (Brownian) fluctuations. Because Brownian motion is random, it alone cannot generate unidirectional motor motion; nevertheless, it plays an important role in the functioning of molecular motors.[7] Two distinct mechano-chemical coupling mechanisms have been proposed that differ in the way the motor utilizes thermal and chemical energies. In the "power-stroke" mechanism, the energy derived from chemical reaction is used directly to drive the motor forward. In the alternative "Brownian ratchet" mechanism, the motor's motion is driven by thermal fluctuations, while the chemical reaction imposes directionality by biasing the motion in a single direction.

The mechano-chemical coupling mechanism employed by RNAP is still under debate (see Chapters 4 and 7). The power-stroke mechanism was proposed based on crystallographic studies of T7 RNAP[8] and suggests that the release of PPi product at the end of the chemical reaction step induces a conformational change in RNAP that makes it forward translocate by 1 bp. However, several other experimental and theoretical studies[9-17] supported a Brownian ratchet mechanism where RNAP can slide back and forth on the DNA template activated by thermal energy and the incorporation of the next nucleotide biases the polymerase forward by one base pair. The most direct evidence for a Brownian ratchet model is the ability of the RNAP to non-catalytically slide backward along the DNA (backtrack) when elongation is blocked *via* NTP starvation or slowed at particular DNA sequences.[11] However, it is not clear whether such thermally-activated sliding occurs at all template positions during active elongation.[18] Recent kinetic modeling combined with single-molecule measurements of RNAP elongation kinetics under various experimental conditions have provided strong support for a Brownian ratchet mechanism.[12,13,15,17]

The second key question that theoretical modeling of transcription must address is the influence of the underlying DNA sequence on RNAP kinetics. Experimentally, it has long been recognized that on a DNA template with varying ATGC content the elongation reaction does not proceed at a uniform rate. In particular, RNAP tends to dwell at some template positions considerably longer than at others, a phenomenon known as transcriptional pausing.[5,6] Bulk biochemical assays have suggested that at least some of the observed pauses were caused by RNAP backtracking.[6,11] At such pause sites transcription is halted until the 3' end of the nascent RNA returns to the active site, either by RNAP forward translocation or by internal cleavage of the RNA at the active site. Single-molecule studies found that pauses could be divided into long/short duration pauses and that the two types of pauses exhibited different sensitivity to external force.[3,19,20] Measurements with improved spatial resolution revealed that RNAP tended to backtrack during the long pauses, but not the short ones, suggesting different pause mechanisms.[21,22] The DNA sequences that are known to induce pauses lack apparent sequence consensus. Theoretical studies suggest that the sequence-dependence of RNAP elongation kinetics is highly correlated with the free energy of the corresponding transcription elongation complex (TEC), which depends strongly on the underlying DNA sequence.[12,17,23,24] These models have been successful in predicting a large portion of experimentally identified pause sites, and provided insight into the mechanism of pausing.

9.2 Background

RNAPs occur as both single- and multiple-subunit enzymes. RNAPs from bacteriophages and mitochondria are representative of the single-subunit family; bacterial, archaeal, and eukaryotic nuclear RNAPs constitute the multiple-subunit family. Although the single-subunit and multi-subunit RNAPs do not likely share a common ancestor, the available biochemical and structural information from representatives of each family shows that these RNAPs share many characteristics,[25,26] a justification for the search of common mechanisms.

Transcription is traditionally divided into three sequential phases: initiation, elongation and termination, although termination can be viewed as an alternate pathway branching from elongation.[27] Here we focus on the elongation stage, during which the RNAP forms a stable transcription elongation complex (TEC) with the template DNA and the nascent RNA, and moves along the DNA incorporating complementary NTPs to the 3' end of the RNA. Each nucleotide addition can be viewed as a competition among active elongation, pausing (a transient conformational state incapable of elongation), arrest (a conformational state incapable of elongation without factor-assisted isomerization back to an active complex) and termination (transcript release and enzyme dissociation from the DNA template) (see Chapters 7 and 10 in the present book).[27]

As depicted in the Structural Atlas and reviewed in Chapter 7, the transcribing RNAP has a crab claw shape with the "jaws" surrounding a central channel, which holds the nucleic acids. The TEC structure contains an open DNA bubble of 12–14 bp and an 8–9 bp RNA-DNA hybrid within the RNAP main internal channel, a smaller secondary channel or pore that likely serves as an entry channel for NTPs, and an RNA exit channel.[26,28,29] The active site, which is the catalytic center of the complex, is located at the junction of the main channel and secondary channels and contains at least one nucleotide binding site and a tightly bound Mg^{2+}. RNAP is a two-metal ion-dependent enzyme and the second active-site Mg^{2+} is thought to be coordinated with the incoming NTP (Chapters 7 and 8).[30]

For a given nascent RNA length, the TEC may exist in slightly different configurations that are due to RNAP translocation, which we will refer to as "translocation states" (see Figure 7.2). The most significant difference among different translocation states is the position of the RNA 3' end relative to the active site of RNAP. NTP incorporation can only occur in the "post-translocation state," where the active site is occupied by the complementary NTP and immediately adjacent to the 3' end of the nascent RNA. Upon NTP incorporation, the 3' end of the RNA occupies the active site and the TEC is in its "pre-translocation state." From this state RNAP must move forward by 1 bp to clear the active site to allow binding by the next incoming NTP and thus to start the next elongation cycle. As mentioned before, it was observed experimentally that TEC can also form "backtracked states" by moving backward non-catalytically and placing the 3' end of RNA into the secondary channel. There is also evidence suggesting that the RNAP can translocate forward beyond the pre-translocation state to "forward-tracked states," which could serve as precursors to transcription termination.[31,32] Such configurations are not catalytically competent and the RNAP must return to the post-translocation state to continue elongation. The pre- and post-translocation states are indispensable for the continuous NTP incorporation cycle, and therefore belong to the "main" elongation pathway; the backtracked and forward-tracked states are part of the "branch" non-essential pathways.

9.3 Mechano-chemical Coupling of Transcription

9.3.1 NTP Incorporation Cycle

A single NTP incorporation cycle must minimally include RNAP translocation from pre- to post-translocation states, NTP binding, NTP hydrolysis and PPi release.[18,33,34] However, the details of the NTP incorporation cycle depend greatly on the specific mechano-chemical coupling mechanism. In a Brownian ratchet model of transcription, prior to NTP incorporation, RNAP can move back and forth on the DNA template, activated solely by thermal energy, so the translocation part of the pathway is largely independent of the chemical reaction part of the pathway. In contrast, in a power-stroke model these two parts are highly correlated: NTP hydrolysis is thought to induce a conformational change of RNAP, which necessarily leads to translocation (as illustrated in Chapter 4 and Figure 7.5A). In other words, the two mechanisms represent "weak" and "strong" coupling between chemical reaction and mechanical translocation, respectively. Consequently, these models make different quantitative predictions of the elongation rate dependence on external load and [NTP], and can be differentiated by kinetic and mechanical measurements.

9.3.2 NTP Incorporation Pathway in a Simple Brownian Ratchet Model

The basic reaction pathway for a Brownian ratchet model involves only one NTP binding site and a minimum number of translocation states as

Kinetic Modeling of Transcription Elongation

shown in:

$$\text{TEC}_{n,\text{pre}} \underset{k_{-1}}{\overset{k_1}{\rightleftharpoons}} \text{TEC}_{n,\text{post}} \underset{k_{-2}}{\overset{\text{NTP } k_2}{\rightleftharpoons}} \text{TEC}_{n,\text{post}} \cdot \text{NTP} \underset{k_{-3}}{\overset{k_3}{\rightleftharpoons}} \text{TEC}_{n+1,\text{pre}} \cdot \text{PPi} \underset{k_{-4}}{\overset{\text{PP}_i \; k_4}{\rightleftharpoons}} \text{TEC}_{n+1,\text{pre}}$$

with $\text{TEC}_{n,\text{forwardtracked}}$ above $\text{TEC}_{n,\text{post}}$ and $\text{TEC}_{n,\text{backtracked}}$ below $\text{TEC}_{n,\text{pre}}$.

(9.1)

where $\text{TEC}_{n,\text{pre(post)}}$ represents the TEC with transcript size n at the pre(post)-translocation state. The main reaction pathway proceeds along the horizontal arrows and includes translocation between pre- and post-translocation state, NTP binding, NTP hydrolysis and PPi release, all of which are potentially reversible.

Reaction pathway (9.1) could be further simplified to reaction pathway (9.2) below certain kinetic limits.[9,12] Under typical experimental conditions where PPi concentration in the solution is very low, the slow pyrophosphorolysis rate makes NTP hydrolysis essentially irreversible ($k_{-4} \sim 0$). Also, there is evidence that NTP hydrolysis rates are much larger than the PPi release rate ($k_{\pm 3} >> k_4$),[35] so the two steps could be combined into one step with a single effective rate. Another simplifying assumption is that both the translocation between the pre- and post- translocation states and NTP binding follow rapid equilibrium kinetics,[9] therefore only the equilibrium constants of these steps need to be considered:

$$\text{TEC}_{n,\text{pre}} \overset{K_1}{\rightleftharpoons} \text{TEC}_{n,\text{post}} \overset{\text{NTP } K_d}{\rightleftharpoons} \text{TEC}_{n,\text{post}} \cdot \text{NTP} \overset{k_{\max}}{\underset{\text{PP}_i}{\longrightarrow}} \text{TEC}_{n+1,\text{pre}}$$

with $\text{TEC}_{n,\text{forwardtracked}}$ above $\text{TEC}_{n,\text{post}}$ and $\text{TEC}_{n,\text{backtracked}}$ below $\text{TEC}_{n,\text{pre}}$.

(9.2)

Notably, in this model, once a complementary NTP binds to the active site, the TEC is locked into the post-translocation state without access to other translocation states until NTP hydrolysis or dissociation. Therefore, it has been suggested that the NTP acts as a pawl in a ratchet and prevents the RNAP from sliding backwards,[14] consistent with the experimental observation that the incoming NTP stabilizes the TEC in its post-translocation state.[14]

9.3.3 NTP Incorporation Pathways in more Elaborate Brownian Ratchet Models

Brownian ratchet models more complex than the simple model described above have also been proposed and involve multiple parallel pathways, additional TEC states, additional NTP binding sites, *etc*. In a quench flow experiment that probed *E. coli* RNAP transient-state kinetics, the rate of single nucleotide incorporation was found to be biphasic: a fraction of the RNAP population exhibited much faster rates of catalysis than the rest.[35,36] Therefore, it was proposed that the elongation complex could exist in either an active (TEC^{act}) or inactive (TEC^{inact}) state, which incorporate NTPs with different rates and are not in fast equilibrium, as shown in reaction pathway (9.3). Based on the dependence of the reaction rate on [NTP], it was also suggested that the transition from an inactive to an active state was facilitated by the binding of the next complementary NTP to an allosteric binding site:

$$\begin{array}{c}
\text{TEC}^{act}_{n,\,post} \xrightleftharpoons{\text{NTP}} \text{TEC}^{act}_{n,\,post} \cdot \text{NTP} \xrightarrow{k_{fast}} \text{TEC}_{n+1,\,pre} + \text{PP}_i \\
\text{NTP} \nearrow \quad \Updownarrow \text{NTP} \\
\text{TEC}^{inact}_{n,\,pre} \rightleftharpoons \text{TEC}^{inact}_{n,\,post} \rightleftharpoons \text{TEC}^{inact}_{n,\,post} \cdot \text{NTP} \xrightarrow{k_{slow}} \text{TEC}_{n+1,\,pre} + \text{PP}_i \\
\Updownarrow \\
\text{TEC}^{inact}_{n,\,backtracked}
\end{array} \quad (9.3)$$

Multiphasic NTP incorporation kinetics were also proposed for human RNA polymerase II (pol II), although the kinetic data were modeled somewhat differently.[37,38] In the pol II model, the fast and slow fractions were proposed to reflect the TEC populations in the post- and pre-translocation states, respectively. The two translocation states interconvert slowly and are not in equilibrium, resulting in biphasic kinetics of NTP incorporation at low [NTP]. Because the two states respond to [NTP] differently, it was proposed that both translocation states could bind to the incoming NTP, and the transition between them was expedited by an NTP-driven "induced fit" mechanism:

$$\begin{array}{c}
\text{TEC}_{n,\,pre} \cdot \text{NTP} \xrightleftharpoons{\text{fast}} \text{TEC}_{n,\,post} \cdot \text{NTP} \longrightarrow \text{TEC}_{n+1,\,pre} \cdot \text{PP}_i \xrightarrow{k_{max}} \text{TCE}_{n+1,\,pre} + \text{PP}_i \\
\text{NTP} \Updownarrow K'_d \qquad \text{NTP} \Updownarrow K_d \\
\text{TEC}_{n,\,pre} \xrightleftharpoons{\text{slow}} \text{TEC}_{n,\,post} \\
\Updownarrow \\
\text{TEC}_{n,\,backtracked}
\end{array} \quad (9.4)$$

A model similar to reaction path (9.4) was proposed in a recent single-molecule study that examined *E. coli* RNAP transcription rate dependence on [NTP] and force. This model also includes a secondary NTP binding site so that NTP binding can take place either before or after translocation.[13] However, unlike reaction path (9.4), the two translocation steps in this model, before and after NTP binding, are assumed to be in equilibrium:

$$\begin{array}{c}
\text{TEC}_{n,\text{pre}} \cdot \text{NTP} \xrightleftharpoons{K_{I2}} \text{TEC}_{n,\text{post}} \cdot \text{NTP} \xrightleftharpoons{} \text{TEC}_{n+1,\text{pre}} \cdot \text{PPi} \xrightarrow{k_{\max}} \text{TCE}_{n+1,\text{pre}} \\
\text{NTP} \updownarrow K'_d \qquad\qquad \text{NTP} \updownarrow K_d \\
\text{TEC}_{n,\text{pre}} \xrightleftharpoons{K_{I1}} \text{TEC}_{n,\text{post}}
\end{array} \qquad (9.5)$$

Compared with the simple Brownian ratchet model (reaction path 9.2), models (9.3)–(9.5) all involve parallel pathways, with the transition between them depending on an NTP binding to a secondary site. This site was postulated to either lie within the streptolydigin-binding region of the *E. coli* RNAP or in the main channel facing the downstream single stranded DNA at $n+1$ and $n+2$ (where n is the RNA length).[36,38] However, so far direct experimental evidence for the extra NTP binding site is still lacking.

An alternative model proposed by Bar-Nahum et al.[14] involves parallel pathways, but with no requirement for an extra NTP binding site. This "dual ratchet mechanism" is based on experimental evidence that the bending of a flexible bridge helix (the F bridge) located at the active site near the 3′ end of the RNA plays an important role in translocation (see Figure 7.1.A).[39–41] According to this model, in addition to the simple Brownian ratchet mechanism in reaction path (9.2), there is a second ratchet imposed by the F bridge, which oscillates back and forth between "straight" and "bent" conformations due to thermal fluctuation. The F helix, the 3′ end of RNA and the incoming NTP substrate interact with each other by competing for the polymerase's active center: an NTP can only bind when the F helix is in its straight form, and the bending of the F helix can force a bound NTP to dissociate, to "push" RNAP forward by one nucleotide, or to cause the 3′ end of RNA to separate from the template DNA into the secondary channel. The reaction pathway is given by pathway (9.6):

$$\begin{array}{c}
\text{TEC}_{n,\text{forwardtracked}} \\
\updownarrow \\
\text{TEC}^S_{n,\text{pre}} \rightleftharpoons \text{TEC}^S_{n,\text{post}} \xrightleftharpoons{K_d} \text{TEC}^S_{n,\text{post}} \cdot \text{NTP} \xrightarrow{k_{\max}} \text{TEC}^S_{n+1,\text{pre}} \\
\updownarrow \qquad\qquad \updownarrow \\
\text{TEC}^B_{n,\text{pre}} \rightleftharpoons \text{TEC}^B_{n,\text{post}} \\
\updownarrow \\
\text{TEC}_{n,\text{backtracked}}
\end{array} \qquad (9.6)$$

where the superscripts S and B denote the straight and bent F bridge conformations, respectively.

9.3.4 NTP Incorporation Pathway in a Power-stroke Model

A power-stroke model has been proposed for the single subunit T7 RNA polymerase based on structural studies where snapshots of the TEC at different stages of the elongation cycle were captured. T7 RNAP was found to undergo a large conformational change from an "open" conformation that could bind the incoming NTP to a "closed" conformation during catalysis (Chapter 4).[8,42] The transition from the "open" to "closed" state occurred upon binding of the NTP to the O helix structure located at the active site, while the pyrophosphate release reversed the transition with a concurrent forward translocation of RNAP by 1 bp. No structure could be obtained of the TEC in a pre-translocation state without the PPi bound within the active site. From these observations Yin and Steitz proposed a power-stroke mechanism in which the chemical energy derived from the NTP hydrolysis reaction directly drives the forward translocation of the RNAP along the DNA template. Because of the tight coupling of the translocation and the PPi release, they can be considered as a single step in the reaction pathway:

$$\text{TEC}_{N,\text{post}} \underset{}{\overset{\text{NTP} \; K_d}{\rightleftharpoons}} \text{TEC}_{N,\text{post}} \cdot \text{NTP} \rightleftharpoons \text{TEC}_{N+1,\text{pre}} \cdot \text{PPi} \xrightarrow{k_{\max}} \text{TEC}_{N+1,\text{post}} \qquad (9.7)$$

In contrast to the Brownian ratchet models, the translocation step is irreversible and occurs only once per NTP incorporation cycle.

9.3.5 Elongation Kinetics

With the simple kinetic scheme in reaction path (9.2), the overall active elongation rate along the main pathway follows Michaelis–Menten kinetics in the presence of competitive inhibitor:[9,12]

$$k_{\text{main}} = \frac{k_{\max}[\text{NTP}]}{K'_d + [\text{NTP}]}, \quad \text{and} \quad K'_d = K_d(1 + K_I) \qquad (9.8)$$

where the effective dissociation constant is larger than the actual NTP dissociation constant K_d, reflecting the competition at the active site between the 3' end of RNA and NTP. Similar arguments apply to the reaction path (9.5), where more inhibition states are present:

$$k_{\text{main}} = \frac{k'_{\max}[\text{NTP}]}{K'_d + [\text{NTP}]}, \quad \text{and} \quad K'_d = K_d \cdot \frac{1 + K_{I1}}{1 + K_{I2}} \qquad (9.9)$$

Importantly, kinetic Equations (9.8) and (9.9) very much rely on the assumption of a rapid equilibrium existing between all states which interconvert during translocation. The importance of this assumption may be illustrated with a comparison of reaction pathways (9.4) and (9.5), which predict different kinetics despite their close resemblance. The model in reaction path (9.5) predicts single-exponential kinetics for the NTP incorporation reaction, while the overall elongation rate changes hyperbolically with varying [NTP]. In contrast, the model reaction path (9.4) predicts double-exponential incorporation kinetics and sigmoidal dependence of overall rate on [NTP].

The kinetics in reaction path models (9.3), (9.4) and (9.6) are more complicated. Despite the apparent difference between reaction paths (9.3) and (9.4), the NTP incorporation kinetics predicted by the two models are similar. In both models the TEC population equilibrates between the pre- and post-translocation states during NTP starvation. Upon NTP addition, the TEC population quickly re-distributes between the NTP bound and unbound states (NTP binding in these models is assumed to be fast), and the distribution depends on the NTP concentration. Different TEC states lead to different NTP incorporation rates, and the resulting kinetics are bi-phasic. Increasing [NTP] plays dual roles: it increases the NTP incorporation rate according to Michaelis–Menten kinetics, and it also shifts the TEC population into an "active" state displaying faster hydrolysis, and thus the overall incorporation rate has a sigmoidal dependence on [NTP]. In contrast, in model reaction path (9.6), the transition between the main and parallel pathways does not depend on [NTP]. Therefore, although the model could account for the biphasic incorporation rate by assuming a slow structural oscillation coupled to polymerization (for instance trigger loop and/or bridge helix oscillations), it could not explain the high sensitivity of the reaction rate to [NTP].

For the power-stroke model in reaction path (9.7), if we assume the NTP binding and hydrolysis reaction are both much faster than the PPi release step, and $[TEC_{n+1,pre} \cdot PPi] = [TEC_{n,post} \cdot NTP] \cdot K$, then the incorporation rate could be expressed as:

$$k = \frac{k'_{max}[NTP]}{K'_d + [NTP]}$$

where the effective NTP dissociation constant $K'_d = K_d/(1+K)$ and $k'_{max} = k_{max}[K/(1+K)]$ is the effective maximum incorporation rate.

9.3.6 Force-dependent Elongation Kinetics

Several single-molecule experiments were carried out to evaluate reaction path models (9.2), (9.5) and (9.7).[13,15,17] Compared with traditional ensemble studies, there are several advantages of single-molecule approaches for studying transcription elongation. First, single-molecule techniques, such as optical and magnetic tweezers, allow application of external force or torque to the TEC, thus selectively perturbing the reaction steps involving RNAP translocation. The nature of this step defines precisely how force affects the overall elongation rate and, therefore, the force-dependence of elongation kinetics is a very critical probe for the elongation mechanism. Second, as mentioned above, elongation has a complicated, multi-branched reaction pathway, and different parts of the pathway may respond differently to perturbations. For instance, active elongation kinetics could have a very different force-dependence than pausing kinetics. By monitoring the RNAP position in real time, especially with high spatial resolution, it is possible to decouple active elongation from pausing, so that their kinetics may be analyzed separately. Third, single-molecule techniques are suitable for measuring properties that are highly unsynchronized among different RNAP molecules (such as pause duration), because they do not require ensemble averaging among a large number of molecules.

Differentiation among reaction path models (9.2), (9.5) and (9.7) is difficult using traditional bulk assays because the predicted kinetics are very similar (see above). However, they predict significantly different force-dependent elongation kinetics. The power-stroke model in reaction path (9.7) contains a set of reversible reaction steps followed by a force-dependent irreversible step, therefore the external force should not affect the NTP dissociation constant but simply the maximal rate constant:

$$k(F) = k(0) \cdot \exp(F\Delta/k_B T) = \frac{k'_{\max}[\text{NTP}]}{K'_d + [\text{NTP}]} \exp(F\Delta/k_B T) \qquad (9.10)$$

where $k(0)$ is the reaction rate at force 0 pN, F is the applied force and Δ is a characteristic distance representing the location of the activation barrier from the pre-translocation state ($0 \leq \Delta \leq 3.4$ nm), and $k_B T$ is the thermal energy (Figure 9.1A). When $\Delta = 0$ bp, there would be no force-dependence on the predicted elongation rate; in other cases, the elongation rate would have the same exponential dependence on F at any NTP concentration. The actual change in the elongation rate at different forces, $k(F_1) - k(F_2)$, would be larger at higher [NTP], making it easier to be detected experimentally.

In the Brownian-ratchet model (reaction path 9.2), the force-affected translocation step is reversible and in rapid equilibrium. The force affects the equilibrium constant K_I and thus the effective NTP dissociation constant by tilting the energy landscape between the pre- and post-translocation states, resulting in an altered elongation rate (see Equation 9.8):

$$k(F) = \frac{k_{\max}[\text{NTP}]}{K_d[1 + K_I(F)] + [\text{NTP}]}$$
$$= \frac{k_{\max}}{1 + \frac{K_d}{[\text{NTP}]}[1 + K_I(0) \cdot \exp(-Fd/k_B T)]} \qquad (9.11)$$

where the characteristic distance d corresponds to 1 bp for transitions between pre- and post-translocation states (Figure 9.1B). Contrary to the previous case, here when [NTP] is infinitely high, the maximal elongation rate would no longer depend on force and the force modulates the effective K_d. Therefore, the sensitivity of the elongation rate to force decreases at higher [NTP]. A similar derivation can be applied to reaction path model (9.5), where applied force shifts the equilibrium distribution between different translocation states. Because in this model translocation could occur either before or after the NTP binding, the force effect contains both NTP-dependent and independent components:

$$k(F) = \frac{k_{\max}}{1 + K_{I2}(0) \cdot \exp(-Fd/k_B T) + \frac{K_d}{[\text{NTP}]}[1 + K_{I1}(0) \cdot \exp(-Fd/k_B T)]} \qquad (9.12)$$

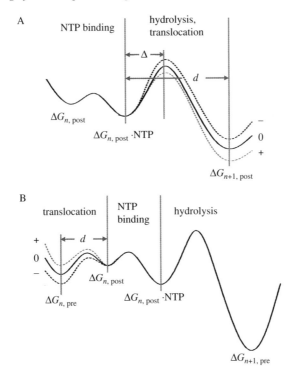

Figure 9.1 Cartoons illustrating the effect of an applied force on the energy landscape corresponding to addition of a single nucleotide during elongation. (A) Power-stroke model. (B) Brownian ratchet model. In both cases, force selectively perturbs the energy landscape in the step involving translocation. Assisting (+) and opposing (−) forces tilt the landscape in opposite directions relative to the zero force case. See text for details.

Therefore, reaction path model (9.5) also predicts a decreasing force-dependent elongation velocity at higher [NTP], but the force-dependence approaches a non-zero limit at very high [NTP].

The Brownian ratchet and power-stroke models predict different trends of elongation velocity force-dependence *vs.* [NTP]; therefore, single-molecule force–velocity measurements can be used to differentiate between the models. It has been shown that *E. coli* RNAP is a powerful motor, and on average ∼25 pN of resisting force was required to stall elongation.[1–3] Under saturating [NTP], elongation velocity remained nearly constant over a wide range of resisting and assisting forces (+25 to −35 pN).[2,3,20,43] Two more recent studies on *E. coli* RNAP found a small but statistically significant dependence of velocity on force, even at high NTP concentrations, and this dependence increased at lower [NTP].[13,17] A study using T7 RNAP found a similar force–velocity relation, arguing for a similar elongation mechanism for the single and multiple subunit polymerases.[15]

Importantly, in reaction path model (9.2), the equilibrium translocation constant K_I can be computed solely based on the sequence-dependent TEC state energy, as will be discussed in detail below. Bai *et al.* showed that once k_{max} and K_d were experimentally determined, the force–velocity curves at various NTP concentrations in reaction path model (9.2) could be predicted without any fitting parameters (Equation 9.11).[17] The agreement between the predicted and the measured force–velocity curves supports the simple Brownian-ratchet mechanism (reaction path 9.2).[17]

Because the power-stroke model as presented in reaction path (9.7) is oversimplified and contains multiple assumptions on reaction kinetics, the power-stroke mechanism cannot be formally ruled out. However, generally speaking, the Brownian ratchet models have found strong support from independent experimental evidence despite some differences in the details of the reaction pathways. In Thomen *et al.*[15] and Bai *et al.*[17] the measured force–velocity relationships of T7 and *E. coli* RNAP at various NTP concentrations were consistent with the simple Brownian ratchet model of reaction path (9.2), while Abbondanzieri *et al.*[13] found that their data could be better explained by the more complicated Brownian ratchet model reaction path (9.5).

9.4 Sequence-dependent RNAP Kinetics

To model sequence-dependent RNAP kinetics, the essential task is to establish the correlation between the kinetic parameters in the NTP incorporation pathway and the sequence of the transcribed DNA. A hint comes from studies of intrinsic transcription termination, where a special sequence known as a terminator destabilizes the TEC and induces dissociation of the RNA and RNAP from the DNA template (Chapter 10).[44] To account for this destabilization, Yager and von Hippel[23] provided a static sequence-dependent thermodynamic analysis of TEC stability of *E. coli* RNAP. This model successfully explained the elevated TEC free energy at intrinsic terminators, and laid out the foundation for kinetic analysis of termination efficiency.[45] Recent theoretical studies extended the thermodynamic model by performing a full kinetic analysis and applying it to active elongation and pausing.[12,17,24] These kinetic models are discussed below.

9.4.1 Thermodynamic Analysis of the TEC

Based on the structure of the TEC, its thermodynamic stability is given by the standard free energy of the complex formation from its components:

$$\Delta G = \Delta G_{\text{DNA bubble}} + \Delta G_{\text{RNA–DNA hybrid}} + \Delta G_{\text{RNAP binding}} \quad (9.13)$$

Here the lower the free energy of a given state the more stable the TEC in that state. The energy cost required for DNA bubble opening $\Delta G_{\text{DNA bubble}}$ is offset by the energy of formation of the RNA-DNA hybrid $\Delta G_{\text{RNA–DNA hybrid}}$ and the stabilizing interactions between the polymerase and the nucleic acids, $\Delta G_{\text{RNA binding}}$. The first two terms of Equation (9.13) can be calculated for a given DNA sequence from the base pairing energy of the nucleic acids. These free energies have been measured in solution only and current models assume the energy values do not change considerably within the TEC. Based on structural data, RNAP interacts primarily with the backbone of the nucleic acids and thus the third term in Equation (9.13) can be treated as a sequence-independent constant.[23]

Another possible energetic contribution that may affect RNAP kinetics comes from the interactions of RNAP with the nascent RNA, ΔG_{RNA}:

$$\Delta G = \Delta G_{\text{DNA bubble}} + \Delta G_{\text{RNA-DNA hybrid}} + \Delta G_{\text{RNAP binding}} + \Delta G_{\text{RNA}} \quad (9.14)$$

The last term represents the change in the folding energy of the free transcript RNA outside the RNAP as transcription proceeds.[24] RNA is known to play a critical role in intrinsic transcription termination and hairpin-dependent transcriptional pausing.[6,46] It has also been proposed that RNA hairpin structures could serve as a barrier to prevent RNAP backtracking and thus influence elongation pausing.[24]

The value of the TEC state energy is highly dependent on the TEC structure used in modeling elongation (size of the DNA bubble, length of the DNA-RNA hybrid, *etc.*). Recent theoretical studies computed the TEC energy either using a fixed TEC structure within the experimentally observed range[12,17,23] or averaging over a range of bubble configurations.[24] For a given TEC structure, the energy depends strongly on the DNA sequence within the complex. On a DNA template of a few hundred basepairs, the TEC energy at different template positions could differ by more than $10k_BT$. This large dynamic range has important implications and naturally leads to the dramatic sequence-dependent kinetic behavior of RNAP.

This overall free energy also depends on the precise translocation state of the TEC. The post-translocation and forward-tracked states tend to be less energetically favorable than the pre-translocation and backtracked states, because the former have fewer base pairs within the RNA-DNA hybrid. Based on the same line of reasoning, the pre-translocation and backtracked states would have approximately the same energy, once averaged over various DNA sequences. Once an incoming complementary NTP binds in the active site and basepairs with the DNA base, this additional interaction energy stabilizes the post-translocation configuration.[8,9]

9.4.2 Sequence-dependent NTP Incorporation Kinetics in Brownian Ratchet Models

Thus far two sequence-dependent elongation models have been formulated, respectively by Bai *et al.*[12,17] and by Tadigotla *et al.*[24] Both models are based on the simple Brownian ratchet mechanism shown in kinetic pathway (9.1).

They are extended to consider sequence-dependent k_{max}, K_d and translocation rates in the branched pathways shown in reaction path (9.2) for all template positions. Since there have been little or no direct measurements of these kinetic parameters, some assumptions must be made to estimate the parameter values. The key differences between the two models are in the assumptions leading to the derivation of the backtracking rates. They lead to different expressions for the elongation rate.

In Bai et al.[12,17] backtracking is assumed to be a slow process, so that during active elongation the backtracked states do not equilibrate with the states within the main pathway. At a majority of template positions and under typical NTP concentrations this model predicts that the probabilities of backtracking are so small that the backtracked states are essentially inaccessible. For these sites, the NTP incorporation follows a single exponential curve with a rate k_{main} as expressed in Equation (9.8). Notably, the equilibrium constant, K_I, is not a free parameter and can be directly calculated from the free energy difference between the pre- and post-translocation states:

$$K_I = \exp[(\Delta G_{post} - \Delta G_{pre})/k_B T] \quad (9.15)$$

As mentioned before, ΔG_{pre} and ΔG_{post} are both strongly sequence-dependent, and thus K'_d as well as k_{main} will be sequence-dependent.

In this model the backtracked rates are calculated based on Arrhenius kinetics, where the transition rate between any two states 1 and 2 is governed by the free energy difference between the peak of the activation barrier for the transition and state 1:

$$k_{1\to 2} = k_0 \exp[-(\Delta G^{\dagger}_{1\leftrightarrow 2} - \Delta G_1)/k_B T] \quad (9.16)$$

where $\Delta G^{\dagger}_{1\leftrightarrow 2}$ is assumed to be the same for all backtracked steps and on all template positions. Under such an assumption, the backtracking rate is solely determined by the TEC free energies. At the few sites displaying a high probability of backtracking, NTP incorporation is predicted to follow a multi-phasic time course.

In contrast, Tadigotla et al.[24] assume that RNAP is capable of rapid backtracking until it encounters the first secondary structure formed by the nascent RNA outside of the RNAP. At this point, further backtracking of RNAP is prevented until the next NTP incorporation. In this model, accessible translocation states both in the main and branch pathways are assumed to be in equilibrium. Thus at each template position the predicted time course of NTP incorporation follows a single exponential with a rate constant:

$$k_{main} = \frac{k_{max}[NTP]}{K'_d + [NTP]} = \frac{k_{max}[NTP]}{K_d \sum_m \exp(\Delta G_{post} - \Delta G_m)/k_B T + [NTP]} \quad (9.17)$$

which depends on the free energy of all accessible states, ΔG_m.

9.4.3 Model Predictions of Pause Locations, Kinetics and Mechanisms

The models of sequence-dependent elongation kinetics described above may be used to predict pause locations and pause kinetics, as well as to provide insights into pause mechanisms. Interestingly, these two models do not always make the same predictions for pausing.

The model in Bai et al. predicts two alternative pausing mechanisms: pausing could be caused either by a slow rate within the main pathway ("on-pathway" pause) or by a relatively fast rate of entry into a nonproductive branch pathway with a slow rate of returning to the main pathway (backtracking pause). Note that the two types of pauses are correlated because different pathways are kinetically competitive: a slow rate in the main pathway increases the probability for the RNAP to enter a branched pathway. Equation (9.15) shows that an on-pathway pause occurs at a template site with a large $\Delta G_{post} - \Delta G_{pre}$, i.e., the pre-translocation state is much more stable than the post-translocation state. Thus the entire RNAP population is expected to follow the same slow kinetics. We refer to this type of pause as a pre-translocation pause. In contrast, a backtracking pause is expected to occur when the pre-translocation state is unstable (high ΔG_{pre}) so that $k_{pre \rightarrow backtracking}$ becomes significant in comparison with k_{main} (Equation 9.8). The dwell time distribution at backtracking pause sites is no longer defined by a single characteristic time but has a long tail due to the slow return of the backtracked RNAP population to the main pathway. Compared to a pre-translocation pause, a backtracking pause tends to occur less frequently but for a longer duration. Because backtracking typically involves translocation over several base pairs (as opposed to 1 bp in the on-pathway pause), a backtracking pause is more sensitive to external force. The two types of pauses therefore provide an explanation for the short and long duration pauses observed in single-molecule experiments.[21,22] This model also correctly predicted prominent pause positions seen in bulk studies, and the force-dependence of pause durations for a known pause sequence detected by single-molecule experiments.[17,47]

In the model of Tadigotla et al., pauses occur when one or more accessible translocation states are significantly more stable than the post-translocation state (Equation 9.17). This model was shown to have successfully predicted a large portion of previously identified pauses.[5] Because most of the pause sites in Bai et al. also have unstable post-translocation states, the two models would have significant overlap in pause sites prediction. However, the pausing kinetics and mechanism of the pauses predicted by the two models are different: most of the pauses predicted by Tadigotla et al. are backtracked, and the NTP incorporation kinetics at these sites all follow a single exponential. Future effort is required to differentiate and evaluate the two models by comparison with more experimental data.

The models mentioned above are mostly based on transcription studies *in vitro*, but have direct implications for cellular processes *in vivo*. It is known that transcription levels of certain genes can exhibit high cell to cell variation (noise), especially when the transcripts are generated in a burst-like fashion.[48-53] Usually, the noise is thought to originate from infrequent transcription initiation.[51-53] A recent model demonstrated that long transcriptional pausing could also lead to bursts of mRNA production and may be a significant contributor to the variability in cellular transcription rates.[54] Further experimental and modeling studies of transcriptional rate, pausing and interactions between elongating RNAPs would shed light on both the fundamental elongation mechanism and the corresponding physiological effects on cell growth and viability.

Acknowledgements

This work is supported by grants to M.D.W. from NSF grant (DMR-0517349), NIH grant (R01 GM059849), and the Keck Foundation's Distinguished Young Scholar Award, as well as postdoctoral fellowships to L.B. from the Rockefeller University and the Damon Runyon Cancer Research Foundation.

References

1. H. Yin, M. Wang, K. Svoboda, R. Landick, S. Block and J. Gelles, *Science*, 1995, **270**, 1653–1657.
2. M. Wang, M. Schnitzer, H. Yin, R. Landick, J. Gelles and S. Block, *Science*, 1998, **282**, 902–907.
3. K. Neuman, E. Abbondanzieri, R. Landick, J. Gelles and S. Block, *Cell*, 2003, **115**, 437–447.
4. J. Gelles and R. Landick, *Cell*, 1998, **93**, 13–16.
5. J. Levin and M. Chamberlin, *J. Mol. Biol.*, 1987, **196**, 61–84.
6. I. Artsimovitch and R. Landick, *Proc. Natl. Acad. Sci. USA*, 2000, **97**, 7090–7095.
7. R. Feynman, R. Leighton and M. Sands, in *The Feynman Lectures on Physics*, Addison-Wesley Pub. Co., Reading, Mass, 1963.
8. Y. Yin and T. Steitz, *Cell*, 2004, **116**, 393–404.
9. R. Guajardo and R. Sousa, *J. Mol. Biol.*, 1997, **265**, 8–19.
10. F. Jülicher and R. Bruinsma, *Biophys. J.*, 1998, **74**, 1169–1185.
11. N. Komissarova and M. Kashlev, *J. Biol. Chem.*, 1997, **272**, 15329–15338.
12. L. Bai, A. Shundrovsky and M. Wang, *J. Mol. Biol.*, 2004, **344**, 335–349.
13. E. Abbondanzieri, W. Greenleaf, J. Shaevitz, R. Landick and S. Block, *Nature*, 2005, **438**, 460–465.
14. G. Bar-Nahum, V. Epshtein, A. Ruckenstein, R. Rafikov, A. Mustaev and E. Nudler, *Cell*, 2005, **120**, 183–193.
15. P. Thomen, P. Lopez and F. Heslot, *Phys. Rev. Lett.*, 2005, **94**, 128102.
16. Q. Guo and R. Sousa, *J. Mol. Biol.*, 2006, **358**, 241–254.
17. L. Bai, R. Fulbright and M. Wang, *Phys. Rev. Lett.*, 2007, **98**, 068103.
18. R. Landick, *Cell*, 2004, **116**, 351–353.

19. K. Adelman, A. La Porta, T. Santangelo, J. Lis, J. Roberts and M. Wang, *Proc. Natl. Acad. Sci. USA*, 2002, **99**, 13538–13543.
20. N. Forde, D. Izhaky, G. Woodcock, G. Wuite and C. Bustamante, *Proc. Natl. Acad. Sci. USA*, 2002, **99**, 11682–11687.
21. J. Shaevitz, E. Abbondanzieri, R. Landick and S. Block, *Nature*, 2003, **426**, 684–687.
22. K. Herbert, A. La Porta, B. Wong, R. Mooney, K. Neuman, R. Landick and S. Block, *Cell*, 2006, **125**, 1083–1094.
23. T. Yager and P. von Hippel, *Biochemistry*, 1991, **30**, 1097–1118.
24. V. Tadigotla, D. O Maoiléidigh, A. Sengupta, V. Epshtein, R. Ebright, E. Nudler and A. Ruckenstein, *Proc. Natl. Acad. Sci. USA*, 2006, **103**, 4439–4444.
25. R. Sousa, *Trends. Biochem. Sci.*, 1996, **21**, 186–90.
26. P. Cramer, *Bioessays*, 2002, **24**, 724–729.
27. P. von Hippel, *Science*, 1998, **281**, 660–665.
28. E. Nudler, I. Gusarov, E. Avetissova, M. Kozlov and A. Goldfarb, *Science*, 1998, **281**, 424–428.
29. S. Darst, *Curr. Opin. Struct. Biol.*, 2001, **11**, 155–162.
30. V. Sosunov, E. Sosunova, A. Mustaev, I. Bass, V. Nikiforov and A. Goldfarb, *EMBO J.*, 2003, **22**, 2234–2244.
31. W. Yarnell and J. Roberts, *Science*, 1999, **284**, 611–615.
32. T. Santangelo and J. Roberts, *Mol. Cell*, 2004, **14**, 117–126.
33. D. Erie, T. Yager and P. von Hippel, *Annu. Rev. Biophys. Biomol. Struct.*, 1992, **21**, 379–415.
34. S. Uptain, C. Kane and M. Chamberlin, *Annu. Rev. Biochem.*, 1997, **66**, 117–172.
35. J. Foster, S. Holmes and D. Erie, *Cell*, 2001, **106**, 243–252.
36. S. Holmes and D. Erie, *J. Biol. Chem.*, 2003, **278**, 35597–35608.
37. Y. Nedialkov, X. Gong, S. Hovde, Y. Yamaguchi, H. Handa, J. Geiger, H. Yan and Z. Burton, *J. Biol. Chem.*, 2003, **278**, 18303–18312.
38. C. Zhang and Z. Burton, *J. Mol. Biol.*, 2004, **342**, 1085–1099.
39. A. Gnatt, P. Cramer, J. Fu, D. Bushnell and R. Kornberg, *Science*, 2001, **292**, 1876–82.
40. D. Vassylyev, S. Sekine, O. Laptenko, J. Lee, M. Vassylyeva, S. Borukhov and S. Yokoyama, *Nature*, 2002, **417**, 712–719.
41. V. Epshtein, A. Mustaev, V. Markovtsov, O. Bereshchenko, V. Nikiforov and A. Goldfarb, *Mol. Cell*, 2002, **10**, 623–634.
42. D. Temiakov, V. Patlan, M. Anaikin, W. McAllister, S. Yokoyama and D. Vassylyev, *Cell*, 2004, **116**, 381–391.
43. R. Davenport, G. Wuite, R. Landick and C. Bustamante, *Science*, 2000, **287**, 2497–2500.
44. K. Wilson and P. von Hippel, *J. Mol. Biol.*, 1994, **244**, 36–51.
45. P. von Hippel and T. Yager, *Proc. Natl. Acad. Sci. USA*, 1991, **88**, 2307–2311.
46. I. Toulokhonov and R. Landick, *Mol. Cell*, 2003, **12**, 1125–1136.

47. A. Shundrovsky, T. Santangelo, J. Roberts and M. Wang, *Biophys. J.*, 2004, **87**, 3945–3953.
48. M. Elowitz, A. Levine, E. Siggia and P. Swain, *Science*, 2002, **297** 1183–1186.
49. E. Ozbudak, M. Thattai, I. Kurtser, A. Grossman and A. van Oudenaarden, *Nat. Genet.*, 2002, **31**, 69–73.
50. W. Blake, M. Kaern, C. Cantor and J. Collins, *Nature*, 2003, **422**, 633–637.
51. L. Cai, N. Friedman and X. Xie, *Nature*, 2006, **440**, 358–362.
52. J. Raser and E. O'Shea, *Science*, 2005, **309**, 2010–2013.
53. J. Yu, J. Xiao, X. Ren, K. Lao and X. Xie, *Science*, 2006, **311**, 1600–1603.
54. M. Voliotis, N. Cohen, C. Molina-Paris and T. Liverpool, *Biophys J.*, 2008, **94**, 334–348.

CHAPTER 10
Mechanics of Transcription Termination

EVGENY NUDLER

Department of Biochemistry, New York University School of Medicine
New York, NY, 10016, USA

10.1 Introduction

Termination signals encoded in DNA that do not require additional protein factors are called intrinsic terminators. Genomic analyses show that in bacteria and bacteriophages intrinsic terminators specify the ends of about half of all annotated protein-encoding transcription units and 70% of non-coding RNA transcription units.[1] Many intrinsic terminators called "attenuators" reside in the 5′ untranslated region (5′ UTR or leader sequence) of an operon and directly control the efficiency of transcription of downstream genes.[2–7]

The typical DNA intrinsic termination signal is composed of a GC-rich palindromic element followed immediately by at least four T bases in a row ("T-stretch").[8] The resulting transcript forms a stable hairpin structure followed by several U residues at the 3′ terminus (Figure 10.1A). Both the hairpin and the U-stretch elements are necessary and sufficient for termination, although sequences just downstream of the T-stretch or far upstream of the terminator can significantly affect the efficiency of the process *in vitro* and *in vivo*.[11–13]

The mechanism of intrinsic termination has been under investigation for the last three decades. Only recently, however, has the reaction been clarified in detail, due principally to an understanding of the structure/function

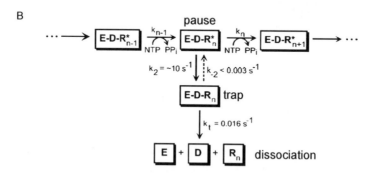

Figure 10.1 Intrinsic termination signal. (A) Depiction of the secondary structure of typical intrinsic terminators. Phage terminators tR2 and t500 have been used for advanced mechanistic studies. (B) A general kinetic scheme for transcription termination at tR2 (ref. 9). E = RNAP, D = DNA template, R = RNA transcript; "n" indicates the length of the transcript in nucleotides. Asterisks (*) mark intermediate states during elongation; E-D-Rn represents the paused state at the termination point, which lasts $\sim$ 1–2 s.[10] Irreversible steps include the formation of the trapped complex and complex dissociation.

organization of the elongation complex at atomic resolution[14–17] and the development of biochemical, protein chemical and single-molecule methods that allow the study of the termination process in real time.[18–20]

10.2 Structure/Function Overview of the Elongation Complex (EC)

As described in the preceding chapters, the catalytically-competent core of bacterial RNAPs has a molecular mass of 400 kDa (Figure 10.2) with subunit composition 2α, β′, β and ω and is evolutionarily conserved in its primary sequence, tertiary structure and function. These properties make bacterial RNAP an excellent model system to understand the basic principles of elongation and termination in all cellular RNAPs.

Three contiguous nucleic acid binding sites in RNAP that hold the EC together have been characterized biochemically and structurally: the DNA duplex binding site (DBS) is located towards the leading edge of the protein, the RNA:DNA hybrid binding site (HBS) is located towards the central portion of the protein, and the RNA binding site (RBS) is located at the trailing edge of the protein (Figure 10.2B).[14–17,21]

The DBS is responsible for the tight binding of ~ 10–12 bp of DNA duplex just downstream from the catalytic site.[22] It is primarily made up of the β′ subunit that forms a clamp which partially encircles the DNA.[14,17] The HBS is defined by its close contacts with the 7–9 bp RNA:DNA hybrid located within a ~ 12 bp stretch of melted DNA known as the transcription "bubble."[21,23] The HBS is also a clamp-like structure that consists of the β′ main channel capped by β lobes 1 and 2.[14–17] Finally, the RBS consists of the β′ rudder, lid, zipper and Zn-finger domains, which, together with the β-flap domain, form the RNA exit channel. This exit channel interacts tightly with a segment of single-stranded RNA corresponding to the 7 to 14 nucleotides located upstream of the RNA 3′ terminus (Figure 10.2).[14,17,24,25] Comparisons of the available high-resolution structures of RNAP indicate that protein domains that participate in the formation of the DBS, HBS and RBS are flexible. They can move apart to accommodate the sigma subunit and DNA during initiation and to release DNA and RNA during termination.[26] RNAP possesses two additional mobile domains that are part of its active center: the bridge helix (or F-bridge) and the trigger loop (or G-loop). As detailed in Chapter 7, these domains play critical roles during catalysis and RNA chain elongation.[27–29] It will be shown below that the bridge helix/trigger loop unit also plays a pivotal role in termination. Protein–nucleic acid interactions at all three nucleic acid binding sites contribute to the overall stability of the EC and render it resistant to high ionic conditions (e.g., 1 M KCl for several hours). Disturbing these contacts at any individual site destabilizes the EC.[22,30] RNA and DNA are rapidly and simultaneously released from the EC during intrinsic termination under physiological salt conditions,[31] implying that the terminator destroys interactions at all three nucleic acid binding sites simultaneously. Our current understanding of this process is described below.

10.3 Mechanism of Intrinsic Termination

The canonical intrinsic terminator is designed so that, at the moment of termination, the U-stretch occupies the entire HBS, generating an unusually unstable ~ 8 bp A:U hybrid,[32] while the inverted repeat encoding the hairpin occupies the single-strand specific RBS. Biochemical analyses have elucidated three major consecutive steps in the pathway leading to termination: transient pausing of RNAP, permanent inactivation (trapping) and, finally, RNA and DNA release (Figure 10.1B). Initially, the EC pauses at the site of termination.[33,34] This pause provides sufficient time for a hairpin structure to form. Once formed, the hairpin permanently inactivates the EC, displaces RNA from the RNA binding channel, and unwinds the upstream half of the hybrid.[34–36] The EC then rapidly dissociates at physiological salt concentrations.

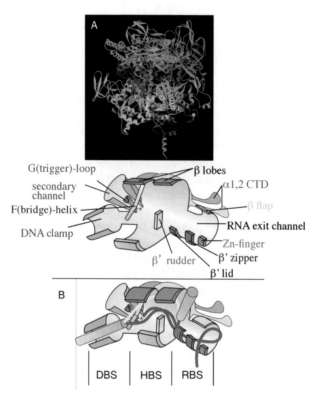

Figure 10.2 Structural and functional organization of the EC. (A) High-resolution crystal structure of *T. aquaticus* RNAP core (top panel) is shown as colored ribbons (αI, light grey; αII, grey; β, light brown; β', light green; ω, dark green; σ, magenta). Mobile domains are shown as follows: β' F-bridge (or bridge helix, residues I705-V835 of Tth β'), purple; β' G-loop (or trigger loop, T1234-L1256), blue; β' zipper (V26-E47), dark blue; β' Zn-finger (G51-S83), green; β' lid (R525-S538), light blue; β' rudder (N584-S602), turquoise; β lobe 1 (A29-D133, G337-S392) and β lobe 2 (R142-N330), orange; and β flap (G757-S789), yellow. The RNAP β' catalytic loop (N737-Q744) is colored red and the Mg^{2+} ion is shown as a small red sphere. The β' N-terminal non-conserved domain (G164-S449) and the CTDs of αI and αII are not shown. (B) Schematic representation of the RNAP structure shown as a simplified cartoon with the same color code as in (A) and all mobile domains indicated. The catalytic center is shown as a small magenta star. (C) Schematic representation of the EC structure. Duplex DNA is shown as flexible green cylinders; nascent RNA, template and non-template DNA strands are shown as red, dark green and yellow worms, respectively. The β flap, β' zipper, Zn-finger and β' lid form the RNA exit channel (RBS). The β' main channel secures the RNA-DNA hybrid in the hybrid binding site (HBS) and clamps the downstream DNA in the DNA duplex binding site (DBS). Transcription goes from right to left.

10.3.1 The Pausing Phase

10.3.1.1 Early Ideas on the Role of Pausing in Termination

Pausing has long been considered as the first and critical step in the termination process.[2,33] Indeed, the early models of termination postulated that the elongation/termination decision is kinetically coupled and hence the longer the EC would stay at the terminator the higher the probability of its dissociation.[37] It was generally believed that the signal responsible for pausing is the termination hairpin itself.[38] Such an assumption was reasonably based on the well-known phenomenon of hairpin-dependent pausing. Hairpin-dependent pausing plays an important role in transcription attenuation, which controls the expression of various metabolic operons in bacteria. For example, the classical attenuation mechanism of the *trp* operon in *E. coli* utilizes a hairpin-dependent pause, which coordinates transcription and translation events, to allow time for the choice of whether or not to transcribe the downstream genes.[5]

The mechanism of hairpin-dependent pausing has been studied in detail. The pause signal relies on a fully folded hairpin that appears ~14 nt upstream of the catalytic site in the RNA exit channel. Once folded the hairpin allosterically modulates the state of the catalytic center of RNAP, presumably *via* the β-flap domain.[39] It reversibly freezes RNAP in the pre-translocated state, *i.e.*, the state in which the RNA 3′ terminus occupies the substrate-binding site (A site). Results obtained by crosslinking the RNA 3′-terminal base to the conserved portion of the trigger loop, in concert with other biochemical data, indicate that in the paused complex, the trigger loop is most likely in the folded, or "closed," conformation (visible in the structures of yeast and *Thermus thermophilus* EC),[40] contacting the 3′-terminal base and blocking access of incoming NTPs to the active center. The hairpin is not the only part of the pause signal. Sequences downstream of the hairpin-encoded palindromic stretch also play an important role in determining the site and efficiency of the pause.[41]

In contrast with hairpin-dependent pausing, at termination the hairpin sequence is located several nucleotides closer to the RNAP catalytic site, a configuration that obviously imposes structural constraints on its folding rate. During termination, the base of the hairpin is just 7–8 nt upstream of the catalytic site, located at the 5′ RNA extremity of the RNA:DNA hybrid in the main channel. Thus, to fully fold, the hairpin must either invade both the RBS and HBS or push the RNAP forward. A third model, which hypothesizes that folding of the hairpin can pull out, or shear, the downstream RNA out of its hybrid with the template DNA, has been recently put forth. This model requires that the huge energy barrier of simultaneous breakage of all the base pairs in the hybrid be overcome. Although no biochemical data has been obtained to support the "shearing" model, recent biophysical data seem to favor this scenario at least at some intrinsic terminators.[42] In all of these cases, the energy barrier opposed to hairpin folding under such circumstances is large.

Therefore, this process is likely to be the rate-limiting step in the termination pathway and some extra time is needed to overcome this barrier. Consequently, pausing at the terminator does indeed favor termination (see below). However, the purpose of pausing is to provide enough time for the hairpin to fold at the right distance from the catalytic site, rather than to increase the probability of complex dissociation.

10.3.1.2 Backtracking-type Pausing at the Terminator

In 1997, the mechanism behind a second class of pauses was established.[23,43] Spontaneous backtracking, *i.e.*, reverse sliding of RNAP along DNA and RNA, was shown, *in vitro*, to drive pauses that occur at numerous sites in natural DNA. During backtracking the catalytic site of RNAP loses its register with the RNA 3' terminus, which is extruded through the secondary channel of the enzyme, and the complex thus becomes temporarily inactivated or paused. Reactivation of the complex usually occurs spontaneously, due to the thermodynamic reversibility of backtracking.[23,43] In some rare instances backtracking may become irreversible. Reactivation of such arrested complexes requires the elongation factor GreB, which stimulates the intrinsic cleavage activity of RNAP and generates a new 3' terminus in register with the catalytic center.[23,44] Alternatively, other motor proteins may push the backtracked RNAP forward and rescue it from the arrested or paused state; examples include trailing ECs or the transcription-coupled repair factor Mfd.[45–47]

One of the major determinants of backtracking-type pausing is the stability of the RNA:DNA hybrid in the EC. A weak hybrid, such as that composed of A:U base pairs, usually induces backtracking.[23] Since backtracking also involves unwinding and rewinding of DNA upstream and downstream of the bubble and reverse threading of RNA through the enzyme, the sequence composition of the whole nucleic acid scaffold and RNA secondary structures are important in determining the efficiency and extent of backtracking.[48]

T-stretches, such as those found in intrinsic terminators, are typical backtracking signals.[49] When the EC reaches the termination site, most of its 8–9 bp hybrid consists of weak A:U base pairs. Regardless of the termination hairpin sequence, the U-stretch induces a brief ($\sim$1–2 s) pause precisely at the termination position.[10,34] This pausing depends strictly on the 3' proximal portion of the U-stretch and can be strongly affected by bases immediately downstream of the catalytic site. Note that the sequence corresponding to the termination hairpin itself does not induce pausing at the termination point. There is a direct correlation between the extent of pausing at the termination point and the efficiency of termination,[10,34] indicating that T-stretch-mediated pausing plays a critical role in the termination mechanism.

As emphasized earlier, the functional role of pausing is therefore to provide additional time for the termination hairpin to fold (Figure 10.1B).

10.3.2 The Termination Phase

Several models are still in play to explain the events that occur after pausing at the termination site. We shall discuss here the two prevalent ones, the "forward translocation" and "allosteric" models (Figure 10.3). According to the forward translocation model, hairpin formation forces the EC to translocate forward 4–5 bp without RNA synthesis (Figure 10.3A).[50,51] This model does not imply any significant conformational changes in RNAP during termination. The allosteric model,[34,36] on the contrary, proposes that hairpin formation leads to extensive conformational changes across the enzyme, leading to termination without forward translocation. In this model, the folding hairpin causes conformational changes in all three binding sites (DBS, HBS and RBS), as well as in the catalytic center, to inactivate and then disrupt the EC. Importantly, the fine details of the mechanism of termination could be partially dependent on the precise sequences of the terminator studied. Details of each model are described below.

10.3.2.1 Forward Translocation Model of Termination

In 1999, Roberts and colleagues proposed a simple mechanism of termination, in which the folding hairpin forces the EC to translocate forward without RNA synthesis so that the template DNA is rewound at the back and unwound at the front of the bubble while the 3' RNA terminus is pulled through the RNA exit channel into solution (Figure 10.3A).[51] The principal experimental observations favoring the forward translocation model come from roadblock experiments performed at modified t500 and tR2 terminators in which it was shown that a site-specific DNA binding protein or interstrand DNA crosslink blocked elongation at the termination site and at the same time inhibited termination.[49,50] These results were interpreted to mean that the roadblocks prohibited forward translocation of RNAP, thus preventing termination.[50]

However, an alternative interpretation is that roadblocks cause rapid backtracking of RNAP before the termination hairpin has a chance to form in the RBS. Such a backtracked EC would be resistant to termination because the portion of the RNA that is normally part of the hairpin would slip back into the enzyme and form a strong GC-rich hybrid in the transcription bubble.

Indeed various protein roadblocks have been found to induce backtracking.[47,52,53] Moreover, backtracking is known to occur when RNAP encounters various lesions in DNA that obstruct its forward propagation,[54] suggesting that interstrand DNA crosslinking adducts used to block forward translocation at the terminator could in fact cause backtracking. This objection has been considered for constructs derived from the t500 and tR2 terminator sites, in particular when the roadblock is the EcoRI Q111 mutant of the EcoRI restriction enzyme.[36,50] This enzyme binds tightly to, but is unable to cleave, it's DNA recognition sequence.[55] When properly placed, this roadblock suppresses termination at the canonical tR2 and t500 sites. However, while there is no evidence for termination suppression *via* backtracking at the t500 hybrid

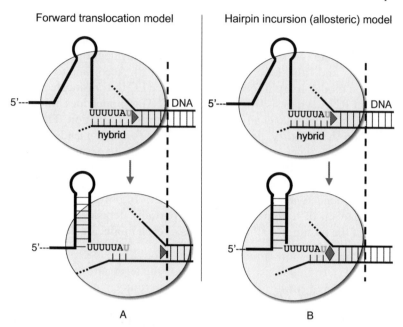

Figure 10.3 Alternative models of termination. The forward translocation model (A) postulates that the hairpin forces the EC to move forward without RNA synthesis. Such movement should lead to downstream DNA duplex unwinding and RNA extraction.[50] The allosteric model (B) postulates that the hairpin does not induce forward translocation, but instead enters the RNA exit channel and main channel to melt the RNA : DNA hybrid and destabilize the EC.[34] Conformational change in the catalytic center (diamond) caused by the invading hairpin leads to irreversible EC inactivation (trapped complex).

constructs, EcoRI Q111 does indeed cause backtracking at the tR2 termination site.[36]

The forward translocation model implies that the RNA 3' terminus is no longer in register with the catalytic center but shifted toward the RNA exit channel (Figure 10.3A).[50] This prediction has recently been tested directly at the tR2 terminator. A crosslinkable RNA probe was used to monitor changes in protein–RNA interactions in the catalytic site during termination. If forward translocation were to occur, the 3' end of the RNA would shift upstream into the main channel and contact the nearby rifampicin-binding region of the β subunit. In the absence of forward translocation, the 3' end should remain in the catalytic site and contact the β' subunit (bridge helix/trigger loop domains).[27,56] The 3' terminal crosslinkable analog of UMP, 4-thio-uridine-5'-monophosphate (sU), was incorporated specifically at the termination site. To prevent immediate dissociation of the termination complex, it was prepared in low salt conditions. (Note that when the resulting "trapped" complex was subjected to higher salt conditions it dissociated immediately.[34,36]) The crosslink was then induced by UV irradiation and mapped to the trigger loop and bridge helix region of the β' subunit in the termination (trapped) and control (active) ECs, respectively. These results excluded any forward translocation of RNAP relative to the RNA at the tR2 terminator, thus providing direct evidence against the forward translocation model of termination here.

A potential caveat of crosslinking experiments is that they were performed in low salt to avoid rapid complex dissociation at the terminator. One can argue that the trapped complex obtained in relatively low salt (e.g., 5 mM KCl and 5 mM MgCl$_2$) is not a ternary complex on the pathway to termination but, rather, a mixture of non-specific complexes between RNAP and RNA and DNA.[57] Experimental data supporting the idea that the trapped complex is a true termination intermediate, in which hairpin-dependent EC inactivation was decoupled from RNA dissociation, are discussed in the next section.

When RNAP translocates the energetics associated with the creation of the DNA bubble varies with the sequence of the single-stranded regions, and this in turn should affect the efficiency of RNA release (see also Chapter 9). The active role of the transcription bubble in termination has indeed been elegantly demonstrated.[58] In this work, RNA release at the terminator was inhibited by non-template strand substitutions that create a permanent mismatched segment at the trailing edge of the bubble. The authors concluded that since the energy associated with DNA rewinding was eliminated, no forward translocation could occur and, hence, termination was less efficient. However, biasing the energetics of bubble rewinding could also affect the efficiency of termination in models that do not require forward translocation. For instance in the competing model presented below, the energy released during bubble rewinding may facilitate DNA:RNA hybrid unwinding during hairpin invasion.

Similar arguments have been made for modifications at the leading edge of the transcription bubble. Since a GC-rich downstream sequence makes DNA unwinding more difficult, an AT or GC rich DNA segment in front of the termination site should either facilitate or suppress termination, respectively, if forward translocation occurs. However, this strong prediction of the model is not consistently supported by experimental evidence. In the case of phage T7 terminators, which have an unusually short U-stretch, the AT-rich downstream sequence indeed potentiated termination.[12] However, in the case of more canonical intrinsic terminators, such as tR2, the results were the opposite in that the termination efficiency was much greater for a GC-rich downstream duplex.[36]

In light of the above discussions forward translocation does not appear to occur at all terminator sequences, a conclusion reinforced by the more recent experiments performed on single molecules.

10.3.2.2 Trapped Intermediate in the Termination Pathway

In low ionic strength buffers, a termination complex can be isolated in which the hairpin has formed, but RNA and DNA have not been released.[34] Such a "trapped" complex is extremely unstable and is characterized by a partially collapsed DNA bubble and a partially unwound RNA:DNA hybrid. It is also irreversibly inactivated, despite the fact that the RNA 3′ terminus is in the catalytic site and maintains the downstream portion of the RNA:DNA hybrid, as revealed by protein–RNA and RNA–DNA crosslinking.[34,36] Mapping of the 3′ RNA contacts in the catalytic site showed that it is in close proximity to the trigger loop, as previously observed in backtracked complexes.[56] However, the trapped complex has been shown to be much more resistant to GreB-mediated cleavage than the backtracked complex.[36] It has therefore been suggested that the hairpin induces a conformational change in the trapped complex that results in a movement of the flexible trigger loop towards the RNA 3′ terminus. This hypothesis is attractive in light of recent structural data demonstrating that the trigger loop can adopt the so-called folded conformation, locking the EC in the pre-translocated state and prohibiting both backtracking and NTP addition (see Chapter 7).[17,28]

Remarkably, not only the 3′ end of the transcript, but also the head of the hairpin (the loop and upper part of the stem) can be crosslinked to the trigger loop in the trapped complex.[36] Such a direct contact between the flexible trigger loop domain and the hairpin could occur only if the hairpin invaded the main

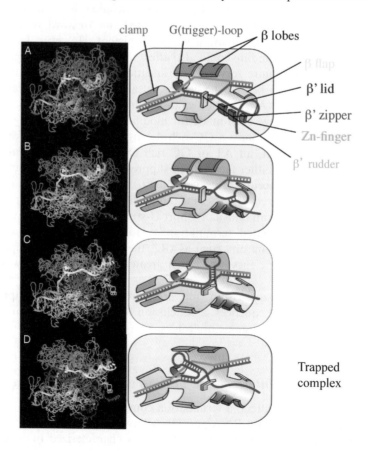

Trapped complex

channel of the EC (*i.e.*, the HBS) (Figure 10.4D below); this could lock the EC in the irreversibly trapped configuration by directly "pushing" or freezing the trigger loop in the folded conformation.

Since RNA–RNAP and DNA–RNAP binary complexes can form in low salt[59] it has been argued that the trapped complex could, in fact, be a mixture of such binary complexes re-associated after termination has taken place.[57] Several observations, however, demonstrate that the trapped complex is indeed a ternary complex with a well-defined trajectory of RNA: (i) In the trapped complex RNA–DNA crosslinking was lost at the −5 position relative to the termination point, but retained at the −3 position.[34] These data show that in the trapped complex RNA and DNA are in the same close proximity as in the normal EC, but the upstream part of the RNA:DNA hybrid is unwound. (ii) The RNA within the trapped EC could be recovered *via* a DNA template immobilized on affinity beads.[36] (iii) In the RNA:RNAP binary complexes, the

Figure 10.4 Allosteric (hairpin invasion) model of intrinsic termination. The RNAP structures with explanatory schemes represent four possible stages of the termination process.[36] Key structural elements are color coded and indicated: gray (green) – non-template (template) strand of DNA; red – nascent RNA; yellow – β flap; violet – β' lid; blue – β' zipper; green – Zn-finger; rose and gold – β lobes 1 and 2; aqua-β rudder; dark blue – trigger loop; teal – DNA clamp and other parts of β'. The star on the scheme denotes the catalytic center. (A) Paused EC. The first step in the termination pathway involves a pause at the end of the U-stretch, the role of which is to provide sufficient time for the hairpin to initiate folding.[34] (B) Hairpin nucleation. Nucleation of the hairpin could be initiated by interactions of the nascent RNA with the Zn-finger and flap domains, which constitute the edge of the exit channel. This induces further motion of the structural elements, which form the exit channel (flap, β-sheet element, zipper and lid), resulting in widening of the channel into a groove and allowing lengthening of the hairpin stem to 2–3 bp. The exit channel funnel may be wide enough to accommodate the hairpin head and a few base pairs of the stem, but to continue its growth the hairpin must open the exit channel into a groove. (C) Hairpin invasion. The hairpin stem grows to 4–5 bp and begins invading the primary channel, unwinding the upstream end of the hybrid and initiating collapse of the upstream edge of the transcription bubble, while simultaneously pushing on the flexible β'-rudder and inducing opening movement of the β lobe 2. This results in widening of the main channel, which allows the hairpin to move toward the downstream DNA and the secondary channel of RNAP. (D) Trapped complex. The hairpin stem grows to its final size of 7–8 bp while its head naturally bends around the downstream edge of the transcription bubble, clashing with the trigger loop. This shortens the hybrid to about 3 bp, further collapsing the bubble. The clashing of the hairpin with the trigger loop may force it to move towards the active site and assume the folded conformation, which would lock the enzyme in the pre-translocated state. Additionally, the hairpin action results in distortion of the hybrid, which could cause complete and irreversible enzyme inactivation. At the same time, trigger loop movement (*i.e.*, folding) induces DNA-binding-clamp opening, resulting in release of the nascent RNA and disengagement of RNAP from the DNA template.

head of the hairpin faces the RNA exit channel, the hairpin stem occupies the main channel (in place of the hybrid), and the RNA 3′ terminus is situated in the secondary channel.[57] It is thus impossible for the observed crosslinking to occur between the trigger loop and the head of the hairpin[36] in the same trapped binary complex. Moreover, rebinding of the same purified RNA transcript to core RNAP (*i.e.*, RNAP–RNA binary complex) gives a totally different and much weaker crosslinking pattern than when RNA is retained in the trapped complex.[36] (iv) Binary complexes formed between RNAP and RNA are more stable than the trapped complex. For example, binary complexes are fully resistant to 40 mM KCl,[57] whereas most trapped complexes rapidly dissociate in such ionic conditions. (v) Finally, inactivated, short-lived intermediates similar in nature to trapped complexes have been detected in real-time single molecule studies.[9] The partial blockage of transcription and DNA replication at two different intrinsic terminators suggests that trapped complexes may even be detected *in vivo*.[47,60]

The process of RNA and DNA release may vary from terminator to terminator and take a much longer time than rapid complex inactivation. The trapped intermediate, therefore, is crucial for efficient termination to occur, as it rapidly and irreversibly halts the complex at the termination point. Once trapped, the EC is destined to dissociate. This "spider" strategy (trap first, kill later) may be characteristic not only of intrinsic terminators but also of factor-dependent termination systems.

10.3.2.3 Allosteric (Hairpin Invasion) Model of Termination

An allosteric model of termination relies on the intrinsic structural plasticity of RNAP to accommodate the hairpin inside the RNA exit and main channels (RBS and HBS). In this model, significant conformational changes across the enzyme occur as the hairpin folds, causing the rapid and simultaneous release of both RNA and DNA. The structures of bacterial and yeast RNAP elongation complexes[15–17,61] in combination with new genetic and biochemical information[36] suggest the following structural model of intrinsic termination:

Hairpin Nucleation Phase. Termination occurs within a short, fixed distance (8 ± 1 bp) downstream from the hairpin base, setting a rigid restriction on the positioning of the hairpin within the enzyme. In the allosteric ("hairpin invasion") model, different mobile parts of the RNAP play specific roles in termination. The flexible flap, zipper and lid domains of the RNA exit channel move slightly apart to accommodate the upper portion of the hairpin (Figure 10.4B). Although the RNA exit channel is wide enough to accommodate the hairpin stem together with a stretch of 5′ proximal RNA,[17] RNA folding within a channel requires more space, suggesting that mild opening is necessary at the stage of hairpin nucleation. In addition, the flap, zipper and Zn-finger (N-terminal Zn-binding-domain of β′) may act as RNA chaperones by actively facilitating hairpin nucleation. The latter hypothesis is supported by the fact that deletion of the flap and Zn-finger, and substitutions in the zipper domain, strongly and specifically inhibit termination.[13,36] Moreover, only those mutations in the RNA exit channel that are predicted to be closest to the upper portion of the folding hairpin have an inhibitory effect on termination. Thus, highly specific interactions between the RNA exit channel elements and the upper portion of the hairpin may facilitate initial RNA folding. Hairpin nucleation is likely to be the most sensitive step in the pathway leading to termination, because it has to compete with the formation of multiple, non-specific alternative RNA structures that occur co-transcriptionally. Such competition helps explain both why RNA-exit channel mutations have the strongest negative impact on termination,[36] and why sequences close to the promoter could have a negative effect on termination despite large distances between them.[11,12] Finally, it is noteworthy that specific alternative RNA secondary structures (*e.g.*, riboswitches) and anti-termination factors readily interfere with hairpin formation to regulate termination.[13,34,62]

Hairpin Invasion Phase. Two other mobile domains of RNAP, the trigger loop and the DNA-clamp, direct the final stages of the termination process. The structural model depicted in Figure 10.4 suggests that once the hairpin grows beyond 4–5 bp it invades the main channel with concomitant melting of the upstream portion of the hybrid. Indeed, unwinding of the upstream portion of the hybrid has been demonstrated directly by RNA:DNA crosslinking experiments[34] and indirectly by the use of RNA:DNA scaffolds.[35] Growth of the hairpin stem to 7–8 bp pulls it deeper into the main channel. Owing to a natural tendency of the RNA hairpin to coil, the head of the hairpin bends around the non-template strand and reaches towards, and directly contacts, the trigger loop. These events were revealed by protein–RNA crosslinking experiments[34,36] that demonstrated the direct contact between the upper portion of the hairpin and the trigger loop in the trapped complex at the tR2 terminator and derivatives. The results described below strongly suggest that a clash between the trigger loop and the hairpin leads both to complex inactivation (trapped state) and destabilization (DNA clamp opening).

Recent structural studies of ECs revealed that the flexible trigger loop can assume at least two conformations, folded or unfolded.[15,16,28,29] The unfolded conformation appears to be necessary for RNA backtracking and transcript cleavage reactions, whereas the folded conformation is required for the catalysis of RNA synthesis. Since the trigger loop is part of the catalytic site, its hairpin-induced displacement can inhibit substrate binding and result in irreversible trapping of RNAP in a pre-translocated position (trapped complex). Indeed, results obtained from crosslinking the RNA 3'-terminal base to the trigger loop in the trapped complex in concert with other biochemical data indicate that (i) the trapped complex is not backtracked and should be in the pre-translocated state as judged from the position of the RNA 3' terminus, and (ii) the trigger loop is most likely in a folded conformation, reaching the

3′-terminal base and blocking access of the active center to NTPs. The trigger loop movement from the unfolded to folded conformation could be directly induced by the invading termination hairpin, which reaches the wall of the secondary channel, "pushes" the trigger loop towards the active center and sterically prevents it from unfolding – hence, the observed crosslinking from the head of the hairpin to the trigger loop. Point mutations in the trigger loop consistently change the efficiency of termination by affecting the rate at which RNAP enters the trapped state.[36]

The forced motion of the trigger loop, which constitutes the bottom of the DNA-binding clamp, may also result in simultaneous opening of the clamp. In turn, opening of the clamp removes all major interactions between RNAP and nucleic acids (Figure 10.4D), thus inducing release of both the transcript and DNA. This model accounts for the fact that G(trigger)-loop specific monoclonal antibodies (G-mAbs) abolish termination:[36] freezing the trigger loop by the action of G-mAbs likely circumvents hairpin-induced trigger loop motion required for clamp opening. Therefore, trigger loop movement must be necessary for hairpin-mediated opening of the enzyme to allow RNA and DNA release. The fact that G-mAbs inhibit elongation, pyrophosphorolysis and termination, but not transcript cleavage,[36] indicates that G-mAbs bind to and stabilize the trigger loop in the unfolded conformation or prevent its movement into a folded conformation. After the trapped EC has formed, the hairpin likely forces the trigger loop to remain in the folded conformation, thus blocking access of the trigger loop to mAbs. This explains why anti-trigger loop mAbs inhibit formation of the trapped complex, but not the release of RNA from an already formed, trapped EC.[36]

10.4 Mechanism of Rho Termination

In addition to intrinsic termination, a second essential mechanism of termination exists in bacteria that serves to separate transcription units, adjust transcriptional yield to translational needs, and suppress activity of some toxic horizontally transferred genes.[63] This mechanism relies on the RNA/DNA helicase Rho.[64,65] Unlike intrinsic termination signals, which are very conserved and highly localized, Rho-dependent terminators are bipartite sites that stretch out over 200 bp of DNA. The upstream portion [Rho-utilization site (RUT)] encompasses a C-rich segment of nascent RNA with little or no secondary structure.[66] The sites where termination actually occurs can be distributed over a large region of DNA, up to 120 bp promoter distal to RUT. In general, Rho termination loci correlate with EC pause sites.[67,68]

The Rho hexamer consists of identical 46-kDa subunits arranged in a ring structure.[69] The actual mechanism by which Rho disrupts the EC is not completely understood. The Rho termination reaction initiates with binding of Rho to RUT. Once bound to RNA, concerted cycles of ATP binding, hydrolysis and release induce conformational changes in the cavity of the hexamer that pull the transcript through and induce Rho to translocate along RNA in a $5' \rightarrow 3'$

direction.[69] The translocation model postulates that Rho catches up to the EC and pulls the transcript away from RNAP and the DNA template. For this to occur, Rho must translocate more rapidly than RNAP, at least at certain DNA locations. Such a kinetic balance may play a significant role in controlling the efficiency of termination.[67,68] However, a sub-population of ECs is resistant to Rho even after the complexes have been stalled by terminating nucleotides or by a protein roadblock (ref. 70 and Epshtein and Nudler, unpublished observations). This suggests that simple kinetic competition is not the only determinant of the site or efficiency of termination.

The key question of how Rho dissociates the EC remains to be addressed. It has been proposed that Rho utilizes its ATP-dependant helicase activity to unwind the RNA:DNA hybrid in the EC.[71] According to the model of hairpin invasion, breaking DNA:RNA interactions at the 5' end of the hybrid destabilizes the EC and is the key step during intrinsic termination.[34-36] However, what is not yet clear in this context is how Rho gains access to the hybrid, which is buried deep in the main channel (HBS) (Figure 10.2). Perhaps specific interactions between Rho and the EC change the conformation of RNAP in a fashion similar to that proposed at an intrinsic terminator, allowing Rho to invade the HBS. Alternatively, Rho might use ATP hydrolysis coupled with the threading motion in its cavity to pull RNA out of the HBS independently of its helicase activity. This latter possibility does not entail specific contacts between Rho and RNAP, which is consistent with the ability of Rho to terminate transcription of non-bacterial RNAP, such as yeast RNAP II (ref. 72) and phage T7 or SP6 RNAPs.[73]

10.5 Summary

Transcription termination is a process of rapid dissociation, at specific sites in the genome, of the elongation complex into the RNA polymerase, DNA template and RNA transcript. This is an integral part of the transcription cycle and one of its key regulatory points. A wealth of information has accumulated over decades of study of transcription termination in various biological systems. Yet, only recently have we begun to understand the mechanism of this process at the molecular level. In bacteria, intrinsic terminators are prototypical termination signals comprising a stable RNA hairpin followed by a stretch of U residues at the 3' terminus. Intrinsic terminators are dominant features of bacterial transcription regulation and play multiple physiological roles: they separate operons into autonomous transcription units, facilitate RNA polymerase recycling, and control expression of downstream genes. This chapter has described the current molecular models of intrinsic termination. Their implications for an understanding of the structure/functional organization of the elongation complex and the processes of elongation and termination have been discussed in general. We also briefly covered extrinsic, or rho-dependent, termination where a helicase known as rho is responsible for potentiating transcription termination.

10.6 Concluding Remarks

The intrinsic terminator signal is seemingly simple. It has, however, taken decades of multidisciplinary research that combined a large body of genetic, biochemical and structural data to come to a general model accounting for its mechanism of action. The allosteric ("hairpin invasion") model of termination described above (Figure 10.4) was the result of searching for a consistent structural representation in which the hairpin invades the exit and main channels of RNAP in the course of termination. The precise allocation of the sequential conformational changes predicted by this model to the various intermediates of the termination pathway is not fully assessed yet. However, this gradual process appears to be consistent with most of the experimental results presented to date.

Interestingly, the positioning of the hairpin in the termination complex, as proposed by the allosteric model, bears a striking resemblance to the positioning of σ-factors within the holoenzyme. Invasion of the primary channel in termination is analogous to invasion of the primary channel by σ region 3.2 in the holoenzyme.[26,74,75] Moreover, σ region 4 acts on the same elements (Zn-finger and flap) of the exit channel and most likely accesses the RNA-exit channel in the same way during initiation as does the nucleating hairpin during termination, by taking advantage of the flexibility of the flap and zipper domains.

The structural organization and principles by which RNA and DNA are retained in the EC are highly conserved among all cellular RNAPs,[15,29,61] suggesting that the fundamental pathway leading to termination is universal. Prokaryotic and eukaryotic termination factors and intrinsic signals likely utilize the same strategy to disrupt the highly processive ECs. Since dismantling the stable EC requires more time than its inactivation, termination factors such as Rho or eukaryotic Pcf11 (ref. 76) may adopt the same strategy as intrinsic terminators, *i.e.*, first to rapidly incapacitate the EC (trap) and then proceed with a slower dissociation event. Future studies on factor-dependent termination will test this prediction.

References

1. E. A. Lesnik, R. Sampath, H. B. Levene, T. J. Henderson, J. A. McNeil and D. J. Ecker, Prediction of rho-independent transcriptional terminators in *Escherichia coli*, *Nucleic Acids Res.*, 2001, **29**, 3583–3594.
2. T. Platt, Transcription termination and the regulation of gene expression, *Annu. Rev. Biochem.*, 1986, **55**, 339–372.
3. D. I. Friedman, M. J. Imperiale and S. L. Adhya, RNA 3′ end formation in the control of gene expression, *Annu. Rev. Genet.*, 1987, **21**, 453–488.
4. C. Condon, C. Squires and C. L. Squires, Control of rRNA transcription in *Escherichia coli*, *Microbiol. Rev.*, 1995, **59**, 623–645.
5. R. Landick, J. R. Turnbough and C. Yanofsky, Transcription attenuation. In *Escherichia coli and Salmonella typhimurium*, F. C. Neidhardt, J. L. Ingraham, K. B. Low, B. Magasanik, M. Schaechter and H. E. Umbarger, eds. (Washington: American Society for Microbiology), 1996, 1263–1286.

6. T. M. Henkin, Transcription termination control in bacteria, *Curr. Opin. Microbiol.*, 2000, **3**, 149–153.
7. E. Nudler and A. S. Mironov, The riboswitch control of bacterial metabolism, *Trends Biochem. Sci.*, 2004, **29**, 11–17.
8. Y. d'Aubenton Carafa, E. Brody and C. Thermes, Prediction of rho-independent *Escherichia coli* transcription terminators. A statistical analysis of their RNA stem-loop structures, *J. Mol. Biol.*, 1990, **216**, 835–58.
9. H. Yin, I. Artsimovitch, R. Landick and J. Gelles, Nonequilibrium mechanism of transcription termination from observations of single RNA polymerase molecules, *Proc. Natl. Acad. Sci. USA.*, 1999, **96** 13124–13129.
10. I. Gusarov and E. Nudler, Control of intrinsic transcription termination by N and NusA: the basic mechanisms, *Cell*, 2001, **107**, 437–449.
11. J. A. Goliger, X. J. Yang, H. C. Guo and J. W. Roberts, Early transcribed sequences affect termination efficiency of *Escherichia coli* RNA polymerase, *J. Mol. Biol.*, 1989, **205**, 331–341.
12. (a) A. P. Telesnitsky and M. J. Chamberlin, Sequences linked to prokaryotic promoters can affect the efficiency of downstream termination sites, *J. Mol. Biol.*, 1989, **205**, 315–330; (b) A. P. Telesnitsky and M. J. Chamberlin, Terminator-distal sequences determine the in vitro efficiency of the early terminators of bacteriophages T3 and T7, *Biochemistry*, 1989, **28**, 5210–5218.
13. R. A. King, D. Markov, R. Sen, K. Severinov and R. A. Weisberg, A conserved zinc binding domain in the largest subunit of DNA-dependent RNA polymerase modulates intrinsic transcription termination and antitermination but does not stabilize the elongation complex, *J. Mol. Biol.*, 2004, **342**, 1143–1154.
14. N. Korzheva, *et al.*, A structural model of transcription elongation, *Science*, 2000, **289**, 619–625.
15. A. L. Gnatt, P. Cramer, J. Fu, D. A. Bushnell and R. D. Kornberg, Structural basis of transcription: an RNA polymerase II elongation complex at 3.3 A resolution, *Science*, 2001, **292**, 1876–1882.
16. H. Kettenberger, K. J. Armache and P. Cramer, Complete RNA polymerase II elongation complex structure and its interactions with NTP and TFIIS, *Mol. Cell*, 2004, **16**, 955–965.
17. D. G. Vassylyev, M. N. Vassylyeva, A. Perederina, T. H. Tahirov and I. Artsimovitch, Structural basis for transcription elongation by bacterial RNA polymerase, *Nature*, 2007, **448**, 157–162.
18. E. Nudler and I. Gusarov, Analysis of the intrinsic transcription termination mechanism and its control, *Methods Enzymol.*, 2003, **371**, 369–382.
19. N. Komissarova, M. L. Kireeva, J. Becker, I. Sidorenkov and M. Kashlev, Engineering of elongation complexes of bacterial and yeast RNA polymerases, *Methods Enzymol.*, 2003, **371**, 233–251.
20. T. J. Santangelo and J. W. Roberts, Formation of long DNA templates containing site-specific alkane-disulfide DNA interstrand cross-links for use in transcription reactions, *Methods Enzymol.*, 2003, **371**, 120–132.

21. E. Nudler, Transcription elongation: structural basis and mechanisms, *J. Mol. Biol.*, 1999, **288**, 1–12.
22. E. Nudler, E. Avetissova, V. Markovtsov and A. Goldfarb, Transcription processivity: protein-DNA interactions holding together the elongation complex, *Science*, 1996, **273**, 211–217.
23. E. Nudler, A. Mustaev, E. Lukhtanov and A. Goldfarb, The RNA-DNA hybrid maintains the register of transcription by preventing backtracking of RNA polymerase, *Cell*, 1997, **89**, 33–41.
24. E. Nudler, I. Gusarov, E. Avetissova, M. Kozlov and A. Goldfarb, Spatial organization of transcription elongation complex in *Escherichia coli*, *Science*, 1998, **281**, 424–428.
25. N. Komissarova and M. Kashlev, Functional topography of nascent RNA in elongation intermediates of RNA polymerase, *Proc. Natl. Acad. Sci. USA*, 1998, **95**, 14699–14704.
26. S. Borukhov and E. Nudler, *RNA polymerase: a vehicle of transcription, Trends Microbiol.*, 2008, **16**, 126–134.
27. G. Bar-Nahum, V. Epshtein, A. E. Ruckenstein, R. Rafikov, A. Mustaev and E. Nudler, A ratchet mechanism of transcription elongation and its control, *Cell*, 2005, **120**, 183–193.
28. D. Wang, D. A. Bushnell, K. D. Westover, C. D. Kaplan and R. D. Kornberg, Structural basis of transcription: role of the trigger loop in substrate specificity and catalysis, *Cell*, 2006, **127**, 941–954.
29. D. G. Vassylyev, M. N. Vassylyeva, J. Zhang, M. Palangat, I. Artsimovitch and R. Landick, Structural basis for substrate loading in bacterial RNA polymerase, *Nature*, 2007, **448**, 163–168.
30. N. Korzheva, A. Mustaev, E. Nudler, V. Nikiforov and A. Goldfarb, Mechanistic model of the elongation complex of *Escherichia coli* RNA polymerase, *Cold Spring Harb. Symp. Quant. Biol.*, 1998, **63**, 337–345.
31. K. M. Arndt and M. J. Chamberlin, Transcription termination in *Escherichia coli*. Measurement of the rate of enzyme release from Rho-independent terminators, *J. Mol. Biol.*, 1988, **202**, 271–285.
32. F. H. Martin and I. Tinoco Jr, DNA-RNA hybrid duplexes containing oligo(dA:rU) sequences are exceptionally unstable and may facilitate termination of transcription, *Nucleic Acids Res.*, 1980, **8**, 2295–2299.
33. J. C. McDowell, J. W. Roberts, D. J. Jin and C. Gross, Determination of intrinsic transcription termination efficiency by RNA polymerase elongation rate, *Science*, 1994, **266**, 822–5.
34. I. Gusarov and E. Nudler, The mechanism of intrinsic transcription termination, *Mol. Cell*, 1999, **3**, 495–504.
35. N. Komissarova, J. Becker, S. Solter, M. Kireeva and M. Kashlev, Shortening of RNA:DNA hybrid in the elongation complex of RNA polymerase is a prerequisite for transcription termination, *Mol. Cell*, 2002, **10**, 1151–1162.
36. V. Epshtein, C. Cardinale, A. E. Ruckenstein, S. Borukhov and E. Nudler, An allosteric path to transcription termination, *Mol. Cell*, 2007, **28**, 991–1001.

37. T. D. Yager and P. H. von Hippel, Transcript elongation and termination in *Escherichia coli*. In *Escherichia coli and Salmonella typhimurium*, F. C. Neidhardt, J. L. Ingraham, K. B. Low, B. Magasanik, M. Schaechter and H. E. Umbarger, eds. (Washington: American Society for Microbiology), 1987, pp. 1241–1275.
38. P. J. Farnham and T. Platt, Rho-independent termination: dyad symmetry in DNA causes RNA polymerase to pause during transcription *in vitro*, *Nucleic Acids Res.*, 1981, **9**, 563–577.
39. I. Toulokhonov, I. Artsimovitch and R. Landick, Allosteric control of RNA polymerase by a site that contacts nascent RNA hairpins, *Science*, 2001, **292**, 730–733.
40. I. Toulokhonov, J. Zhang, M. Palangat and R. Landick, A central role of the RNA polymerase trigger loop in active-site rearrangement during transcriptional pausing, *Mol. Cell*, 2007, **27**, 406–419.
41. C. L. Chan, D. Wang and R. Landick, Multiple interactions stabilize a single paused transcription intermediate in which hairpin to 3′ end spacing distinguishes pause and termination pathways, *J. Mol. Biol.*, 1997, **268**, 54–68.
42. M. H. Larson, W. J. Greenleaf, R. Landick and S. M. Block, Applied force reveals mechanistic and energetic details of transcription termination, *Cell*, 2008, **132**, 971–982.
43. N. Komissarova and M. Kashlev, RNA polymerase switches between inactivated and activated states by translocating back and forth along the DNA and the RNA, *J. Biol. Chem.*, 1997, **272**, 15329–15338.
44. S. Borukhov, V. Sagitov and A. Goldfarb, Transcript cleavage factors from *E. coli*, *Cell*, 1993, **72**, 459–466.
45. J. S. Park, M. T. Marr and J. W. Roberts, *E. coli* Transcription repair coupling factor (Mfd protein) rescues arrested complexes by promoting forward translocation, *Cell*, 2002, **109**, 757–767.
46. V. Epshtein and E. Nudler, Cooperation between RNA polymerase molecules in transcription elongation, *Science*, 2003, **300**, 801–805.
47. V. Epshtein, F. Toulmé, A. R. Rahmouni, S. Borukhov and E. Nudler, Transcription through the roadblocks: the role of RNA polymerase cooperation, *EMBO J.*, 2003, **22**, 4719–4727.
48. V. R. Tadigotla, D. O'Maoileidigh, M. A. Sengupta, V. Epshtein, R. H. Ebright, E. Nudler and A. E. Ruckenstein, Thermodynamics and kinetics-based identification of transcriptional pauses, *Proc. Natl. Acad. Sci. USA*, 2006, **103**, 4439–4444.
49. E. Nudler, M. Kashlev, V. Nikiforov and A. Goldfarb, Coupling between transcription termination and RNA polymerase inchworming, *Cell*, 1995, **81**, 351–357.
50. T. J. Santangelo and J. W. Roberts, Forward translocation is the natural pathway of RNA release at an intrinsic terminator, *Mol. Cell*, 2004, **14**, 117–126.
51. W. S. Yarnell and J. W. Roberts, Mechanism of intrinsic transcription termination and antitermination, *Science*, 1999, **284**, 611–615.

52. P. A. Pavco and D. A. Steege, Elongation by *Escherichia coli* RNA polymerase is blocked in vitro by a site-specific DNA binding protein, *J. Biol. Chem.*, 1990, **265**, 9960–9969.
53. D. Reines and J. Mote Jr, Elongation factor SII-dependent transcription by RNA polymerase II through a sequence-specific DNA-binding protein, *Proc. Natl. Acad. Sci. USA*, 1993, **90**, 1917–1921.
54. V. S. Kalogeraki, S. Tornaletti, P. K. Cooper and P. C. Hanawalt, Comparative TFIIS-mediated transcript cleavage by mammalian RNA polymerase II arrested at a lesion in different transcription systems, *DNA Repair (Amst.)*, 2005, **4**, 1075–1087.
55. D. J. Wright, K. King and P. Modrich, The negative charge of Glu-111 is required to activate the cleavage center of EcoRI endonuclease, *J. Biol. Chem.*, 1989, **264**, 11816–11821.
56. V. Markovtsov, A. Mustaev and A. Goldfarb, Protein-RNA interactions in the active center of transcription elongation complex, *Proc. Natl. Acad. Sci. USA*, 1996, **93**, 3221–3226.
57. M. Kashlev and N. Komissarova, Transcription termination: primary intermediates and secondary adducts, *J. Biol. Chem.*, 2002, **277**, 14501–14508.
58. A. M. Ryder and J. W. Roberts, Role of the non-template strand of the elongation bubble in intrinsic transcription termination, *J. Mol. Biol.*, 2003, **334**, 205–213.
59. C. R. Altmann, D. E. Solow-Cordero and M. J. Chamberlin, RNA cleavage and chain elongation by *Escherichia coli* DNA-dependent RNA polymerase in a binary enzyme.RNA complex, *Proc. Natl. Acad. Sci. USA.*, 1994, **91**, 3784–3788.
60. E. V. Mirkin, D. C. Roa, E. Nudler and S. M. Mirkin, Transcription regulatory elements are punctuation marks for DNA replication, *Proc. Natl. Acad. Sci. USA*, 2006, **103**, 7276–7281.
61. K. D. Westover, D. A. Bushnell and R. D. Kornberg, Structural basis of transcription: nucleotide selection by rotation in the RNA polymerase II active center, *Cell*, 2004, **119**, 481–489.
62. A. S. Mironov, I. Gusarov, R. Rafikov, L. E. Lopez, R. Kreneva, D. Perumov and E. Nudler, Sensing small molecules by nascent RNA: a mechanism to control transcription in bacteria, *Cell*, 2002, **111**, 747–756.
63. C. J. Cardinale, R. S. Washburn, V. R. Tadigotla, L. M. Brown, M. E. Gottesman and E. Nudler Termination factor Rho and its cofactors NusA and NusG silence foreign DNA in *E. coli*, *Science* 2008, **320**, 935–938.
64. J. W. Roberts, Termination factor for RNA synthesis, *Nature*, 1969, **224**, 1168–1174.
65. J. P. Richardson, Structural organization of transcription termination factor Rho, *J. Biol. Chem.*, 1996, **271**, 1251–1254.
66. C. M. Hart and J. W. Roberts, Rho-dependent transcription termination. Characterization of the requirement for cytidine in the nascent transcript, *J. Biol. Chem.*, 1991, **266**, 24140–24148.

67. L. F. Lau, J. W. Roberts and R. Wu, RNA polymerase pausing and transcript release at the lambda tR1 terminator in vitro, *J. Biol. Chem.*, 1983, **258**, 9391–9397.
68. D. J. Jin, R. R. Burgess, J. P. Richardson and C. A. Gross, Termination efficiency at rho-dependent terminators depends on kinetic coupling between RNA polymerase and rho, *Proc. Natl. Acad. Sci. USA*, 1992, **89**, 1453–1457.
69. E. Skordalakes and J. M. Berger, Structural insights into RNA-dependent ring closure and ATPase activation by the Rho termination factor, *Cell*, 2006, **127**, 553–364.
70. K. W. Nehrke, F. Zalatan and T. Platt, NusG alters rho-dependent termination of transcription in vitro independent of kinetic coupling, *Gene Expr.*, 1993, **3**, 119–133.
71. C. A. Brennan, A. J. Dombroski and T. Platt, Transcription termination factor rho is an RNA-DNA helicase, *Cell*, 1987, **48**, 945–952.
72. W. H. Lang, T. Platt and R. H. Reeder, *Escherichia coli* rho factor induces release of yeast RNA polymerase II but not polymerase I or III, *Proc. Natl. Acad. Sci. USA*, 1998, **95**, 4900–4905.
73. Z. Pasman and P. H. von Hippel, Regulation of rho-dependent transcription termination by NusG is specific to the *Escherichia coli* elongation complex, *Biochemistry*, 2000, **39**, 5573–5585.
74. D. G. Vassylyev, *et al.*, Crystal structure of a bacterial RNA polymerase holoenzyme at 2.6 A resolution, *Nature*, 2002, **417**, 712–719.
75. K. S. Murakami, S. Masuda and S. A. Darst, Structural basis of transcription initiation: RNA polymerase holoenzyme at 4 A resolution, *Science*, 2002, **296**, 1280–1284.
76. Z. Zhang and D. S. Gilmour, Pcf11 is a termination factor in *Drosophila* that dismantles the elongation complex by bridging the CTD of RNA polymerase II to the nascent transcript, *Mol. Cell*, 2006, **21**, 65–74.

CONCLUSION
Past, Present, and Future of Single-molecule Studies of Transcription

CARLOS BUSTAMANTE[a,b] AND JEFFREY R. MOFFITT[a]

[a] Department of Physics, University of California, Berkeley CA 94720-7300, USA; [b] Departments of Chemistry, Molecular and Cell Biology, and Howard Hughes Medical Institute, University of California, Berkeley CA 94720-3220, USA

C.1 Introduction

In the last two and a half decades, molecular biophysics has experienced something close to a revolution. The advent during this period of gene cloning and single point mutagenesis, on the one hand, and of powerful methods of structural determination, on the other, have had an enormous impact on our understanding of biological function at the molecular level. Unfortunately, the great expansion of structural information that began to occur around that time was not accompanied by an equivalent increase in our understanding of the dynamics and functionality of these structures. The reason for this asymmetry is not difficult to understand. Methods of structural determination such as NMR or X-ray crystallography can provide atomic scale detail because all the molecules in the crystal are in a single state or very close to a single state. Thus, the resulting picture is a macroscopic average of the individual structures of each molecule in the sample, an accurate but static picture of the ensemble.

Function, on the other hand, necessarily involves a change of structure in time. Even if the ensemble is initially synchronized, once that change has been triggered, the molecules will soon find themselves in various different states because each of them will follow a distinct trajectory, evolving differently in response to local fluctuations that vary from molecule to molecule. Thus, within a short time the individual molecules in the ensemble lose memory of their initial phase and become asynchronous with each other. Following the evolution of an ensemble of molecules involves averaging the asynchronous signals of the individual molecules. Such averaging provides only a "kinetic" description of the process, a sort of idealized, "mean dynamics" of the molecular population that no longer describes the true dynamics of any one given molecule in the ensemble. Moreover, ensemble studies tend to mask rare but potentially important molecular trajectories in a process or short-lived intermediate states. Finally, ensemble kinetic studies tend to smear out the inherent thermally-induced fluctuations that dominate the dynamics of the individual molecules. Great progress has been made with techniques to synchronize molecular ensembles, such as stopped-flow instruments with increasingly small dead-times or the use of photo-activated processes; however, these techniques remain limited by these shortfalls of ensemble averaging.

One way to circumvent the problem of synchronization is to follow the individual trajectories of single molecules as they undergo their dynamic processes.[1-12] In the end the final analysis will still be an ensemble one, average rates and average fluctuations will be the quantities measured. But because averages are built one molecule at a time, out of the individual molecular trajectories, it is possible to extract much more information, *e.g.*, the degree of static or dynamic heterogeneity in a system, the probabilities of rare events, and even the full probability distributions that govern fluctuations. It has now been almost two decades since single-molecule methods were first introduced in biochemical research. In the intervening time, single-molecule methods have come of age and important new results have emerged in a broad range of biological problems. Today, as testified from the many examples given in this book, we can say that single-molecule methods are often the methods of choice to investigate dynamics problems in biochemistry.

C.2 RNA Polymerase as a Molecular Machine: Past and Present

Fifty years ago, biochemists described cells as small vessels that contain a complex mixture of chemical species undergoing reactions through diffusion and random collision. This description was satisfactory inasmuch as the intricate pathways of metabolism and, later, the basic mechanisms of gene regulation and signal transduction were still being unraveled. Gradually, and in part as a result of the parallel growth in the structural understanding of the molecular components of the cell, biologists came to understand that the cell resembles more a mechanical factory in which many of the processes are performed by specialized and often localized machinery rather than in small reaction vessels. Many of these functions involve highly specialized molecular complexes whose function is often mechanical in nature. In particular, many of these molecular machines operate cyclically and by coupling a thermodynamically favorable chemical reaction with a mechanical task of some sort, functioning effectively as energy transducers or molecular motors. The need of such specialized molecular entities is clear. Cells have polarity and they must, therefore, carry out several directional processes. For example, metabolites must be transported across membranes and against chemical gradients, molecules and even organelles must be transported from one part of the cell to another, and the linear information encoded in DNA must be read in a sequential fashion. None of these processes can be performed by thermally induced random diffusion and collision, as these can lead only to non-directional transport, low reaction rates and inefficient reaction yields. Molecular motors are evolution's solution to this problem. Molecular motors are unlike their macroscopic counterparts in that the energies at which they operate are comparable to that of the thermal bath; therefore, molecular motors are intrinsically open systems that constantly exchange energy with the thermal bath that surrounds them and, as a result, their dynamics are dominated by fluctuations. Thus, sitting astride the boundary between deterministic and random processes, the function of molecular motors is to tame the stochastic nature of Brownian processes, generating biased, directional motion.

Transcription is an example of a directional processes in which the linear information encoded in the DNA duplex must be read sequentially to be converted into a facsimile version that contains the same information in the form of a messenger RNA molecule. RNA polymerase, the molecule responsible for catalyzing this process, must perform this task as a directional machine and, indeed, as a molecular motor. Thus, studies of RNA polymerase, as a dynamic, fluctuating, molecular motor capable of producing force and torque, promise to reveal not only important details on the biologically fundamental process of transcription but also general design features and properties of molecular machines.

As detailed in the Part II of this book, the application of single-molecule methods to transcription started with the work of Yin et al.,[13] who used optical tweezers to characterize the force–velocity relationship for this enzyme. These studies and succeeding ones,[14,15] established the stalling force of the motor in the range 15–25 pN, depending on the experimental conditions. Subsequently, single-molecule manipulation experiments revealed important information about the dynamics of transcription elongation. It was found that during elongation the enzyme does not move in a continuous manner but in a discontinuous fashion, pausing often at different locations along the template and in some cases even arresting in an irreversible fashion. The detailed dynamics of transcription elongation are important because they determine the overall rate of transcription and the mRNA throughput, both of which play an important regulatory role in gene expression. The existence of pauses was first inferred from bulk studies, but using these methods it was not possible to separate the pause frequency (*i.e.*, the efficiency of a particular pause site) from the pause length or life-time. Single-molecule experiments allowed for the first time to investigate systematically these complementary aspects of elongation dynamics.[16] These studies revealed that RNAP pauses in a sequence-dependent manner, and that these pauses can sometimes lead to irreversible arrest of the enzyme. It was shown that the probability of entering a pause depends on the rate of transcription. The faster the transcription, the less likely that the molecule will pause, confirming as suggested earlier from bulk studies that there exists a kinetic competition between the on-pathway transcription and pausing. A pause is therefore an "off-main-elongation-pathway" state. Moreover, it was shown that the off-pathway event that leads to a pause is required to enter a permanent pause or arrest of the enzyme.

Bulk studies were used to establish the ability of RNA polymerase to move backwards along the template so that the 3′ end of the nascent transcript becomes disengaged from the active site of the enzyme, a mechanism termed "backtracking."[17–19] This process has been recently confirmed by the use of optical tweezers with improved spatial resolution.[20–22] The process of backtracking plays an important role in the editing capabilities of the enzyme[23] and leads to elongation pausing and arrest. The origin and the nature of pauses as observed in single-molecule experiments is still a matter of debate, as discussed in Chapter 9. Some authors postulate the existence of two distinct types of pauses: (i) intrinsic, short or "ubiquitous" ones that would be sequence dependent – and to which the application of forces had little effect[23b] – and (ii) longer pauses that these authors associate with backtracking.[24] Conversely, a recent single-molecule study of yeast RNA polymerase[20] revealed that the distribution of residence times of this enzyme in pauses follows not a multi-exponential dependence but a $t^{-3/2}$ power law that extends to all pause lifetimes. The existence of a power law for the pauses with long durations, known to be backtracked enzymes, suggests that backtracking could occur through simple diffusion of the enzyme over the template, as the first passage time to a barrier by a random-walk mechanism should follow such a power law. Moreover, the fact that the power law extends to short pause durations, previously associated with the "ubiquitous" pause class, suggests that these pauses may in fact be governed by the same diffusive behavior. This suggestion is particularly provocative since it unifies the multiple types of pauses under one single mechanism, diffusive backtracking. In fact, a similar power law appears to apply for the data obtained previously for the *E. coli* enzyme,[53] suggesting that this is a general property of polymerases. Despite the appeal of a single pause mechanism, whether all pauses involve backtracking to a greater or lesser degree is still a matter of discussion. It is clear, however, that single-molecule experiments performed on specific sequences and with the use of high-resolution methods should shed more light on the nature of pauses in transcription elongation.

One way to address the nature of the various dynamical processes during elongation is to perform experiments at different temperatures. An optical tweezers experiment in which the heating caused by the laser light was used to vary the temperature of the medium revealed interesting information on transcription elongation by *E. coli* RNAP.[25] This study reported that transcription elongation speeds up by as much as $1.5\,\text{bp}\,\text{s}^{-1}$ per degree but, surprisingly, it found that neither the pause frequency nor the pause lifetimes varied with temperature. While the independence of life-times on temperature is consistent with a one-dimensional diffusion postulated above as the mechanism of backtracking, the independence of pause frequencies with respect to temperature is more difficult to rationalize. Presumably, there should be an activation energy that must be overcome for the enzyme to leave on-pathway elongation and generate a backtracked state. It is possible that the observed temperature independence of the pause frequency is the result of the limited range of temperature in which these studies were performed (22–35 °C). Future studies on larger temperature ranges may clarify this point. Investigated in ref. 53; see *Note in Proof* that follows this chapter.

Most of the studies of transcription elongation have been carried out on the prokaryotic enzyme. Recently, methods of factor-less initiation first utilized in bulk were adapted to develop a single-molecule assay of transcription elongation by yeast RNAP.[20] This study compared the motor properties of the eukaryotic and the prokaryotic enzyme. It was found that the former is a significantly weaker motor, capable of generating forces of only 7 pN on average. This smaller stall force for the eukaryotic enzyme appears at odds with the fact that during its normal function in the cell this enzyme is expected to negotiate transcription through a DNA template organized in the form of chromatin. Significantly, this same study reported that the transcription elongation factor TFIIS is capable of increasing the stall force of the enzyme almost three-fold. Thus, it appears that the low stall force of the enzyme may be part of a regulatory strategy of the cell in which trans factors play an important role in the ability of the enzyme to overcome the barriers imposed by the presence of nucleosomes on the template. Moreover, the discovery that the pause durations were best described with a power law, which properly predicts the relatively large fraction of extremely long pauses observed, also sheds specific insight into the nature of the terminal stall. In particular, it was found that the terminal stall corresponds not to a thermodynamic stall due to the applied tension, but rather a kinetic stall due to entry into an extremely long-lived backtracked state. In this fashion, trans acting factors such as TFIIS can increase the effective "strength" of RNAP polymerase not through thermodynamic means but rather by rescuing the enzyme from long-lived off-pathway states.[20]

Although transcription elongation was the first aspect of transcription studied by single-molecule approaches, more recently several important results have emerged in the study of transcription initiation. They are reviewed in Chapters 5 and 6. Specifically, magnetic optical tweezers were used to investigate the formation of open promoter complexes with *E. coli* RNAP.[26] It was found that during open promoter complex formation the DNA must unwind 1.2 ± 0.1 turns. Significantly, it was also found in this study that strong promoters are more easily melted than weak promoters, suggesting a possible mechanism for their strength.

Another important question recently addressed by single-molecule methods is the mechanism of abortive initiation. As detailed in Chapter 5, Fluorescence Resonance Energy Transfer (FRET) was used by Kapanidis et al. to distinguish between different mechanistic alternatives[27–30] by labeling the leading and trailing ends of RNAP, the DNA both upstream and downstream of the promoter, or by labeling two different points on the polymerase to detect any contraction or expansion of the protein itself. These experiments clearly demonstrated a contraction between upstream and downstream DNA and a concomitant contraction between the leading edge of the polymerase and the downstream DNA, which is consistent with the scrunching model.[31,32]

Single-molecule FRET experiments were also used by Margeat et al.[33] to demonstrate that the sigma subunit is not obligatorily released upon binding to the promoter. This group also showed that elongation complexes harboring transcripts several tens of nucleotides long retain more than 50% sigma, with half-lifetimes close to 1 h.

Atomic Force Microscopy (AFM) has also been used to study initiation and elongation complexes at the single-molecule level.[34,35] In particular, by comparing the DNA contour length of free DNA with that of open promoter complexes via direct imaging of RNAP bound to DNA, it was found that RNAP wraps DNA around the protein surface. However, the extent of permanent DNA wrapping, which reflects the extent of protein-upstream-DNA interaction, is not a common feature of all promoters but strongly correlates with the upstream sequence of a particular promoter. For instance, RNAP bound to λ P_R induces a full turn of DNA wrapping whereas no DNA wrapping was observed when the same RNAP was bound to a lacUV5 promoter.[36] Although these elements seem to have little effect on promoter function, the particular geometry of the lambda operator region suggests that such distal upstream interactions may be used by a P_R-bound RNAP to downregulate the activity of the antagonist P_{RM} promoter, an effect known as the lambda P_R-P_{RM} interference (C. Rivetti, C. Bustamante unpublished results).

C.3 Technical Developments in Optical Tweezers

While the ability to follow individual RNAP molecules as they transcribe mRNA has revealed extensive information on the mechanism of not just elongation but also pausing, many features of the kinetic mechanism of RNAP still remain elusive. The recent development of high-resolution optical tweezers[21,37–39] – instruments capable of observing the discrete steps of nucleic acid motors – promise to provide a direct method by which these details can be probed. With the resolution necessary to resolve the discrete, single-nucleotide steps of RNAP, such instruments allow not just the measurement of the step size of this motor[21] but, perhaps more importantly, the full distribution of dwell times before each of these steps has been completed. These distributions and their dependence on nucleotide concentration and force fully characterize not just the mean rate of elongation but also the natural stochastic fluctuations in this rate. Thus, measurement of these distributions promises to reveal not only the number of kinetic transitions per cycle but also the pathway by which the motor moves through these kinetic states.[40–42] To accomplish this task, it is not sufficient to have only high spatial resolution; significant temporal resolution is also required. For RNAP, in particular, the average rise of B-form DNA, just 3.4 Å bp^{-1}, imposes the need to resolve DNA length changes of only a few ångströms. In addition, RNAP has been reported to transcribe RNA as fast as $\sim$20 bp s^{-1} in the case of *E. coli.* RNAP[21] and $\sim$15 bp s^{-1} for Eukaryotic Pol II,[20] suggesting that a temporal resolution of at least 50 ms is also required. In recent years, both technical and theoretical advances have not just suggested that such resolution is possible: ångström resolution on a time scale of the order of a fraction of a second has already been achieved.[21,38,39]

The development of high-resolution optical trapping techniques has required a detailed understanding of the noise that limits these measurements. This noise can be divided into two broad categories: environmental noise and Brownian noise. Environmental noise not only stems from thermal, acoustic and mechanical fluctuations in the environment of the instrument, but from the noise associated with components of the instrument as well, such as laser position, power and polarization. This noise has been successfully addressed in two ways. Environmental fluctuations can be monitored and subtracted from the final signal, *e.g.*, by monitoring the movements of a fiducial mark[43,44] placed on the sample chamber surface. Alternatively, the instrument can be designed to isolate the measurement from many of these noise sources, *e.g.*, by adding a second optical trap so that the resulting dual trap optical tweezers effectively levitates the experiment off the surface of the chamber.[21,22,38,39] By addressing these sources of noise, it is now possible to follow motions of the trapped bead on the sub-ångström scale over a wide range of time scales.[21,38,43,44]

However, the temporal and spatial resolution relevant to most experiments is the ability to detect changes in the tether length, not in the motion of the beads. It is in attempts to better resolve this tether length that the second source of noise, the Brownian fluctuations of the system, becomes a significant obstacle. Even in the perfectly isolated instrument, fluctuations induced by the stochastic Brownian force establish a fundamental limit to spatial and temporal resolution. Fortunately, such noise is amenable to theoretical description.[38,39,45] Such description has shown that the effect of the Brownian noise can be reduced through proper choice of experimental design, *i.e.*, smaller beads, shorter DNA tethers, and increased signal averaging. Remarkably, increasing the trap stiffness, thus lowering the fluctuations of the beads, does nothing to improve this limit. This counter-intuitive result stems from the fact that it is the tension fluctuations induced on the DNA tether, not the positional fluctuations of the beads, that limit the ability to estimate the length of the DNA molecule. In a dual trap optical tweezers, the fluctuations of the two beads can be thought of as the combination of correlated and anti-correlated motions. Because only the anti-correlated movement of the beads induces a fluctuating tension on the molecule, the noise associated with correlated motions is irrelevant. Removing this correlated noise by tracking the position of both beads simultaneously and monitoring the length of the tether with a differential measurement provides a significant improvement in the spatial and temporal resolution of the experiment.[39] Surprisingly, the ability to remove a portion of the noise by differential measurement not only lowers the Brownian limit to spatial and temporal resolution for dual trap tweezers in which both traps are monitored as compared to dual trap tweezers in which only one trap is monitored, this limit is also lowered compared to that of a single trap instrument with only one bead subject to fluctuations. In combination with the increased isolation provided by the dual traps, this correlation analysis makes the dual trap optical tweezers the preferred choice for high-resolution optical trapping. Such an instrument holds the current record for resolution, 3.4 Å steps of a DNA tether with a temporal resolution of 250 ms.[38]

Another important technical development, the design of hybrid optical tweezers that combine optical manipulation with single-molecule fluorescence detection, holds great promise for deciphering additional features of the motor mechanism. In particular, while statistical analysis of motor dwells can yield detailed information on the kinetic transitions that occur within a motor's mechanochemical cycle, this analysis provides little information on the physical conformational changes that correspond to these kinetic transitions. Optical tweezers, because of their projection of all biomolecular motions onto a single common axis – the molecular extension – can only measure conformational changes that affect the length of the nucleic acid tether. On the other hand, single-molecule fluorescence measurements, because of the flexibility in fluorophore location, can yield dynamic information on almost any internal conformational change. Moreover, these techniques can now do so with both sub-nm and 10s of ms scale resolution.[8,12] Unfortunately, the combination of these two techniques has proven to be quite difficult, due to both the added complexity of instruments and reduced dye lifetimes from the simultaneous exposure of fluorophores to fluorescent excitation light and near-infrared trapping light. Despite these difficulties, excellent proof-of-principle experiments have been demonstrated,[37] and application of these techniques should follow in the next few years.

C.4 A Look into the Future

What implications do these advancements have for the study of RNAP? The ability to observe ångström-scale motions on the second time scale is sufficient for the detection of the discrete steps of RNAP and has allowed measurements that confirm the expected single-base-pair step size.[21] Unfortunately, the temporal resolution of these measurements has been limited to the second scale. This restricts observations to conditions of limiting nucleotide concentration, under which the observed dwell time distributions are dominated by a single rate-limiting, kinetic transition, the binding of a single nucleotide.[21] Extracting additional information on the kinetic cycle will require the ability to observe the discrete steps of RNAP over the entire range of nucleotide concentration, in particular where binding is not rate-limiting. If limited only by Brownian noise, it should be possible to increase the temporal resolution of measurements while maintaining ångström-scale spatial resolution. Achieving this goal will certainly require additional attention to residual experimental noise sources. However, the ability to test predicted models of the mechanochemical cycle of the polymerase with direct observation of the dwell time distribution at all nucleotide concentrations makes this a worthwhile pursuit. Moreover, direct observation of the displacements in both short and long pauses promises to settle the debate over pause mechanism.

With the combination of optical tweezers and single-molecule fluorescence, it should be possible to monitor conformational dynamics within a single protein domain while simultaneously following the discrete movements along its molecular track.[46] With such a measurement, internal conformational changes and the mechanical stepping can be ordered in time, and, more interestingly, inter-event dwell time distributions can be formed, yielding the same detailed kinetic information as discussed above but now between two events in the same kinetic cycle. By varying the location of the fluorophores, it should be possible to systematically identify the exact conformational changes that correspond to each kinetic event.

We have argued above that single-molecule methods, far from being merely confirmatory of the main results obtained by traditional bulk methods, have contributed significant new insight and details about the transcription machinery and its dynamics. Yet, our picture of the transcription process is patchy and incomplete. Although the main features of each process, initiation, elongation and termination are basically known, several important questions must be answered before a complete picture is available. Here, we will just mention a few of the issues that should become clarified through the use of single-molecule methodology.

The first has to do with the actual mechanism of translocation during processive elongation. Several bulk assays suggest that RNA polymerase is likely to operate not as a power-stroke generating machine but as a Brownian ratchet.[47] Although this conjecture appears correct, the answer still awaits a definitive experiment in this area. Moreover, as discussed by Wang and collaborators in Chapter 9, several kinetics schemes can be postulated to account for some of the bulk kinetic and single-molecule dynamics observations. However, most of these models remain untested. Clearly, the advent of ultrahigh-resolution optical tweezers instruments such as the ones discussed above will play an important role in discriminating between linear *vs.* branched pathways and in identifying possible new intermediates of the reaction. Moreover, as in other mechanochemical processes in the cell, force is one of the products of the transcription reaction. Consequently, external forces can be used to aid or inhibit the reaction. Plots of the effect of this external mechanical parameter as a function of different reagent conditions on the rate of transcription can provide information about where in the mechano-chemical cycle of the motor the force is actually generated.[48]

Other aspects of elongation dynamics that will likely be probed in more detail include the nature of pausing and the backtracking process. The ability to follow the diffusion of RNAP along the template and its eventual return to the active site using low-drift high-resolution instruments will give us an unprecedented description of the mechanism of this important regulatory process. Furthermore, it will be possible to understand the interplay between backtracking and the effect of formation of secondary structures in the RNA transcript.

Another central question that remains unanswered is what are the limits of torque generation in RNAP? In both prokaryotes and eukaryotes, RNAP most likely transcribes not free but torsionally constrained DNA. Under these circumstances, as transcription proceeds, RNAP creates positive supercoils ahead of itself and negative supercoils in its wake.[49] The presence of these strained configurations implies the generation of a reaction torque on the enzyme that accordingly must be able to work against an externally applied torque. Systematic experiments in which the transcription elongation rate is monitored as a function of the applied torque should provide an important insight into this aspect of the polymerase motor.

Another fertile area of research lying ahead involves the mechanisms of transcription termination. Here, the ability to directly manipulate the nascent transcript as transcription proceeds will make it possible to address several questions regarding the nature of the termination process itself. In particular, it will be possible to answer the question of whether the contact of the RNA with the polymerase leads to termination *via* the induction of an allosteric conformational change or through a novel mechanical dislodging of the hybrid.[50]

As mentioned above, it has been found that there are significant differences in the mechanical abilities of prokaryotic and eukaryotic enzymes. Yeast RNAP II has been shown to be able to generate forces only one-third those generated by *E. coli* RNAP but that can increase significantly in the presence of elongation factors added in trans. Yet, the question still remains as to what the dynamics of transcription elongation of RNAP II are in the presence of its natural substrate, chromatin. Little is known about how RNAP overcomes the presence of the nucleosome. The crucial difference in mechanical properties between the prokaryotic and eukaryotic enzyme strongly suggests that there is much to be gained by studying transcription elongation by the RNAP II over modified or unmodified (by acetylation or phosphorylation) nucleosomes and in the presence or absence of chromatin remodeling factors.[51,52]

Finally, all this complex intermolecular dynamics must be accompanied by the corresponding changes in the internal degrees of freedom in the enzyme itself. These intramolecular changes can be best monitored by single-molecule fluorescence assays such as FRET. The advent of robust combined optical or magnetic tweezers/single-molecule microscopy instruments capable of simultaneously monitoring force, displacement and changes in the emission from single fluorophores will greatly speed up the development of a detailed moving picture of the transcribing polymerase.

In the last 20 years, structural biologists have put forth an unspoken challenge to the rest of the scientific community. The most intricate and detailed crystal structures, such as those of RNAP polymerase, simply beg to be animated, to be complemented with dynamic information on the ångström scale. Single-molecule techniques with the ability to resolve ångström scale and millisecond dynamics, such as the ultrahigh-resolution methods described above, are the answer to this call. The promise of these emerging methods is that they will not only complement but complete the exquisite atomic pictures that structural techniques can now routinely yield.

References

1. C. Bustamante, Y. R. Chemla, N. R. Forde and D. Izhaky, *Annu. Rev. Biochem.*, 2004, **73**, 705.
2. T. R. Strick, J. F. Allemand, V. Croquette and D. Bensimon, *Physics Today*, 2001, **54**, 46.
3. T. R. Strick, J. F. Allemand, D. Bensimon, A. Bensimon and V. Croquette, *Science*, 1996, **271**, 1835.
4. S. B. Smith, L. Finzi and C. Bustamante, *Science*, 1992, **258**, 1122.
5. C. Bustamante, L. Finzi, P. E. Sebring and S. B. Smith, paper presented at the Optical Methods for Ultrasensitive Detection and Analysis: Techniques and Applications 1991.
6. C. Bustamante, Z. Bryant and S. B. Smith, *Nature*, 2003, **421**, 423.
7. C. Bustamante, J. C. Macosko and G. J. L. Wuite, *Nature Reviews Molecular Cell Biology*, 2000, **1**, 130.
8. W. J. Greenleaf, M. T. Woodside and S. M. Block, *Annu. Rev. Biophys. Biomol. Struct.*, 2007, **36**, 171.
9. E. Betzig and R. J. Chichester, *Science*, 1993, **262**, 1422.
10. M. Orrit and J. Bernard, *Phys. Rev. Lett.*, 1990, **65**, 2716.
11. W. E. Moerner and L. Kador, *Phys. Rev. Lett.*, 1989, **62**, 2535.
12. P. V. Cornish and T. Ha, *ACS Chemical Biology*, 2007, **2**, 53.
13. H. Yin *et al.*, *Science*, 1995, **270**, 1653.
14. R. J. Davenport, G. J. L. Wuite, R. Landick and C. Bustamante, *Science*, 2000, **287**, 2497.
15. M. D. Wang *et al.*, *Science*, 1998, **282**, 902.
16. N. R. Forde, D. Izhaky, G. R. Woodcock, G. J. L. Wuite and C. Bustamante, *Proc. Natl. Acad. Sci. USA*, 2002, **99**, 11682.
17. N. Komissarova and M. Kashlev, *J. Biol. Chem.*, 1997, **272**, 15329.
18. E. Nudler, A. Mustaev, A. Goldfarb and E. Lukhtanov, *Cell*, 1997, **89**, 33.
19. N. Komissarova and M. Kashlev, *Proc. Natl. Acad. Sci.*, 1997, **94**, 1755.
20. E. A. Galburt *et al.*, *Nature*, 2007, **446**, 820.
21. E. A. Abbondanzieri, W. J. Greenleaf, J. W. Shaevitz, R. Landick and S. M. Block, *Nature*, 2005, **438**, 460.
22. J. W. Shaevitz, E. A. Abbondanzieri, R. Landick and S. M. Block, *Nature*, 2003, **426**, 684.

23. (a) M. T. Marr and J. W. Roberts, *Mol. Cell*, 2000, **6**, 1275; (b) K. C. Neuman *et al.*, *Cell*, 2003, **115**, 437.
24. K. M. Herbert *et al. Cell*, 2006, **125**, 1083.
25. E. A. Abbondanzieri, J. W. Shaevitz and S. M. Block, *Biophys. J.*, 2005, **89**, L61.
26. A. Revyakin, R. H. Ebright and T. R. Strick, *Proc. Natl. Acad. Sci. USA*, 2004, **101**, 4776.
27. J. D. Gralla, A. J. Carpousis and J. E. Stefano, *Biochemistry*, 1980, **19**, 5864.
28. A. J. Carpousis and J. D. Gralla, *J. Mol. Biol.*, 1985, **183**, 165.
29. B. Krummel and M. J. Chamberlin, *J. Mol. Biol.*, 1992, **225**, 239.
30. B. Krummel and M. J. Chamberlin, *J. Mol. Biol.*, 1992, **225**, 221.
31. A. N. Kapanidis *et al.*, *Science*, 2006, **314**, 1144.
32. N. K. Lee *et al.*, *Biophys. J.*, 2007, **92**, 303.
33. A. N. Kapanidis *et al.*, *Mol. Cell*, 2005, **20**, 347.
34. C. Rivetti, M. Guthold and C. Bustamante, *EMBO J.*, 1999, **18**, 4464.
35. C. Rivetti, S. Codeluppi, G. Dieci and C. Bustamante, *J. Mol. Biol.*, 2003, **326**, 1413.
36. S. Cellai *et al.*, *EMBO Reports*, 2007, **8**, 271.
37. J. R. Moffitt, Y. R. Chemla, S. B. Smith and C. Bustamante, *Ann. Rev. Biochem*, 2008, **77**, 205.
38. C. Bustamante, Y. R. Chemla and J. R. Moffitt, In: *Single-Molecule Techniques: A Laboratory Manual.*, eds. P. R.. Selvin and T. Ha, Cold Spring Harbor Laboratories, Cold Spring Harbor, New York, 2008, pp. 297–324.
39. J. R. Moffitt, Y. R. Chemla, D. Izhaky and C. Bustamante, *Proc. Natl. Acad. Sci.*, 2006, **103**, 9006.
40. Y. R. Chemla, J. R. Moffitt and C. Bustamante, *J. Phys. Chem. B* (2008), **112**, 6025.
41. J. W. Shaevitz, S. M. Block and M. J. Schnitzer, *Biophys. J.*, 2005, **89**, 2277.
42. M. J. Schnitzer and S. M. Block, *Cold Spring Harbor Symposia on Quantitative Biology*, 1995, **60**, 793.
43. A. R. Carter *et al.*, *Applied Optics*, 2007, **46**, 421.
44. L. Nugent-Glandorf and T. T. Perkins, *Opt. Lett.*, 2004, **29**, 2611.
45. F. Gittes and C. F. Schmidt, *Eur. Biophys. J. Biophys. Lett.*, 1998, **27**, 75.
46. A. Ishijima *et al.*, *Cell*, 1998, **92**, 161.
47. G. Bar-Nahum *et al.*, *Cell*, 2005, **120**, 183.
48. D. Keller and C. Bustamante, *Biophys. J.*, 2000, **78**, 541.
49. L. F. Liu and J. C. Wang, *Proc. Natl. Acad. Sci.*, 1987, **84**, 7024.
50. M. H. Larson *et al.*, *Cell*, 2008, **132**, 971.
51. Y. L. Zhang *et al.*, *Biophys. J.*, 2007, **360A**.
52. A. Shundrovsky, C. L. Smith, J. T. Lis, C. L. Peterson and M. D. Wang, *Nat. Struct. Mol. Biol.*, 2006, **13**, 549.
53. Y. X. Meija, H. Mao, N. R. Forde and C. Bustamante, *J. Mol. Biol.*, 2008, **382**, 628.

Note in Proof: Ref. 53 was published during the revision of this book. This work investigates the pausing behavior of prokaryotic RNAP over a wide temperature range, 7 °C–45 °C, and shows that the pause density is very sensitive to temperature, particularly below ~ 20 °C. In addition, the authors show that the pause durations are well described by a $t^{-3/2}$ power law, consistent with the eukaryotic enzyme. Moreover, counter to what is predicted for a simple multi-exponential process, these distributions are remarkably insensitive to temperature—exactly as is expected for a diffusive mechanism.

Subject Index

'A' site (aka IS site)
  allosteric NTP binding model, 223
  E site distinction from, 242–4
  in nucleotide addition cycle, 207, 208, *256*
  nucleotide selection intermediates, 242
  RNA occupancy and abortive initiation, 176
  role in transcript mediated proofreading, *256*
  role in two-pawl ratchet model, 226
  steric clash in bridge-helix model, 215
abortive cycling, 21
abortive initiation
  ALEX studies, 133–42
  DNA scrunching in, 173–6, 307
  FRET and magnetic trap studies, 142–5
  FRET studies, 126, 135
  structural transitions in, *7, 98*
  three models of, 133–5, *136,* 139–40
accurate transcription
  error rate, 192, 241, 354
  other factors affecting, 257
  possible role for pyrophosphorolysis, 209–10
  possible role for transcript cleavage, 211
  possible role for trigger loop, 219
  proofreading contribution, 241, 254–8
  RNA self-cleavage and, 257
activators
  promoter co-dependence, *28*
  promoter regulation and, 22, 24

active center
  see also catalytic sites
  ability to accommodate scrunched DNA, 140
  bacterial and eukaryotic, 258
  domains of, 283
  features of, elongation complex, 207, 265
  introduction of a third Mg^{2+} ion, 248, 252
  NTP access routes, 213, 238–41
  nucleotide addition models, 213–19
  ppGpp and coordination of a third Mg^{2+} ion, 249
  proofreading and, 254–8
  protein regulators, 250–4
  regulatory checkpoints, 237–44
  remote control by ω subunit, 254
  role of Mg^{2+}, 237–8, 248
  small molecule regulators, 244–50
AFM. *see* atomic force microscopy
Alberts, Bruce, 203
ALEX (alternating-laser excitation) spectroscopy
  abortive initiation studies, 134–5, 142, 180
  biomolecule concentrations for, 145, 147
  combination with TIRF, 142
  ensemble FRET compared with, 132
  fate of σ^{70} in elongation, 126–32
  methodology, 121–4
  millisecond-ALEX, 124, 142, *143,* 145
  σ release assay, 129–30
  temporal resolution, 124
  three-color alternating-laser excitation (3c-ALEX), 144–6
  transcription mechanism studies, 124–6

allosterism
  allosteric NTP binding model, 223–6
  allosteric termination model, 287, *288, 291,* 292–4
  communication between DksA and ω, 254
  hairpin-dependent pausing and, 285
  proposed for NTP binding site, 222
  streptolydigin regulation and, 245
  TFIIS regulation and, 252
α-subunit C-terminal domains (CTDs)
  base pairs targeted by, 18
  modification by T4 bacteriophages, 32
  σ factor interaction, 50, 84
  transcription activators and, 24, 29
  wrapping of upstream DNA, 44
α-subunit domains, 15
amino acid stacking, 52
AMP-cPP, 242, 246
antibiotics. *see* Microcin J25; rifampicin; streptolydigin; tagetitoxin
arrested complexes
  backtracking of, 238
  as a conformational state, 265
  reactivation by GreB, 250, 286
  rescue by transcript cleavage, 211
Arrhenius law/kinetics, 181, 276
atomic force microscopy (AFM)
  DNA wrapping, 307
  open complex structure and, 45, 77–8, 168
ATP
  as an energy source, 193–4
  promoter regulation by ppGpp and, 30–1
attenuation, 281, 285
A:U hybrids, 202, 283, 286
auxiliary subunits, 82–4

backtracking
  see also forward-tracking
  of arrested complexes, 238
  effect on leading-edge FRET, 128
  evidence for 'Brownian ratchet' mechanism, 264, 276
  free energy and, 275

induced by protein roadblocks, 287
pause length and, 264–5
questions remaining over, 197, 201
resolvable by Gre factors, 27, 237
role of the secondary channel, 250
termination and, 202–3
two-pawl ratchet model and, 226
backtracking pauses, 211–12, 227, 277, 286, 305
bacterial nucleoid. *see* nucleoid
bacteriophage subversion of host RNAP mechanisms, 32–3
bacteriophage T5, T7. *see* T5, T7
base pairing energy, 275
basic patch, Gre factors, 251
BDloopI and II, 213
  see also fork loop 2
bend deformations, supercoiled DNA, 163, 168
β subunit and DNA melting, 58–9
β' subunit
  see also bridge helix; trigger loop
  intermediate destabilization by, 57
biophysical models of elongation, 227–8
biophysical techniques
  elongation phase investigation, 228
  limitations for dynamic systems, 302–3
  termination phase investigation, 202
braking, termination viewed as, 191
bridge helix (F-bridge)
  conformational change and translocation, 109–11
  nucleotide addition models based on, 213–16
  power stroke model and, 271
  and RNAP catalytic activity, 213
  role in allosteric NTP binding model, 225
  role in dual ratchet mechanism, 269
  role in trigger loop catalysis, 219
  role in two-pawl ratchet model, 226
  streptolydigin effects, 244–5
Brownian noise, 264, 304, 308–10
'Brownian ratchet' mechanism, 98, 219–22
'dual ratchet mechanism', 269

Subject Index 317

energetics of, 264
equilibration between translocational states, 107, 197
more elaborate model, 267–79
sequence dependent kinetics, 275–6
simple model, 266–7, 272–4, 275–6

calibration of magnetic trapping sensors, 161–6
CAP (Catabolite Activator Protein)
number and concentration in *E. coli*, 73
as a transcription activator, 24
catalytic sites
see also active center
mechano-chemical coupling, 194–200, 198–9
proportion in the pre-translocated register, 197, 199
trigger loop role, 217–19
cation involvement. *see* Mg^{2+} ions
challenges for the future. *see* research opportunities
channels
see also main channel; RNA exit channel; secondary channel
NTP delivery to the active center, 213, 238–41
TEC active site and, 265
'chaseable complexes', 129
ChIP-chip, 228
chromatin
DNA binding in bacteria, 69
elongation dynamics and, 306, 311
chromatin immunoprecipitation followed by whole-genome microarrays (ChIP-chip), 228
chromosomes
intracellular environment and, 26–7, 31
topology and modeling, 227–8
circular dichroism studies, 84
clamp
in core RNAP structural overview, 3
hairpin invasion termination model, 293–4
clamp, β′
downstream duplex binding at, 283

FRET studies of, 126
hairpin invasion termination model, *290*
closed complexes (RP$_c$)
formation, 49–51
proposed structure, 46–7
transition to open, *46,* 116
co-activation, 28–9
conformational changes
see also deformations; DNA melting; plectonemic form; promoter unwinding; toroidal form
allosteric termination model, 292
at the catalytic site, 199
on DNA binding, *44,* 45
and duplex translocation, 98–9, 106–8
evidence from kinetics and, 40, 310
FRET and ALEX studies of, 118, 124–6, *127*
nucleotide loading, 212
open and closed state, 106, 270
ppGpp binding sites and, 250
protein, and promoter unwinding, 171
T7 elongation complex, 102, 369–270
transcription activation by, 24–6, 60–1
in transcription initiation, 7
transient intermediates and, 116
translocational models, 221
trapped termination complexes, 290
trigger loop folding, 219, 293–4
conformational state, arrest as, 265
consensus promoters
closed complex formation from, 50
consensus elements of, 17–19
constraints
in magnetic trapping, 161
supercoiled DNA, 70
copying fidelity. *see* accurate transcription
core enzyme
DNA binding effects, 45
holoenzyme distinguished from, 14
interactions with σ subunit, 42–3
relationship of subunits, 5
structural overview, *2–4*

counterion effects and promoter
  recognition, 48–9
coupling
  mechano-chemical, 194–200, 198–9,
    263–4, 266–74
  translocation and topology, 200–1
crosslinking experiments, 55–8
  initiation to elongation changes,
    101–2, 207, 215–17
  open complexes, 20
  superhelical DNA, 71
  termination phase, 287–94
  time-resolved, 56
CRP (Cyclic AMP Receptor Protein)
  number and concentration in *E. coli*, 73
  as a transcription activator, 24
crystallography
  see also power stroke mechanism
  achievements and limitations, 40–1,
    115–16, 302
  of active sites, 207, 212–13, 249–50
  of holoenzymes, 42–7, 84
  of intermediates, in open complex
    formation, 40
CTDs. *see* α-subunit C-terminal domains
  (at alpha)

DBS (DNA duplex binding site), 283, *284*
defocusing optical imaging, 119
deformations of DNA
  linear and torsional, 161
  loop, twist and bend, 163
diffusion, pausing and, 305, 314
dissociation constants. *see* kinetics
distance restrained docking, 147
distances, intramolecular, *200*
  see also FRET
DksA cofactor
  as an auxiliary subunit, 82
  promoter regulation and, 83–5
  response to ppGpp and STP and, 31
  RNAP regulation and, 250–1, 253–4
DNA chaperones, 74
DNA melting
  see also torsional deformation
  intermediates and mechanism of,
    42, 47, 52–3

main channel entry model, 240
sequential linear pathways, 54–7
by specific subunits, 57–9
DNA repair processes, 27
DNA supercoiling. *see* superhelicity
DNA topoisomerase interactions, 85
DNA–DNA distances, FRET
  analysis, 126, *127*
DNaseI hypersensitivity, 44, 54, *56*
  see also footprinting methods
docking, 16, 147
donor–acceptor pairs (FRET), 117–18
double psi-β barrels (DPBB), 207
Dps (DNA Protection during
  Starvation) protein, *73*
dumbbell configuration, optical
  trapping, 195, 196
dynamic heterogeneity, 120

E. coli (Escherichia coli)
  determination of subunit structure,
    14–17
  K-12 as a model system, 13–14
  number and concentration of DNA
    binding proteins, *73*
  pre- and post-translocated, 197
  secondary NTP binding site, 268
  stall force for elongation, 196, 273
  streptolydigin effects, 245
E (FRET efficiency) function, 117–18,
  122, 125, 146
  simplified form as mean, E^*, 129,
    130, 135
E site (NTP entry site), 213, 217
  distinction from IS, 242–4, 246
  misincorporation and proofreading,
    255–7
  nucleotide binding, 111
  secondary channel NTP access and,
    239–40
  tagetitoxin and, 248–9
electrostatic interactions, 48–9
elemental pauses, 212
elongation complex (EC, RD_e or
  TEC)
  'activated' and 'inactivated' states,
    223, 271

Subject Index

active site features of, 207
cartoon schematics of, 6
'chaseable complexes', 129
first stable complex, $RD_{e,11}$, 129, 131
free energy and kinetics, 265, 274–5
inactivation prior to termination, 283
leading-edge FRET/ALEX results, 129, 130
open (RPo) and closed (RPc) complexes, 7
structure and function overview, 282–3, 284
von Hippel's model, 21
elongation factors
  eukaryotic RNAP force generation and, 311
  nusA and the σ-cycle, 19
elongation phase
  basic mechanism, 206–12
  biophysical models, 227–8
  DNA topological changes in, 75–6
  dynamics of, 304
  fate of σ^{70} initiation factor, 126–32
  involves translocation, not scrunching, 140
  kinetics and kinetic modeling of, 263–78, 270–4, 305
  negative superhelicity effects, 78
  processivity, 101, 112–13
  regulation by pausing, 211–12, 305
  RNAP as an engine, 191, 193–203
  single-unit and multi-subunit RNAPs, 112–13
  structural transitions, 7, 97, 98–103
  technologies for investigating, 228
  torsional-translational coupling, 158
  transcription bubble fate, 181–2
  translocation and strand separation, 103–12
energetic barriers
  to complex formation, 39
  elongation models, 227–8
  intermediate re-equilibration, 62
  pausing and, 212
  to promoter escape, 173, 178–80

energy, mechanical
  efficiency of conversion, 194, 197–8
  molecular complexes as transducers, 304
  stressed transcription complexes, 158
energy landscape
  see also free energy
  DNA supercoiling, 70, 87–8
  elongation, 273
  nucleotide incorporation, 193–4
  thermal environment and, 304
  transcriptionally active complex formation, 61–2
  translocation, 107, 289
engine, elongating RNAP as, 191, 193–203
ensemble studies
  ensemble FRET, 119–20, 128
  FRET compared with ALEX, 132
  shortcomings of, 303
entropic and enthalpic regimes, 162
entry site, NTP. see E site
environment, intracellular
  DNA compaction in, 26–7
  the global RNAP economy, 30–2
  transcription response to changes, 27–30
environmental stress, 80–2, 87
error rate. see accurate transcription
Escherichia coli. see E. coli
evanescent-wave microscopy, 182
experimental techniques. see research opportunities; technologies for RNAP investigation
extensibility of DNA, 161–6
extension-supercoiling curve, 167, 171
extension-time traces, 166–7
extrinsic termination (rho termination), 294–5

F-bridge. see bridge helix
Fe-BABE cleavage studies, 102, 223
fidelity. see accurate transcription
'fingers' domain and NTP incorporation, 102, 104
  see also Zn^{2+} finger region

FIS (Factor for Inversion Stimulation)
  DNA topology and, in exponential growth, 87
  number and concentration in *E. coli*, 73
  topological forms stabilized by, 75
5' UTR (untranslated region), 281
flap
  in core RNAP structural overview, *3*
  hairpin invasion termination model, *290*, 292
  in RNA exit channel, 283
fluorescence approaches
  see also FRET
  elongation studies using, 228
  single-molecule fluorescence studies, 119, 309–10
  TIRF microscopy combinations, 142, 144
  translocation studies in real time, 223
fluorescent microscopy, 86, 142, 144
footprinting methods, 20
  abortive initiation, 173
  closed complex formation, 50–1, 55
  of elongation complexes, 222
  initial transcription, 133
  nucleation stage of DNA melting, 52
  open complex intermediates, 41, 44–6
  time resolved X-ray, 61, 116
  toroidal HU-DNA complex, 74
  trigger loop mutations, 215
force-dependent elongation kinetics, 271–4
force *vs.* extension response curve, 161–2
forces
  derived from chemical free energy, 194
  exerted in magnetic trapping, 161
  point of generation, 310
  stall force for elongation, 196, 273, 304, 306
force–velocity relationships, 194, 197, 222, 228, 273–4, 304
fork-junction DNA
  binary complex with RNAP, 43–5
  evidence on nucleation-propagation, 52–3
  σ region 2-interactions, 55

fork loop 2 (BDloopI), 224, 245
Förster radius (R_o), 118
forward-tracking, 266–7, 269, 275
free energy
  see also energy landscape; mechanical forces
  nucleotide incorporation, 193–203
  pauses as energy barriers, 212
  sequence dependent kinetics, 265, 274–6
FRET efficiency. see E (FRET efficiency)
FRET (fluorescence/Förster resonance energy transfer) spectroscopy
  see also ALEX spectroscopy
  analysis of protein-protein distances, 124–5
  complementarity with magnetic traps, 142–5, 182
  ensemble FRET, 119–20
  evaluation of initial transcription models, 135–40
  leading-edge and trailing-edge assays, 126–31
  methodology, 117–24
  single molecule FRET (smFRET), 120–2, 124–6, 143
future challenges. see research opportunities

G-loop. see trigger loop
G-mAbs (G-loop specific monoclonal antibodies), 294
gel electrophoresis, abortive RNA, 177
genes
  encoding RNAPs, identification, 15
  number of RNAP molecules compared to, 30
  varying expression frequency, 17
Gfh1 transcription inhibitor, 252–3
Gibbs free energy. see free energy
global regulators, 29
Gre binding site
  arrested complex reactivation, 286
  elongation complex schematic, *6*
  secondary channel binding, 250

Subject Index

Gre factors (GreA and GreB)
  as auxiliary subunits, 82, 85
  effect on $Mg^{2+}II$, 237
  promoter escape facilitated by GreA, 257
  as transcription cleavage factors, 211, 237
  transcription effects of, 83–4
  trapped termination complex resistance to GreB, 290
growth phases
  DNA superhelicity and, 71, 83
  FIS effects in exponential phase, 87
  holoenzyme response to altered template topology, 80–2
  nucleoid protein composition and, 75
  open complex overstabilization and, 60
  RNAP binding state and, 30–1
  stationary phase, 31–2

H-NS (Histone-like Nucleoid Structuring protein, aka H1), 73, 74, 75, 78–9
hairpin invasion termination model, 287, 288, 291, 292–5
  hairpin invasion phase, 293–4
  hairpin nucleation phase, 292–3
hairpin-stabilized pauses, 212, 275, 285
hairpin structure
  formation in intrinsic termination, 283, 286, 287
  termination sequence, 201, 281
HBS (RNA:DNA hybrid binding site), 283, 284, 285, 291, 295
helicase function of polymerases, 97, 108
helicase (rho) termination, 294–5
helixes. see bridge helix; O helix; superhelicity, DNA
heparin resistance, 14, 20, 39
highly transcribed regions, 26–7, 30
history of RNAP research, 13–18, 191–2, 302–4
holoenzyme
  distinguished from core enzyme, 14
  modules moving in formation of, 43
  processes carried out, 38–9
  response to altered template topology, 80–2

structural characterization, 2–4, 7, 42–7
transcription in vitro, 19
Hopfield Criterion, 209, 255
HU (Histone-like Unwinding protein), 71, 72–4
human RNAPs, 213, 223, 226, 240–1, 268
hydrolysis. see pyrophosphate dissociation

i site. see 'P' site
i+1 site. see 'A' site
IHF (Integration Host Factor), 73
immobilized molecules
  FRET and ALEX studies of, 124–6, 136, 142, 143
  thermodynamic force measurements, 194, 196
in vitro experiments
  natural environment differences, 26
  transcription by the E. Coli holoenzyme, 19–20
inchworm model, 98, 133, 137–9
indirect readout, 49
'induced fit' mechanism, 268
inhibitors. see regulators
initial transcribing complex (RP_{itc})
  formation, 49, 133, 159
  FRET studies on RP_o and, 135
initiation factors. see σ subunits (under sigma)
initiation phase
  DNA melting preceding, 47
  importance of σ factors, 14
  intermediates between promoter recognition and, 39
  investigation by magnetic trapping, 159
  kinetic analysis of promoter escape and, 141–2
  mechanism of, 133–41
  $σ^{70}$ initiation pathways, 84–5
  transition to elongation, 7, 97, 98–112, 131
intermediate complex (RP_i), 116
intermediates
  characterization of, in open complex formation, 40–1, 47–8
  destabilization of, by specific subunits, 57–9

nucleotide selection, 242
open, closed and intermediate complexes, 116
between promoter recognition and transcription initiation, 39
prospects for identifying new, 310
stressed intermediates, 141
between transcription initiation and elongation, 102
trapped termination complexes, 289–92
intrinsic pauses, 305
intrinsic repeats, 70
intrinsic termination, 275, 281, *282*, 283–95
IS site. *see* 'A' site
isomerization
closed complex formation, 50, 51–60
initially-transcribing complex formation, 159, 180
σ domain 1.1 and open complex formation, 58

jaws
in closed complex formation, 51
in core RNAP structural overview, *3*
DNA binding and distances between, 45
in elongation complex, 265
Jencks, W. P., 198, 203

kinetics
biphasic and multiphasic, of nucleotide incorporation, 268, 271
conformational change evidence from, 40
of elongation, 263–78
of initial transcription and promoter escape, 141–2
kinetic models of nucleotide addition, 223–8
of open complex formation, 39, 159, 168–72
of σ^{70} retention, 131
of streptolydigin inhibition, 246
of termination, 201, 295
transcription bubble and, 179–82

labeling reactions
ALEX and, 147
ensemble FRET and, 120, 129
radiolabeling, 223
single-molecule FRET and, 124–5, 145–6
lac promoter
CRP-dependence, 24
DNA wrapping, 307
early investigations, 17, 22
RP_o characterization at consensus, 166–72
superhelicity investigations, 77, 80
lac repressor, 22, 201
*lac*UV5-CONS, 135
*lac*UV5 mutation, 47, 50–1, 54–8, 129–32, 135–7
λ P_R region, 50–1, 54–5, 60, 307
lasers. *see* ALEX
leader sequence, 281
leading-edge-FRET assay, 126–31
lid
in core RNAP structural overview, *3*
hairpin invasion termination model, *290, 292*
in open complex model, 45
in *T. aquaticus* RBS, 283
in T7 RNAP elongation complex, 101
ligand dependent modulation, 29
linear reaction pathway
promoter complexes and intermediates, 39, 47
sequential DNA melting pathways, 54–7
Lineweaver–Burk plot, 169–70
linker, σ3.2, 181
linking number, *Lk,* 70, 162, 167

magnesium. *see* Mg^{2+} ions
magnetic optical tweezers, 306
magnetic trapping/magnetic tweezers *see also* optical trapping
calibrating the DNA sensor, 161–6
elongation complex investigations, 271
FRET and, 125, 140, 142–5
future directions, 182–3
general features of, 159–61
temporal resolution of, 144, 160, 161

Subject Index

main channel
  in hairpin-dependent pausing, 285
  hairpin invasion of, 290–2
  mechanism of repeated translocation via, 228
  NTP active site access via, 213, *239*, 240–1
  Thermus aquaticus, 238
mechanical forces
  Bruce Alberts' view, 203
  DNA twisting in transcription regulation, 158
  molecular motor analogy, 191–204, 303–7
  RNAP as an engine, 191, 193–203
  sewing machine analogy, 198
  terminator sequence as a brake, 201–2
  tethering strategies, 195
mechano-chemical coupling, 194–200, 198–9, 263–4, 266–74
Mg^{2+} ions
  in bacterial RNAP structural overview, 2–5
  DksA coordination of, 253
  in elongation complex, *6*, 106, 207, 265
  involvement in reverse reactions, 211
  MgI and MgII, *6*, 237
  regulator introduction of a third cation, 248, 249
  regulator stabilization of Mg^{2+}II, 251, 253
  role in nucleation-propagation, 53
  single-unit and multi-subunit RNAPs, 112–13
  trigger loop folding and, 216
  universality of in RNAP catalysis, 207, *208*
Michaelis complex, 199
Michaelis constant, 107, 193
Michaelis–Menten kinetics, 39, 169–70, 270–1
Microcin J25 (McJ) inhibition, 238–9
millisecond-ALEX, 124, 142, *143,* 145
molecular biophysics limitations, 302–3
molecular motor analogy. *see* mechanical forces

molecular structures. *see* structural atlas
movies, of structural transitions, 97, 101–2, 104, 106, 108
multi-subunit RNA polymerases
  common ancestral core, 207
  initiation to elongation, 103
  sensitivity to regulation, 237
  single-unit and, 97–9, 112–13, 237, 265
  structure elucidation, 14–15, 238
mutations
  EcoRI Q111, 287–8
  *lac*UV5 mutation, 47, 50–1, 54–8, 129–32, 135–7
  P266L mutant, 102
  RNAP, effects on DNA supercoiling, 85
  'stringent RNA polymerase' mutant, 86

N-terminal domain
  initiation and elongation, 99–101
  σ^{70} region 1.1, 58, 116, 119, 126
N25 early promoter (phage T5), 173–6, 182
NAC. *see* nucleotide addition cycle
nanomanipulation, 144, 157–8
  see also magnetic trapping
  detecting promoter unwinding with, 164
nanosecond-ALEX, 124
NAPs. *see* nucleoid associated proteins
negative supercoiling
  see also superhelicity, DNA
  nucleoid and nuclear DNA, 70
  promoter unwinding and, 164, *165*
  promotion of transcription by, 76–7, 78, *79*
noise
  Brownian noise, 264, 304, 308–10
  instrumental noise, 308, 310
NTP-driven translocation model, *224,* 226
NTP (nucleotide triphophate) concentration
  abortive initiation studies using, 174, *178*
  velocity-force trends and, 273–4

nucleation-propagation stage, DNA melting, 52–3
nucleoid
  DNA compaction within, 26–7, 69–70
  structure of, 72–5
  transcription foci in, 86
nucleoid associated proteins (NAPs), 73, 74
  co-activation and, 29–30
  DNA binding proteins in *E. coli*, 73
  DNA compaction and, 26–7
  RNAP cooperation with, 86–7
  supercoiled DNA constraint by, 72
nucleosomes, 306, 311
nucleotide addition cycle (NAC), 206, 207–8, 210
  active site models, 238–241, 213–19
  inhibition by small molecules, 243
  kinetic models, 223–8, 268
  secondary channel involvement, 239–40
nucleotide/NTP binding
  allosteric site proposed, 222
  as driving mechanism for translocation, 97–8, 111–12
  E site involvement and, 239
  elongation complex schematic, 6
  mismatched binding, 110, 111
  secondary site in *E. coli*, 268–9
nucleotide/NTP incorporation
  biphasic and multiphasic kinetics, 268
  elongation complex stages, 104, 105
  free energy of, 193–203
  misincorporated, 210, 211, 219
  models of, 266–70
  selectivity and error correction, 241–4
nucleotide/NTP loading
  active center access routes, 213, 228–9, 238–41
  structural basis of, 209, 212–19
nucleotide sequences
  most-studied promoters, 48
  and RNAP kinetics, 264–5
  sequence dependent kinetics, 265, 274–8
  termination sequence, 201–2, 274
NusA (N utilization substance A) elongation factor, 19

O helix, 104, 106, 221, 270
ω subunit
  active center remote control, 254
  σ factor balance and, 81, 83–4
open complex formation
  characterization of intermediates, 40–1, 47–8, 116
  energy barriers to, 56
  footprinting of *lacUV5* complex, 77
  kinetics of, 168–72
  leading-edge FRET/ALEX results, 129, 130
  σ^{70} retention, 131
  σ domain 1.1 and, 58–9
  σ region 2-DNA interactions, 54–5
  single-molecule studies in, 41
  transcriptionally active, 60–1
open complexes (RP_o)
  characterization at two canonical promoters, 166–72
  with fork-junction DNA, 43–5
  FRET studies on RP_{itc} and, 135
  isomerization to RP_{itc}, 159
  lifetime and waiting time, 169–72
  location of scrunching within, 140
  promoter unwinding, 159, 164, 168, 171
  regulation by overstabilization, 59–60
  sigma-core, core-DNA and sigma-DNA interactions, 53
  σ^{70} release, 132
  structural model of, 45–7
  transition from closed to, 46
  use in *in vitro* experiments, 20–1
optical trapping techniques/optical tweezers
  see also magnetic trapping
  ALEX and, 147
  dual trap, 308–9
  elongation complex investigations, 271, 304–6
  elongation rates, 181–2
  force–velocity curves, 194
  FRET and, 125
  mechanical analysis, 195, 196
  recent developments, 307–9
  single-molecule fluorescence combined with, 309–10

Subject Index 325

supercoiled DNA, 164
translocation mechanisms, 112
orientation factor, k^2, FRET studies, 118–19

'P' site
  in nucleotide addition cycle, 207, *208*
  RNA occupancy and abortive initiation, 176
pausing
  see also arrested complexes
  backtracking as suggested cause, 264
  as DNA encoded, 207
  elongation phase regulation by, 211–12, 305
  force dependence and, 271
  hairpin-dependent transcriptional pausing, 275
  locations, kinetics and mechanisms, 277, 305
  pause prevention by transcript cleavage factors, 250, 252, 253
  power law governing durations, 305–6, 314
  response to misincorporation, 255
  role in termination, 283, 285–6
  tagetitoxin and intrinsic pausing, 248
  temperature effects on, 306, 314
  trigger loop role in, 216–17
  two-pawl ratchet model and, 226–7
  types of pause, 212, 264, 277, 305
pawl and ratchet, *267*
  see also two-pawl ratchet model
PCf11 termination factor, 286
persistence length (of DNA), 74, 162, 168
phage T5, T7. *see* T5, T7
phasing process, spacer region and isomerization step, 54
placeholders, σ^{70} region 1.1, 116, 119
plasmid DNA, 70–1, 78, 85, 162–4
plectonemic form
  bacterial DNA, 71–2, 78
  promoter unwinding and, 164–5
  topological amplification and, 165
  torque and, 164
  writhe component and, 162

power law, pause durations, 305–6, 314
'power stroke' mechanism, 97–8, 220–1, 264, 269–70, 272–4
ppGpp (guanosine tetraphosphate)
  DksA potentiation of, 253
  effect on open complex lifetime, 172
  mechanism of transcription initiation effects, 84–5
  promoter regulation by ATP and, 30–1
  σ factor competition and, 81, 82–3
  transcription regulation, 249–50
PPi. *see* pyrophosphate dissociation
pre-insertion site (PS), 104, 109, 111–12, 242
  streptolydigin inhibition of transition to IS, 246
pre-insertion states, *209*, 217–19
pre-translocation pauses, 277
pre-translocation state. *see* registers
primary channel. *see* main channel
probe-stoichiometry histograms, ALEX, 145–6
promoter attachment, multi-subunit RNAP, *16*
promoter clearance/escape
  DNA scrunching and, 172–3, 176–82
  DNA twisting and translocation in, 158
  elongation complex structure and, 101
  facilitated by GreA, 257
  FRET studies, 126, 141
  kinetic analysis of initial transcription and, 141–2
  mechanisms proposed for, 173
  sigma-core interactions and, 43
  stressed intermediates in, 141, 178–9
  structural transitions in, 7
  transcription bubble in, 176–7, 179–81
promoter recognition
  and initiation, intermediates, *20*
  intermediates between, and transcription initiation, 39
  stages of, 48–9
promoter regulation
  see also activators; repressors
  by ATP and ppGpp, 30–1
  biochemistry of, 22–6
  nucleoid associated protein role, 26

promoter unwinding
  detection using nanomanipulation, 164–5
  kinetics, 168–70
  open and initially-transcribing complexes, 159
  structural analysis, 176–9
promoters
  activator co-dependent, *28*
  activator-dependent, 24
  binding and open complex formation, 47, 59
  characterization and core elements, 17–19
  closed complex dependence on, 49–50
  interaction with protein transcription factors, 19
  malT promoter, 60
  open complex characterization at canonical, 166–72
  open complex regulation by overstabilization, 59–60
  phage T7, 99
  sequences of most-studied, *48*
  structure of, and DNA supercoiling, 78–80
  twist deformations at, 158–9
  ubiquity of linear pathways, 39
proofreading, 254–8
  contribution to accurate transcription, 241
  and streptolydigin inhibition, 246
propagation stage, 52–3
protein–DNA distances, FRET, 126, *127*
protein–protein distances, FRET, 124–5, *127*
proteins
  E. coli DNA binding proteins, *73*
  interaction with promoters, 19
  regulation by, 250–4
  supercoil constraining, 71
proximity constraint, smFRET, 121
PS site. *see* pre-insertion site
pulse-chase experiments, single-molecule, 176–9
pulsed interleaved excitation, 124, 144

purification requirements, 120
pyrophosphate (PPi) dissociation
  as driving mechanism for translocation, 97–8, 106, 221
  release as completion of nucleotide addition cycle, 207
pyrophosphorolysis, 209–10, 255

questions remaining. *see* research opportunities

R (donor–acceptor distance), 117–18
rate constants. *see* kinetics
rate-limiting step, complex formation, 39, 60–1
RBS (RNA binding site), 283, *284,* 285
RD$_e$. *see* elongation complex
reaction kinetics. *see* kinetics
reaction pathways
  elongation complex, 266–74
  parallel pathways, 269
  promoter complexes and intermediates, 39
  sequential linear, 54–7
recruitment
  efficiency of, 30–1
  transcription activation by, 24–6, 29
regimes, DNA extension curve, 163
registers
  detecting, 222–3
  interconversion, and translocation models, 221, 225
  multiphasic incorporation kinetics and, 269
  pre- and post-translocated, 266, 275
  pre-translocated, 207, 209, 212, 293
regulation by pausing, 237, 311
regulator-dependent pauses, 212
regulators
  see also promoter regulation; transcription factors
  active center sensitivity to, 236–7
  likely mode of action, 62
  MerR family, 54
  proteins as, 250–4
  small molecules as, 244–50

Subject Index

relative probe stoichiometry ratio, *S*, 123–4, *125*
relaxed DNA
  superhelicity and, 70–1, *76,* 77, 80–2, 87–8
  twist and writhe deformations, 162–3
repressors
  biochemistry of bacterial, 22, *23*
  rarity of multiple, 27
  and supercoiled DNA, 75
research opportunities
  elongation phase, 228–9
  from initiation to elongation, 116–17
  questions not addressed elsewhere, 33
  single molecule studies, 145–7, 182–3, 309–12
  translocation mechanisms, 200, 223
reverse reactions
  pyrophosphorolysis and transcript cleavage as, 208–11
  reverse translocation, 221
rho termination, 294–5
ribosomal RNA, rrnB P1 promoter, 166
riboswitches, 293
ribozymes, 257–8
rifampicin binding region
  role in allosteric NTP binding model, 225
  role in forward translocation model, 288
rifampicin inhibition, 14–15, 136, 139, 174
rifamycin binding site, *6*
RNA chaperones, 292–3
RNA exit channel
  blocking by σ^{70} domain 4, 181
  hairpin invasion model of termination, 292
  RBS and β-flap domain as, 283
roadblock experiments, 287
rotational freedom, FRET studies, 119
RP$_c$. *see* closed complexes
RP$_{itc}$. *see* initial transcribing complex
RP$_o$. *see* open complexes
*rpo*B, C and D genes, 15
rrnB P1 promoter, 166

rudder
  in core RNAP structural overview, *3*
  hairpin invasion termination model, *290*
  in open complex model, 45
  in RBS, 283
RUT (rho-utilization site), 294

S (relative probe stoichiometry ratio), 123–4, *125*
scrunching
  in abortive initiation, 173–6, 307
  in promoter escape, 172–3, 176–82
scrunching model, initial transcription, 98–9, 159
  evaluation using FRET, 133–4, *138,* 139–40
  not applicable to elongation, 140
secondary channel
  NTP active site access via, 213, 239–40
  position in core and elongation complex, *5, 6*
  regulation by proteins binding in, 250–4
  tagetitoxin binding, 248
  Thermus aquaticus, 238
selectivity regulation, model, *83,* 84
sensor-kinases, 27
sequence dependent kinetics, 265, 274–8
sequestration of excess RNAP, 32
'shearing' model of termination, 285
σ cycle, 19
σ^{54} subunit
  promoters dependent on, 29
  unusual transcription activation and, 24–6
σ^{70} subunit, 17
  closed complex formation, 50
  domain 4 in promoter escape, 181
  initiation pathways, 84–5
  interaction with fork-junction DNA, 44, *46*
  modification by T4 bacteriophages, 32
  open complex formation, 39–42, 45
  promoter activation, *25*
  in promoter escape, 142

rearrangement, on holoenzyme
    formation, 42–3
  in transcription elongation, 126–32
σ^S subunit
  DksA and, 85
  in non-growing *E. coli,* 81–2
σ subunits
  ALEX studies of, 147
  ensemble-FRET study, 119
  exchange of, as stress response, 81–2, 83
  initiation and elongation differences, 103, 140
  multiplicity of, in bacterial RNAP, 15
  region 1.1, 57–8, 116, 119–20, 126
  region R2, 129
  region R4, 129, 296
  release studies, 129–30, 132, 307
  RNAP structural atlas, *2–4*
  role in isomerization, 51–60
  σ 3.2 linker, 103, 181
  structure and protein domains, 16–17
  termination complex hairpin similarities, 296
  as transcription initiation factor, 14
σ^{38} subunit, 32
single-molecule fluorescence techniques, 119, 309–10
single-molecule force-clamp transcription assays, 222, 228
single-molecule nanomanipulation analysis, 157–9
  see also magnetic trapping
single-molecule pulse-chase experiments, 176–9
single-molecule studies
  see also ALEX; FRET; magnetic trapping
  history, status and prospects of, 302–12, 314
  structure of open complex intermediates, 41
  transcription elongation reaction paths, 271–2
  transcription mechanisms, 124–6, *196*
  of transient intermediates, 117, 120–1

translocation mechanisms, 112
trapped complexes and, 292
solenoidal wraps. *see* plectonemic form
sorangicin binding site, 6
spacer regions, 54–5
spatio-temporal resolution
  magnetic trapping, 165–6
  optical tweezers, 308–9
specificity loops, 99, *100*
spectroscopy. *see* ALEX; FRET
static heterogeneity, 120
stationary phase, 31–2
  see also growth phases
stoichiometry histograms, fluorophore, 121–2
strain
  energy and promoter escape, 173
  in ternary complexes, 158
streptolydigin-binding loop (fork loop 2), 224, 245
streptolydigin (Stl)
  allosteric inhibition and translocation mechanism, 111
  bridge-helix model and, 215
  as eubacterial RNAP inhibitor, 244–7
  mechanism of inhibition, *243,* 245–6
  trigger loop model and, 216
streptolydigin (Stl) binding site
  crystallography of complex, 244–5
  elongation complex schematic, *6*
  as secondary NTP binding site, 269
stress, environmental, 80–2, 87
stressed intermediates, 141
stringent response
  DNA superhelicity and, 80
  ppGpp as mediator, 249
  transcription foci and, 86
stringent response factors. *see* DksA
'stringent RNA polymerase' mutant, 86
structural atlas of RNAP, 1–7
  active center in, 207
  Stl binding site, 245
  text references to, 42–3, 115–16, 181, 238
structural basis of transcription phases, 96–8

Subject Index

structural flexibility, active complexes, 61–2
structural transitions. *see* conformational changes
subdomains, T7 RNAP N-terminal domain, 101
subunits
  see also α,β subunit; multi-subunit; σ subunit
  assembly of, 15
  auxiliary, 82
  bacterial core RNAP, *2–3, 5*
  Eschericia coli RNAP, 14–17
  single-unit and multi-subunit RNAPs, 97–9, 112–13, 237, 265
supercoiled DNA
  see also plasmid DNA
  intracellular compaction by, 26–7
  linking numbers and, 162
  negative supercoliing, 70, 76–9, 164–5
  and promoter structure, 78–80
  and promoter unwinding, 164
  supercoiling utilization, 75–8
superhelical density, 70–1
superhelicity, DNA, 69–72
  causal role for RNAP in, 85–6
  effect on promoter activity, 78–80
  energy conversion, 87–8
  extension as a function of supercoiling, *160,* 166, *167*
  generation by relaxed DNA template transcription, *76*
  open complex stability and, 171–2
  topological changes in transcription, 75, 78
  topological forms stabilized by H-NS and HU, *74*
  toroidal and plectonemic configurations, 70–1, 74
  transcription model, 82–5
  twist and writhe components, 70
surface helical repeat, 70
synchronization problem, 303

T5 N25 early promoter, 173–6
T7 A1 promoter, 52, 54

T7 RNA polymerase
  initiation and elongation phases, 98–102
  pre- and post-translocated, 197
  structural studies, 96–8
  termination behavior, 289
  translocation and strand separation, 103–8, 199
T-stretches, 281, 286
tagetitoxin, *243,* 247–9
'tau plots', 169–70
TEC (transcription elongation complex). *see* elongation complex
technologies for RNAP investigation, 33, 309–12
  see also fluorescence approaches; research opportunities; single-molecule studies
temperature effects
  see also noise
  closed complexes, 51
  DNA melting, 54
  on pausing, 306, 314
  transcription activation rates, 60–1
temporal resolution, 309
  ALEX, 124, 145
  ALEX and FRET, 144
  magnetic trapping, 160, 161
  magnetic trapping and FRET, 182
  optical tweezers, 308
termination factors, 286
termination phase
  allosteric (hairpin invasion) model, 287, *288, 291,* 292–4
  biophysical investigations, 202
  DNA topological changes in, 76
  forward translocation model, 287–9
  intrinsic termination, 275, 283–94
  pausing phase preceding, 202, 285–6
  physiological roles of intrinsic terminators, 295
  rho termination, 294–5
  RNAP 'braking' during, 191, 201–2
  torsional-translational coupling, 158
  transcription bubble fate, 181–2
  trapped intermediates, 289–92
  trigger loop alteration effects on, 217

termination sequences, 201–2, 274, 281
tether length, 144, 308–9
Tethered Particle Motion (TPM)
  analysis, 195
TFIIB domain, 103
TFIIS transcription factor
  RNAP regulation by, 252
  stall force in yeast RNAP and, 306
  as transcription cleavage factor, 211
thermodynamic force, 194, 196
thermodynamics
  of the elongation complex, 274–5
  of intermediates, in open complex
    formation, 40–1
  model of transcript elongation, 227–8
Thermus aquaticus
  main and secondary channel, 238
  open complex studies, 159
  RNA self-cleavage in, 257
  structural overview of core RNAP, *2–3*
  structural overview of the elongation
    complex, *284*
Thermus sp.
  bridge helix folding in, 213
  Gfh1 inhibitor, 250, 252–3
  streptolydigin inhibition, 244–7
Thermus thermophilus
  evidence for pre-insertion sites, 242
  initiation to elongation, 103, 108–9,
    112
  ppGpp binding site, 249–50
  tagetitoxin inhibition, 248
three-color alternating-laser
  excitation (3c-ALEX), 144–6
time. *see* temporal resolution
TIRF (total-internal-reflection
  fluorescence) microscopy
  ALEX spectroscopy combined
    with, 142
  single-molecule FRET combined
    with, 144
topological amplification, 165
topological changes coupled to
  translocation, 200–1
'topological differentiation', 88
topological domains, 71

topology
  see also conformational changes;
    torsional deformation
  combined with enzymology, 192
  DNA twist and writhe, 162
toroidal form, 70–1, 73–5, 78
  see also plectonemic form
torque
  DNA supercoiling and, 163–4, 171
  generation limits, 311
  promoter escape energy barrier and,
    181, 182
torsional deformation, DNA
  magnetic trapping investigations,
    160–1
  transcription stages and, 157
  translational coupling, 158
TPM (Tethered Particle Motion)
  analysis, 195
trailing-edge-FRET assay, 126–9, 131
transcript-assisted proofreading, 257–8
transcript cleavage, 210–11, 217
transcript cleavage factors, 237, 250, 256
  see also Gre factors; TFIIS
  proofreading role, 256–7
  structural comparison, 251
transcript elongation. *see* elongation
  phase
transcription bubbles
  constraining supercoiled DNA, 70, 75
  during elongation and termination,
    181–2
  HBS close contact with, 283
  location of scrunching in open
    complexes and, 140–1
  during promoter escape, 176–7, 179–81
  role in termination, 289
  T7 RNAP, *100*
transcription factors
  see also activators; regulators;
    repressors; TFIIS
  multiplicity of and environmental
    changes, 27
  promoter regulation by, 22
transcription fidelity. *see* accurate
  transcription

Subject Index

transcription foci, 86
transcription initiation. *see* initiation phase
transcription process
  analysis of intermediates, 116
  cartoon of, *18*
  highly transcribed regions, 26–7
  significance of, 13–14
  single-molecule FRET and ALEX studies, 124–6
  supercoiling influence, model, 82–6
transcription pulses
  efficiency of termination, 182
  pulse-chase experiments, 176, *177*
transcription regulators. *see* regulators
transcription termination. *see* termination phase
transient binding, single-stranded DNA, 58
transient-excursions model, 133, *136,* 137
transient upstream contacts, RP_c, 50
translocation and strand separation
  multi-subunit cellular RNAPs, 108–12
  T7 RNAP, 103–8, 199
translocation mechanism, 97–8, 107, 109–12
  see also Brownian ratchet; power stroke
  coupling to topological changes, 200–1
  FRET analysis, *146*
  models of, 219–23
translocation register/states. *see* registers
trapped termination complexes, 289–93
trigger loop (G-loop)
  bridge-helix-centric models, 213, 215–16
  in hairpin-dependent pausing, 285
  hairpin invasion termination model, 293–4
  power stroke model and, 271
  and RNAP catalytic activity, 213
  role in nucleotide addition and pausing, 216–17, 227
  role in substrate loading and catalysis, 217–19
  streptolydigin effects, 244–5, 247
  trapped termination complexes and, 290, 293
twist deformation *(Tw)*
  see also torsional deformation
  at the promoter, 158–9, 165
  superhelical DNA, 70, 162

two-pawl ratchet model, *224,* 226–7
tyrT promoter
  AFM of holoenzyme complex, *77*
  supercoiled DNA interaction, 77–8

U-stretch, 202, 283, 286, 289
ubiquitous pauses, 305
upstream activating sequence (UAS), 77
upstream contacts, RP_c formation, 50–2
UV crosslinking. *see* crosslinking experiments

velocity-force relationships, 194, 197, 222, 228, 273–4, 304

wrapping, DNA around RNAP, 44–5, 49–51, 56, 62
  AFM evidence, 45, 307
  footprinting evidence, 20, 44–5, 56
  supercoiling and, 70, 75, 77
writhe component *(Wr),* 70, 162, 165
  see also plectonemic form

X-ray crystallography. *see* crystallography
X-ray footprinting methods, 61, 116

yeast RNAP, 15
  bridge-helix-centric models and, 213
  elongation complex, 108–9, 111–13
  initiation to transcription, 97, 103
  single-molecule studies of elongation, 306

zbd (zinc-binding domain), *3*
zipper
  in core RNAP structural overview, *3*
  hairpin invasion termination model, *290,* 292
  in the RBS, 283
Zn^{2+} finger region
  DNA interaction and, 44–5
  hairpin invasion termination model, *290,* 292
  in the RBS, 283
zone sedimentation studies, 84